DETERGENTS
IN THE
ENVIRONMENT

SURFACTANT SCIENCE SERIES

1. Nonionic Surfactants, *edited by Martin J. Schick* (see also Volumes 19, 23, and 60)
2. Solvent Properties of Surfactant Solutions, *edited by Kozo Shinoda* (see Volume 55)
3. Surfactant Biodegradation, *R. D. Swisher* (see Volume 18)
4. Cationic Surfactants, *edited by Eric Jungermann* (see also Volumes 34, 37, and 53)
5. Detergency: Theory and Test Methods (in three parts), *edited by W. G. Cutler and R. C. Davis* (see also Volume 20)
6. Emulsions and Emulsion Technology (in three parts), *edited by Kenneth J. Lissant*
7. Anionic Surfactants (in two parts), *edited by Warner M. Linfield* (see Volume 56)
8. Anionic Surfactants: Chemical Analysis, *edited by John Cross* (out of print)
9. Stabilization of Colloidal Dispersions by Polymer Adsorption, *Tatsuo Sato and Richard Ruch* (out of print)
10. Anionic Surfactants: Biochemistry, Toxicology, Dermatology, *edited by Christian Gloxhuber* (see Volume 43)
11. Anionic Surfactants: Physical Chemistry of Surfactant Action, *edited by E. H. Lucassen-Reynders* (out of print)
12. Amphoteric Surfactants, *edited by B. R. Bluestein and Clifford L. Hilton* (see Volume 59)
13. Demulsification: Industrial Applications, *Kenneth J. Lissant* (out of print)
14. Surfactants in Textile Processing, *Arved Datyner*
15. Electrical Phenomena at Interfaces: Fundamentals, Measurements, and Applications, *edited by Ayao Kitahara and Akira Watanabe*
16. Surfactants in Cosmetics, *edited by Martin M. Rieger* (out of print)
17. Interfacial Phenomena: Equilibrium and Dynamic Effects, *Clarence A. Miller and P. Neogi*
18. Surfactant Biodegradation: Second Edition, Revised and Expanded, *R. D. Swisher*
19. Nonionic Surfactants: Chemical Analysis, *edited by John Cross*
20. Detergency: Theory and Technology, *edited by W. Gale Cutler and Erik Kissa*

ADDITIONAL VOLUMES IN PREPARATION

DETERGENTS IN THE ENVIRONMENT

edited by

Milan Johann Schwuger

Institute of Applied Physical Chemistry
Research Center Jülich GmbH
Jülich, Germany

CRC Press
Taylor & Francis Group
Boca Raton London New York

CRC Press is an imprint of the
Taylor & Francis Group, an informa business

First published 1997 by Transaction Publishers

Published 2018 by CRC Press
Taylor & Francis Group
6000 Broken Sound Parkway NW, Suite 300
Boca Raton, FL 33487-2742

© 1997 by Taylor & Francis Group, LLC
CRC Press is an imprint of Taylor & Francis Group, an Informa business

No claim to original U.S. Government works

ISBN 13: 978-0-8247-9396-8 (hbk)

Visit the Taylor & Francis Web site at
http://www.taylorandfrancis.com

and the CRC Press Web site at
http://www.crcpress.com

Library of Congress Cataloging–in–Publication Data

Detergents in the environment / edited by Milan Johann Schwuger.
    p. cm.— (Surfactant science series ; v. 65)
    Includes index.
    ISBN 0-8247-9396-X (hardcover : alk. paper)
    1. Detergents—Environmental aspects. 2. Soil pollution.
  I. Schwuger, M. J. (Milan J.) II. Series.
TD196.D48D48 1996
  628.5'2—dc20   96–31584
                                          CIP

Preface

In the past, research on the environmental behavior of detergents was focused on biodegradability. Two books in this series, Volumes 3 and 18, explore surfactant biodegradation. However, as important as this subject is, it does not provide a sufficient basis on which to determine the behavior of detergents completely. For example it does not consider the migration of substances into soil and sediments, and all the ensuing interactions. Even readily biodegradable surfactants experience synergistic or antagonistic effects as they interact with different soil contaminants or components of organic and inorganic origin. The result may be a change in mobility and adsorption equilibrium.

Major changes in the composition of detergents during the past 20 years began with the introduction of zeolites. Later, new polymers and surfactants were introduced and salt content was reduced. Current detergents have new properties and new qualities of environmental behavior. Some recently developed ingredients are not eliminated by biodegradation. Moreover, processes of sedimentation, precipitation, and adsorption are of crucial importance. Therefore, it is absolutely necessary to examine this behavior extensively from the physicochemical point of view. For this reason, the present book is devoted to these interactions. It is the first book available based on this conception, thus closing a gap in consideration of the environmental influence of detergents.

I hope this new approach will be of interest and will be well received as a contribution to the extensive literature on biodegradability.

Milan Johann Schwuger

Contents

IV. Complexing Acids

Contributors

Angelika Bartelt Development OEM Spray Coatings, BASF L + F AG, Münster, Germany

Fritz H. Frimmel Engler-Bunte-Institute, University of Karlsruhe, Karlsruhe, Germany

Dieter Horn Polymer Physics, BASF AG, Ludwigshafen, Germany

Ulrich Kaluza Product Safety and Environmental Protection, BASF AG, Ludwigshafen, Germany

Dieter Kiessling Specialty Chemicals, BASF AG, Ludwigshafen, Germany

Gerd Kloster[†] Institute of Applied Physical Chemistry, Research Center Jülich GmbH, Jülich, Germany

Erwin Klumpp Institute of Applied Physical Chemistry, Research Center Jülich GmbH, Jülich, Germany

Peter Kuhm R & D Automotive Industry, Henkel KGaA, Düsseldorf, Germany

Claus Peter Kurzendörfer R & D Physical Chemistry, Henkel KGaA, Düsseldorf, Germany

Inge Langbein BASF AG, Ludwigshafen, Germany

[†]Deceased.

Franz Malz   Consultant, Essen, Germany

Ulrich Schröder   Specialty Chemicals, BASF AG, Ludwigshafen, Germany

Milan Johann Schwuger   Institute of Applied Physical Chemistry, Research Center Jülich GmbH, Jülich, Germany

Jean-Marie Séquaris   Institute of Applied Physical Chemistry, Research Center Jülich GmbH, Jülich, Germany

Josef Steber   Department of Ecology, Henkel KGaA, Düsseldorf, Germany

DETERGENTS IN THE ENVIRONMENT

I
Surfactants

1

Loading Surface Waters
with Surfactants

FRANZ MALZ   Consultant, Essen, Germany

I. THE PROBLEM OF SURFACTANTS (1959–1964)

In 1959, German rivers were foaming. After the long dry summer, the MBAS
level of drinking water from the Ruhr River increased at domestic taps to as
much as 1.73 mg/liter in the city of Essen on December 21, 1959. Downstream

of dams massive and stable foam layers formed (Fig. 1). In canal locks barges disappeared in the foam. Long foam trails like condensation formed behind barges on the Rhine. The (few) biological sewage treatment plants with activated sludge processes produced foam barriers a couple of meters tall, which were blown by the wind into the surrounding water (Fig. 2).

In the decade between 1950 and 1960, synthetic surface-active substances replaced "natural" soap products in many application areas in household and industry. These new surfactants offered extreme technical benefits but they had substantial disadvantages for water management. Because they showed only a very poor biodegradation profile, surfactants could be found very soon in increasing concentrations in sewage waters and rivers, but also in ground water for drinking water supply as a result of soil infiltration. This led occasionally to extreme impairment of water quality management.

This problem also occurred in other Western industrialized states at the same time. In Great Britain, the increasing consumption of surfactants was characterized by two substantial increases during the years 1951–1954 and 1958–1960. Thus, similar difficulties in water management emerged as in Germany. In Germany, the consumption of a few thousand tons per year of surfactants in 1950 increased to nearly 80,000 ton per year in 1960 with an increasing trend. In the United States, the market figures developed in the same way but 2 years ahead.

FIG. 1 Foam layers on the Lippe River downstream the Buddenburg barrier.

FIG. 2 Foam layers above the aeration tank of a biological sewage treatment plant.

Among the surfactants used, the anionic surfactant tetrapropylene benzene sulfonate (TPS) counted for more than 80% of the market volume, with the remainder 15% nonionic and 5% cationic surfactants. TPS is considered "biologically hard." The elimination rate in the few existing biological sewage treatment plants achieved a maximum of 25%. The sewage treatment plants acted more as a reservoir for peak loadings on washing days. Adsorption by the sludge in the pretreatment stage accounted for the largest part of the elimination rate. The typical influent concentration at municipal sewage treatment plants was around 20 mg/liter MBAS on washing days during the peak time. Concentrations of about 16 mg/liter have been measured in the effluent that reached the surface waters. In the Ruhr River, the most important river for the drinking water supply of the Rhinish-Westphalian industrial area, MBAS concentrations around 0.7 mg/liter were measured.

Substantial concentrations of MBAS have been found occasionally in the river or surface waters used for drinking and industrial process water supply.

For the Ruhr area, this had dramatic effects (Fig. 3) during the last weeks of the dry year 1959. An emergency situation occurred for the water supply in the city of Essen. Apart from the quantitative supply situation, the quality of the drinking water was especially affected. The drinking water for Essen is Ruhr River water artificially filtered through soil. At the time of the extreme drought,

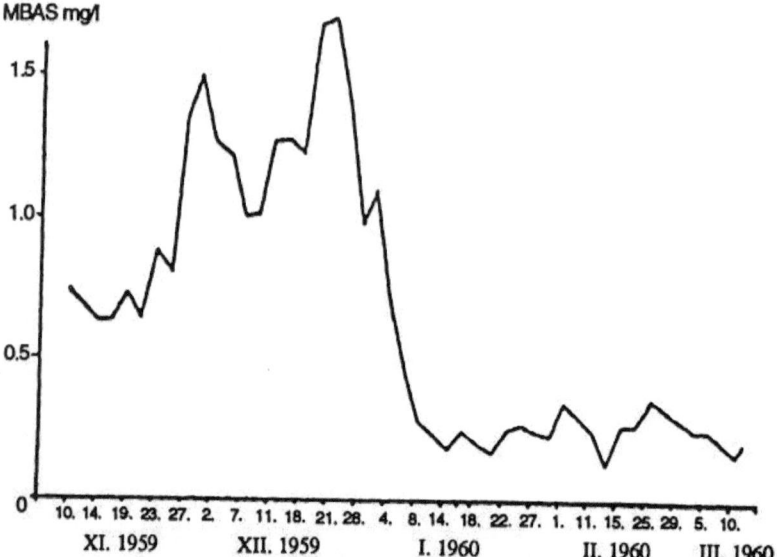

FIG. 3 Development of the surfactant concentration (MBAS) in the drinking water of the city of Essen in 1959.

the standby reservoir water was not available and the water from the Rhine could not be used. Over the course of a few weeks, the drinking water became increasingly contaminated by sodium chloride and water hardening substances but also with surfactants, up to 1.73 mg/liter analyzed as MBAS, by the need to recirculate the water via sewage treatment plants, the Ruhr River, infiltration processes, and the drinking water preparation plant. The existence of the surface-active substances was obvious through strong foaming of the drinking water at domestic taps.

In the Ruhr, Lippe, and Neckar rivers, high foam barriers formed—up to several meters high and more than a hundred meters long. During sewage treatment, substantial dispersion of the suspended solids was observed with increasing concentrations of surfactants. This impacted on the efficiency of the sedimentation process. In biological sewage treatment plants with aeration tanks, the efficiency of the oxygen uptake of the medium and fine bubbly aggregates was reduced by more than 80% during the activated sludge stage. Foam flotation of the activated sludge substantially disturbed the functioning of the sewage treatment unit. Foam layers several meters high formed above the aeration tank. Not only aerobic biological stages were influenced, but also anaerobic sludge treatment in the digesters was hampered to a major extent by surfactants adsorbed on the sludge. Swimming sludge layers formed in the reactors; foam and sludge entered the draining system of the digester up to the gas engines. The anaerobic reactions of the fatty acid degradation were interrupted at MBAS concentrations above 500 mg/liter.

There is a long history of coping with the many technical problems of sewage treatment and drinking water supply to achieve improvement on short notice, such as antifoaming by mechanical and chemical techniques or by charcoal filtration during drinking or waste water treatment. There was only one way to solve the increasingly serious problem of biologically hard TPS, however:

Conversion from poorly biodegradable surfactants to biologically weak i.e., good and fast (that is, thoroughly and rapidly biodegradable molecules)
Systematic implementation of biological waste water treatment

If today the surfactant concentration of large rivers, such as the Rhine, is as follows:

Anionic surfactants around 0.05 mg/liter MBAS
Anionic surfactants about <0.01 mg/liter LAS
Nonionic surfactants about <0.01 mg/liter BiAS
Cationic surfactants about <0.01 mg/liter DSBAS

Highly sensitive analytical methods must be developed to identify the industrial and other human-generated sources of surfactants against the background of biogenic surface-active substances. All this can be assessed properly only if the environmental situation 35 years ago is considered—when German rivers foamed.

II. CORRELATION OF SURFACTANTS WITH WASTE WATERS AND SURFACE WATERS

The phrase ''Surfactants load surface waters'' has gained symbolic meaning in ecological thinking. To enter the very complex area of surfactants and water protection, it is very useful to show the real correlation between surfactants and detergents and cleaning products usage, waste water, drainage systems, sewage treatment plants, surface waters, and soil as existing today in a country with developed environmental protection (Fig. 4):

Surfactants and other ingredients are used to produce detergents and cleaning products.
They are used in household as well as in institutional and industrial areas.
To do the job surfactants and other ingredients must meet clearly defined technical requirements.
After use the surfactants and ingredients become part of the waste water unless they are already ''used up'' and chemically modified.

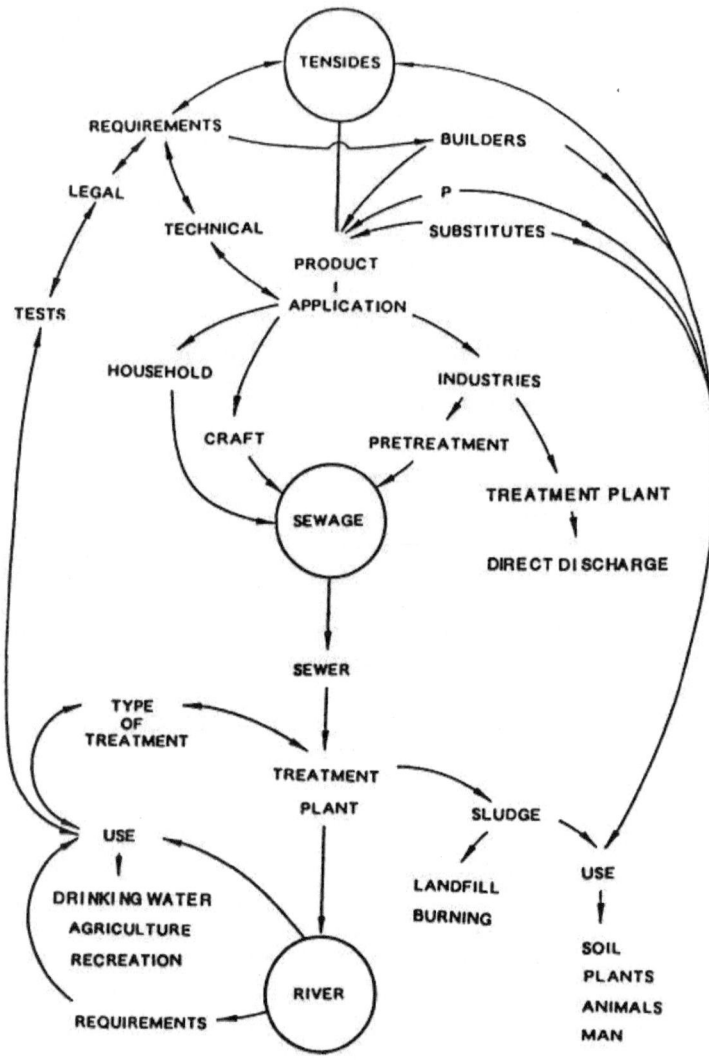

FIG. 4 Correlation of surfactants with waste water and surface waters.

In Germany, domestic and about 60% of industrial waste waters are collected in municipal sewage treatment plants and, together, are biologically treated. Pretreatment may be necessary for "origin-related" waste water streams from industry and commercial practices, which can also affect surfactant levels.

Intensive biochemical and physical degradation and elimination reactions, even of surfactants, begin in the sewer system. Depending on the length and aeration of the drainage system, this can account for up to 30%.

Degradation and elimination reactions occur in biological sewage treatment plants. The extent depends upon the sludge loading and the residence time, respectively. Apart from the metabolic reactions of bacteria, physicochemical elimination processes occur, including adsorption on sewage sludge.

Sewage treatment effluent is discharged into surface waters. Requirements for the quality of sewage treatment effluents are determined by the intended use of the surface waters. This impacts on the intensity of the waste water treatment process and the technical conception of the sewage treatment plant, including further cleaning steps, for example in view of phosphorus and nitrogen elimination.

Sewage treatment effluent is diluted in surface waters. The dilution factor varies substantially from place to place. In surface waters, naturally and occurring human-generated substances, including some remaining surfactant traces, are further intensively biologically metabolized by a "self-cleaning" process.

Depending upon the intended use of the surface waters and the performance profile of sewage treatment plants, there are legal requirements for surfactants and ingredients of detergents and cleaning products. Test systems have been developed that until now have been guided by the technical description of a biological sewage treatment plant. In the future development of requirements for surfactants and ingredients, these test systems will have to consider several areas, such as sewage treatment plants and aquatic ecosystems.

For managing sewage sludge, agricultural use has precedence, and, thus, a direct correlation with soil protection is provided.

All activities and consequences in the area of surfactants and environmental protection can be correlated and starting points for analysis can be gained.

III. OBJECTIVES OF WATER PROTECTION

In the Rhine Water Quality Report 1988 (*Rheingütebericht*, 1988), the State Office for Water and Waste of Northrhine-Westfalia concluded the following:

The Rhine water quality is by far better than its reputation. The public picture of a "dead Rhine River" is absurd. The river has recovered from the heavy loading of the late 1960s and early 1970s. In the area of Northrhine-Westfalia, the Rhine River has achieved water quality Class 2 in most areas. Also, the pollution with organic substances is substantially lower than at the beginning of the 1980s.

This clear position should be seen as the interim result of the "environmental protection program" of the German federal government for the interdisciplinary task in the period preceding the turn of the century. The pragmatic decision of the early 1960s for a way to solve the surfactant issue served as a model and influenced the environmental protection program announced in 1971, with its defined goals for "waters":

Restoration and maintenance of ecological equilibriums

Assurance of a high-quality water supply for population and industry

Water quality class 2 for all surface waters (quality class 2–3 as a minimum)

Purification of nearly all municipal, industrial, and agricultural waste waters

Development of "environmentally acceptable" production processes and "environmental chemicals" (e.g., detergents) that have no negative impact on surface waters.

The activity field for Water in the area of environmental protection was thus defined. The program also serves to define objectives for detergents and cleaning products.

The effect of this environmental protection program was reflected in substantial improvement in water quality by the end of the 1970s, for example in the Rhine River and the rivers of Northrhine-Westfalia. For example, the water quality of the Lippe River provides good evidence of the effect of waste water treatment in cities and industries. Figure 5 shows the percentage distribution of the various water quality classes along a length of river of about 150 km, from Lippborg to the mouth of the Rhine at Wesel, encompassed by the Lippeverband. Although around 1970 major parts of this river still showed water quality class 4 and 3–4, the biological situation of the Lippe improved substantially in the 1970s. Even in 1978 the originally dominating quality classes 4 and 3–4 had disappeared, and by 1982 classes 2–3 (with a trend to 2) became the most visible.

For lowland rivers loaded with intensively cleaned municipal and industrial waste waters, it is assumed that water quality class 2–3 represents a high degree

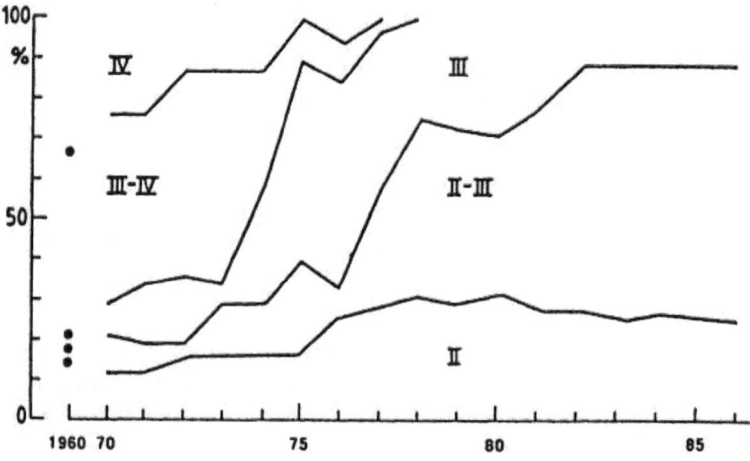

FIG. 5 Development of the water quality of the Lippe River.

of purity. This was confirmed by the increase in good fishing results from 1976 to 1985, with a quota from about 10,000 kg fish to more than 30,000 kg fish per year. Also, trout reappeared.

This effect of a general improvement in the water quality situation in the Rhine is also reflected in the level of surfactants at the sampling station at Bimmen (on the border of Germany and the Netherlands). In Fig. 6 the concentrations and loadings of anionic surfactants (MBAS) during the years 1974–1984 are listed. The downward trend continues. Currently, the levels are below 0.05 mg/liter, which according to Hellmann not only include surfactant residues

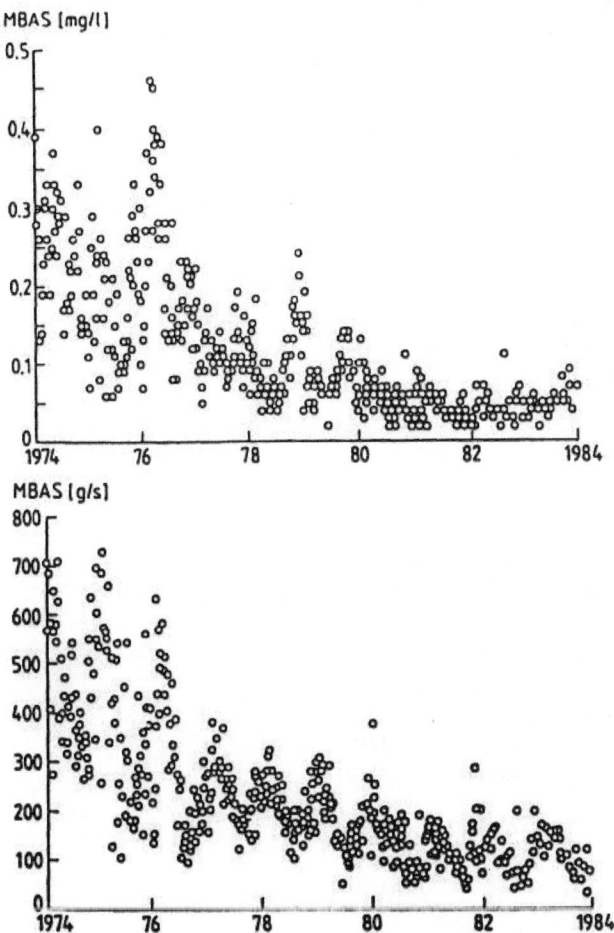

FIG. 6 Surfactants (MBAS) in the Rhine River (according to Hellmann).

caused by human activity, but also include many surface-active ''natural'' components. Both Figs. 5 and 6 provide important evidence for surfactants and water protection, they prove the close correlation between the following:

Construction of full biological sewage treatment plants
Biodegradable surfactants

Despite higher consumption of surfactants, the surfactant concentration and loading of the Rhine are continuously decreasing. The way chosen to solve the problem appears to have been correct.

IV. WATER-RELEVANT SURFACTANT CONSUMPTION (1991)

To assess the surface water situation, it is necessary to break down the consumption figures of surfactants. According to thorough investigations by Noll, conducted in 1991, the consumption of surfactants in the Federal Republic of Germany (FRG) was of the order of about 383,000 ton active material:

Anionics	211,000 t/a	=	55%
Nonionics	150,000 t/a	=	39%
Cationics	15,000 t/a	=	4%
Amphoterics	7,000 t/a	=	22%
	383,000 t/a	=	100%

From these, consumption volumes for use in detergent and cleaning products according to the German Detergent Law (WRMG) were as follows:

Anionics	211,000 t/a	→	131,000 t	=	62%
Nonionics	150,000 t/a	→	61,000 t	=	41%
Cationics	15,000 t/a	→	4,000 t	=	57%
	383,000 t/a	→	206,000 t	=	54%

From the consumption of surfactants that were not used in products belonging to the WRMG, that is, $383,000 - 206,000 = 177,000$ t/a, were the following:

83,000 t/a	not relevant for surface waters
31,000 t/a	relevant for surface waters but not regulated by the WRMG
83,000 t/a	relevant for surface waters, difference between export and import of preparations containing surfactants

Summing the total consumption under the category of relevance to surface waters or not, the distribution results were as follows:

Not water relevant	83,000 t/a	=	22%
Water relevant	63,000 t/a	=	16%
Export-import			
Not WRMG relevant	31,000 t/a	=	8%
WRMG relevant	206,000 t/a	=	54%
	383,000 t/a	=	100%

Glues, colors, varnishes, plastics, and construction materials are examples of application areas of surfactants not relevant to surface water. Areas of industrial use of surfactants containing preparations not covered by the WRMG are the textile industry, paper production, the oil industries, and fire fighting, among others.

The distribution of the various types of surfactants is important given the ecological considerations of surfactants in preparations regulated by the WRMG:

Anionic surfactants			
LAS	47,000 t/a	=	36%
Alkylsulfonates	22,000 t/a	=	17%
Alkyl sulfates	14,000 t/a	=	11%
Alkyl ether sulfates	33,000 t/a	=	25%
Others	15,000 t/a	=	11%
	131,000 t/a	=	100%
Nonionic surfactants			
Alcohol ethoxylates	42,000 t/a	=	69%
Alkanolamides	4,000 t/a	=	6%
Alkoxylates	6,000 t/a	=	10%
Others (<1,000 t)	9,000 t/a	=	15%
APEO	0 t/a	=	0%
	61,000 t/a	=	100%

where APEO = alkylphenolethoxylate. Among the cationic surfactants used in preparations regulated by the WMRG, the alkylammonium compounds outweigh the imidazolinium compounds by several times.

Comparing these figures with previous statistics, it becomes obvious that there is a general shift of surfactants being used in detergent and cleaning products.

	1988	1991
Anionics	51%	63%
Nonionics	37%	29%
Cationics	10%	5%
Amphoterics	2%	2%

Regarding the main surfactant groups, the various surfactants show the following shift:

Anionics	1988	1991
LAS	52%	36%
Alkylsulfonates	16%	17%
Alcoholsulfates	9%	11%
Alcohol ether sulfates	19%	25%
Others	4%	11%
Nonionics	1988	1991
Alcohol ethoxylates	64%	69%
Alkanolamides	8%	6%
Alkoxylates	7%	10%
Others	20%	15%
APEO	1%	—

This survey shows the following:

A substantial decrease in LAS use.
An increase in alkyl ether sulfates.
An increase in alcohol ethoxylates.
APEOs are no longer used.

Generally, the total consumption of surfactants has stabilized. On the other hand, a substantial shift can be noticed within the main groups. The most welcome change is the elimination of APEOs. In total, fewer cationic surfactants are used compared with previous years. The introduction of DEQ (ditallow ester of 2,3-dihydroxypropyltrimethylammonium chloride) instead of DTDMAC (ditallow dimethylammonium chloride) can be seen as the solution to a problem because DEQ biodegrades very well.

To obtain complete clarification of surfactant consumption, the analogous breakdown of surface water-relevant application areas within an industry covering products that until now have not been regulated by the WRMG is absolutely necessary. However, assuming that more than 70% of the industrial waste waters are discharged into municipal sewage treatment plants and are treated together with domestic waste water, this question becomes less relevant if the effluent concentrations of the municipal sewage treatment plants are examined. Still, it is absolutely necessary to require the greatest possible elimination of surfactants by chemical means or the use of ready biodegradable surfactants from direct or indirect industrial contributors.

V. SURFACTANTS IN SURFACE WATERS

A. Current Situation (1987–1992)

To assess the impact of surfactants on water quality, the Rhine River is a good example. Comprehensive and well-documented data exist for the Rhine River as well as its main branches. The Rhine can be also used as an example because more than 42 million people live in and important big industries are located in its heavily industrialized catchment area. About 60% of all waste waters of Germany (before 1989) reach the Rhine. The international observation point Bimmen at the German-Dutch border is used to describe Rhine water quality but not exclusively in view of surfactants. Furthermore, extremely comprehensive data about the Rhine in the area of Duesseldorf exist as result of investigations by Fischer, Gerike, and Steber.

Currently, the following surfactant concentrations in the Rhine River at Bimmen are reported:

Anionic surfactants, 0.05 mg/liter of MBAS
Anionic surfactants, <0.01 mg/liter of LAS (linear alkyl benzene sulfonate, Na salt)
Nonionic surfactants, <0.01 mg/liter of BiAS (bismuth active substance)
Cationic surfactants, <<0.01 mg/liter of DSBAS (disulfine blue active substance)

The following surfactant concentrations are assumed to be of human origin:

Anionic surfactants, <0.01 mg/liter
Nonionic surfactants, <0.005 mg/liter
Cationic surfactants, <0.005 mg/liter

In the area of Duesseldorf-Himmelgeist, the following figures are reported for the Rhine:

Anionic surfactants, 0.08 mg/liter of MBAS
Anionic surfactants, 0.006 mg/liter of LAS
Nonionic surfactants, 0.01 mg/liter of BiAS

It is reported later that it is absolutely necessary to bring perspective to these results, especially for anionic surface-active substances that might be of human-biogenic origin. In general, it can be assumed that in surface waters the correlation between MBAS and LAS is 10:1.

Reports about Ruhr River water quality mention an average concentrations as follows:

Anionic surfactants, 0.07 mg/liter of MBAS
Nonionic surfactants, 0.015 mg/liter of BiAS

at the Ruhr River mouth into the Rhine. The Ruhr River is of special interest for many quality considerations and analyses because it carries a high level of treated waste water but is also used for the drinking water supply of the highly populated industrialized area between the Ruhr and the Lippe rivers along the Emscher River.

Well-documented data exist from the investigations of Steber, Gerike, and Fischer for the catchment area of the Rhine River, especially for the Neckar, Main, and Ruhr, with 140 measuring points for anionic surfactants and 20 measuring points for nonionic surfactants. The average concentrations are as follows:

Anionic surfactants, ≤ 0.07 mg/liter of MBAS
Anionic surfactants, ≤ 0.02 mg/liter of LAS
Nonionic surfactants, ≤ 0.01 mg/liter of BiAS

Bavarian rivers are reported to show the following concentrations:

Isar	LAS	<0.01 mg/liter
	downstream of sewage treatment effluent	
		0.015–0.03 mg/liter
Iller	LAS	<0.002 mg/liter
Inn	LAS	<0.002 mg/liter
Salzach	LAS	<0.002 mg/liter
Lech	LAS	<0.003 mg/liter
Occasionally		0.006 mg/liter

The BiAS values are of the same order of magnitude.

Measurements at Asten on the Danube in 1987 showed the following results:

Anionic surfactants, 0.01 mg/liter of MBAS
Anionic surfactants, 0.002 mg/liter of LAS
Nonionic surfactants, <0.006 mg/liter of BiAS

Summarizing all these results, the average concentrations in the large rivers are as follows:

Anionic surfactants, 0.05 mg/liter of MBAS
Anionic surfactants, <0.01 mg/liter of LAS
Nonionic surfactants, <0.01 mg/liter of BiAS
Cationic surfactants, <<0.01 mg/liter of DSBAS

Currently, a nationwide monitoring program is in progress that is intended to identify the actual concentration of surfactants in many small and medium-sized rivers.

There are no data from the state official monitoring programs of surface waters because surfactants are no longer listed as criteria in the very comprehensive catalog of parameters to be monitored. This may be because the measured concentration in surface waters is significantly below the guideline of 0.2 mg/liter drinking water and that a dying fish has never been correlated with water loading by surfactants because the NOEC values have never been exceeded.

In many analyses it has been investigated and evidence has been provided that these concentration ranges in surface waters are plausible.

Gerike calculated the following concentrations from his comprehensive measurements on the Rhine, Neckar, Main, and Ruhr for 1985, that is, when the concentration had stabilized at a very low level:

	MBAS (mg/liter)		BiAS (mg/liter)	
	Expected	Measured	Expected	Measured
Rhine	1.3	0.03	0.6	0.02
Neckar	2.2	0.01	1.0	0.02
Main	2.2	0.02	1.1	0.02
Ruhr	2.5	0.01	1.2	0.01

This resulted in 99% degradation and elimination in sewage treatment plants as well as by self-purification of the waters.

A very simplified calculation for the Rhine considering the relatively dry year

1991 with 1.780 m³/s MQ is about 22% below the average of many years and, based on the preceding mentioned consumption figures leads to the following results under the following assumptions:

MQ = 1.780 m³/s
Water-relevant consumption of surfactants:

Anionics	165,000 t/a (WRMG 131,000 t/a)	
Nonionics	118,000 t/a (WRMG 61,000 t/a)	
Cationics	12,000 t/a (WRMG 10,000 t/a)	

60% consumption in the Rhine river catchment area
>90% treated by secondary (biological) sewage treatment plants
>95% degradation/elimination
No degradation/elimination in surface waters

Anionic surfactants		
Mass concentration	1.76	mg/liter MBAS
Exposition	0.25	mg/liter MBAS
Measured	0.08	mg/liter MBAS
Measured	<0.01	mg/liter LAS
Nonionic surfactants		
Mass concentration	1.26	mg/liter BiAS
Exposition	0.18	mg/liter BiAS
Measured	<0.01	mg/liter BiAS
Cationic surfactants		
Mass concentration	0.13	mg/liter DSBAS
Exposition	0.018	mg/liter DSBAS
Measured	<<0.01	mg/liter DSBAS

Calculations do not represent reality, especially not in view of the degradation and elimination properties of surfactants in sewage treatment plants and surface waters, as well as in view of the problems caused by biogenic substances detected by the analytical methods as well. However, the figures confirm that the human-generated rest concentration is expected to be less than 0.5% of the consumption. In this calculation for the Rhine, the contributions from Switzerland and France are not yet included.

B.   The Rhine River: Development 1958–1992

Today's knowledge about concentrations and loadings can only be assessed if they are correlated with the values of 35 years ago. The Rhine at point 729 km at Duesseldorf-Himmelgeist is the most intensively sampled observation point worldwide. In 1958, Fischer began with detailed investigations continued by Gerike and Steber today. The development of the surfactant concentration in the Rhine river is documented in the best possible way by these measurements.

The MBAS load of the Rhine doubled from 300 to 600 g/s from 1958 to 1964. During this period, poorly biodegradable TBS was used. After conversion to ready biodegradable surfactants in October 1964 and despite the continued substantial increase in surfactant consumption, the load of the Rhine decreased from 700 to 500 g/s from 1965 to 1974. As a result of the construction of secondary sewage treatment plants in the mid-1970s, the loading decreased continuously from 1975 to 1982. Since then it has been about 100 g/s measured as MBAS.

The 100 g/s MBAS corresponds to about 40 g ABS (alkyl benzene sulfonate) according to the methods of Waters et al. (1991) and to <10 g/s LAS by HPLC. The latter value represents the human-generated rest concentration of anionic surfactants (Fig. 7).

Parallel to the reduction in loading with anionic surfactants, loading by nonionic surfactants also decreased (Fig. 8). Between 1974 and 1987 the loading decreased from 80 to 30 g/s and is now (1990–1992) about 15 g/s. Having considered surfactant consumption in the Rhine River catchment area up to Duesseldorf-Himmelgeist, these loadings correspond to a degradation/elimina-

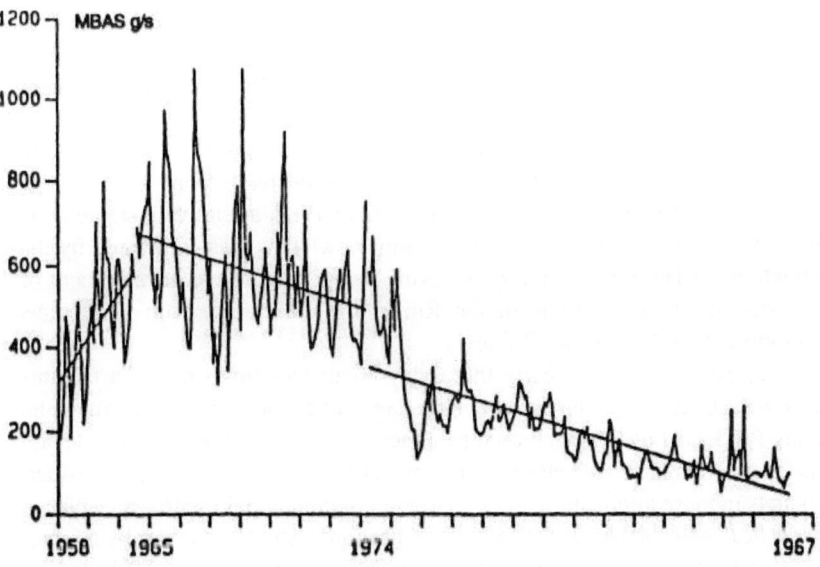

FIG. 7 Trend of anionic surfactants (MBAS) in the Rhine River (according to Gerike).

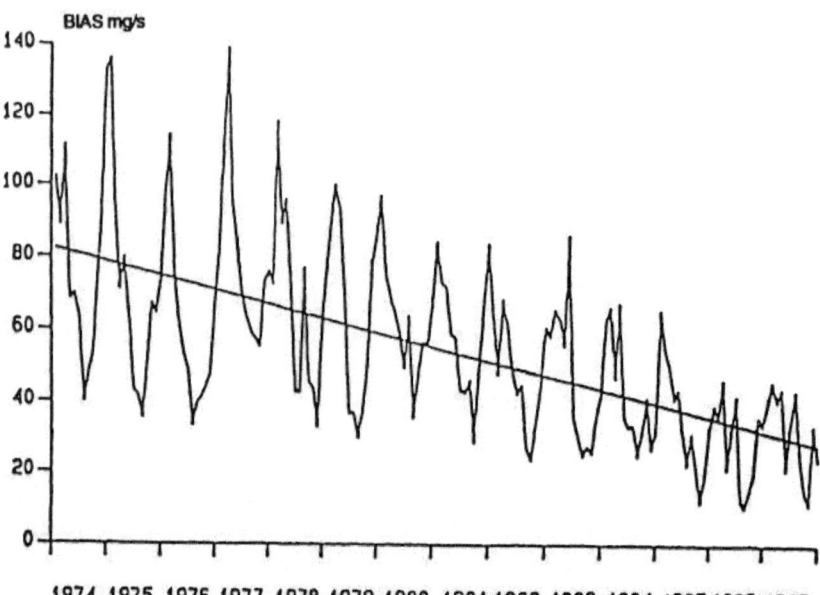

BIAS mg/s

1974 1975 1976 1977 1978 1979 1980 1981 1982 1983 1984 1985 1986 1987

FIG. 8   Trend of nonionic surfactants (BiAS) in the Rhine River (according to Gerike).

tion rate of significantly above 99% for anionic surfactants and more than 98% for nonionic surfactants, according to Gerike.

Using the MBAS values for the observation station Bimmen at the German-Dutch border for surfactant considerations as published by the International Commission for Protecting the Rhine River and by the International Working Group of the Water Plants at the Rhine (IAWR), a decrease from 0.152 (1976) to 0.05 mg/liter MBAS (1982) can be noted. As of 1982, a relatively stable level of about 0.05 mg/liter has been observed until now. This was confirmed by the investigations of Hellmann (Fig. 9). To bring perspective to the development of these surfactant concentrations in the Rhine, the correlation with surfactants consumption should be made (Table 1).

This comparison shows clearly that today about four times more surfactants are being used and that the use of surfactants relevant to the Detergent and Cleaning Products Law has increased by a factor of 3.5. The applications relevant for surface waters have increased by a factor of about 10; by far the largest part of products in these application areas are treated in municipal waste water plants. This contrasts with MBAS and BiAS concentrations, which are lower by a factor of 10, and LAS concentrations, which are lower by a factor of 50 based on

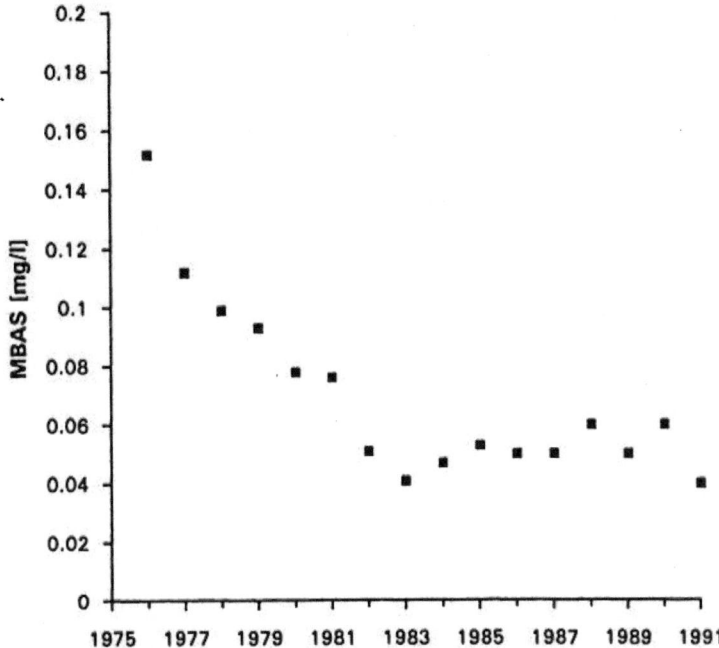

FIG. 9 Trend of anionic surfactants (MBAS) in the Rhine River at Bimmen (according to Irk and Iawr).

MBAS concentrations in 1960. Considering the pseudosurfactants (according to Hellmann) in these concentrations, it can be said that, in general, the concentration of anionic surfactants has decreased by a factor of 50 and that of nonionic surfactants by a factor of 20.

C. The "Conversion" (1964–1965)

In view of the solution of the surfactant problem, the relatively short period between mid-1964 and 1965 was the most interesting period—the "conversion" period. The First Detergent Law of September 5, 1961 and the corresponding ordinance of September 1, 1962 required that as of October 1, 1964 only those surfactants are allowed to be marketed in detergents that are biodegradable by more than 80%, measured in a dynamic test according to a specified test protocol of 1962.

The conversion in October 1964 showed effects after only a few months in the few biological sewage treatment plants: only 30% of municipal waste waters

TABLE 1 Development of Surfactant Consumption and Surfactant Concentrations in the Rhine River

Surfactant consumption (1000 ton/a)	1960	1991	x Factor
Surfactants total	100	383	3.8
Anionics	80	211	2.6
Nonionics	15	150	10
Cationics	5	15	3
Amphoteric		7	
Surfactants relevant to WRMG	60	206	3.4
Anionics	48	131	2.7
Nonionics	9	61	6.8
Cationics	3	10	3.3
Amphoterics		4	
Relevant to surface waters	10	94	9.4
Not relevant according to WRMG		4	
Concentration in the Rhine river, mg/liter			
MBAS	0.5	0.05	-10^{-1}
LAS		<0.01	-5^{-1}
BiAS			with respect to MBAS
DSBAS	0.1	<0.01	-10^{-1}
		<<0.01	

were treated. Until the First Detergent Law came into effect, the degradation/ elimination rate in sewage treatment plants of the Lippe-Verband was about 19% according to Malz; it increased to 60% as of end October 1964 until early 1965. Although the effluent concentrations were about 1 mg/liter of MBAS before the surfactant boom, they increased to 8 mg/liter in 1964; in the years 1965–1966 they decreased again to about 1 mg/liter.

Similar results have been reported by the sewage treatment plants of the Ruhr-Verband, from Hamburg and Bavaria. In only a few sewage treatment plants of the Niers-Verband have lower degradation rates been observed. This was thought to be because these sewage treatment plants were located in the catchment area of the German-Dutch border and many people bought their detergents in the Netherlands.

The degradation of the biologically "soft" surfactants began in the drainage system on the way to the sewage treatment plant. It reached degradation rates between 20 and 40% depending on the flow time; in the Emscher River, degradation rates of up to 50% have been measured. This "predegradation" was not considered in the mass balance calculations of surfactant elimination in sewage treatment plants: in those days a degradation of "only" 75% was measured although more than 80% was required by law.

The effect of the conversion on surface waters also became evident very quickly. At equal water flow of the Ruhr, Booksteeg measured the following MBAS concentrations:

1,390 kg/day MBAS before conversion (1964)
435 kg/day MBAS after conversion (1965)

This corresponded to a decrease of 70%. Figure 10 shows that the MBAS concentration in Ruhr waters continuously increased before the conversion between the city of Hagen to the mouth of the Rhine River. After the introduction of biodegradable surfactants, not only did the concentration in river waters decrease substantially in 1965 but there was also a decreasing trend downward to the mouth that was not caused by dilution but by degradation in the river.

This degradation in river waters was also observed in the Emscher River, consisting of about 80% waste water. At the confluence with the Rhine River, MBAS concentrations of about 6 mg/liter were measured in 1964 but only 2–3 mg/liter in 1965. This means that the biodegradation process begins before processing at the waste water treatment plant. According to the investigations of Huber of Bavarian surface waters, the surfactant concentration decreased by 40–50% shortly after the conversion.

Fischer et al. have monitored the conversion phase in many rivers. In 1964, they found in the Neckar over a distance of about 200 km and at 30 sampling

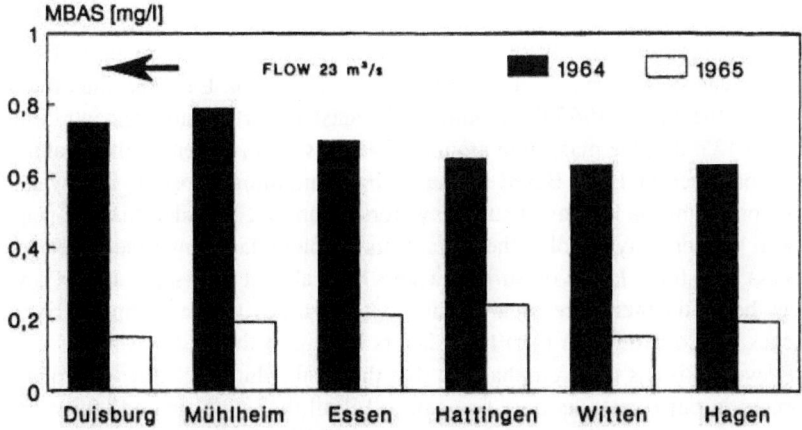

FIG. 10 Changes in the surfactant concentration (MBAS) in the Ruhr River after conversion to biodegradable surfactants, 1964–1965 (according to Bucksteeg, 1966).

TABLE 2   Effects of the Conversion to Biodegradable Surfactants on
Surface Waters

River	1964 (mg/liter MBAS)	1965 (mg/liter MBAS)	Decrease (%)
Rhine D-Himm.	0.38	0.26	30
Neckar (reach)	0.64	0.32	50
Main (reach)	0.55	0.23	58
Ruhr (mouth)	0.75	0.15	80
Mosel (reach)	0.34	0.25	26
Saar	0.82	0.59	28
Sauer	0.35	0.21	40
Agger	0.18	0.12	33
Ahr	0.30	0.13	57
Lahn	0.55	0.21	62
Lippe	0.88	0.39	56
Sieg	0.33	0.17	48
Wupper	2.25	1.11	51

points an average MBAS concentration of 0.64 mg/liter MBAS; in 1965 this
concentration was only 0.32 mg/liter, corresponding to a decrease of 50%.

In the Rhine River, decreases of up to 30% have been found. In the Main
River, the values have been up to 65% lower. Table 2 provides a survey of the
situation in other surface waters.

According to the report by Husmann and Malz to the federal government in
1967,

> Summarizing the results of these investigations of the various rivers in Ger-
> many, it can be said that from 1965 before the Detergent Law became effec-
> tive until the end of 1965 the minimum decrease in surfactants concentration
> is about 13% and the maximum around 77%. This positive trend will certainly
> have continued in 1966. Based on the available monitoring results in sewage
> treatment plants as well as in surface waters it can be concluded that the way
> chosen by Germany to solve the surfactants problem has shown the expected
> success. The foam layers on surface waters have almost all disappeared. They
> occur here and there, are local events, but the times where complete river
> reaches are covered with high foam layers belong to the past.
>
> However, it has to be emphasized that the final solution of the surfactants
> problem in our rivers can only be achieved if all the domestic and industrial
> waste waters containing surfactants are treated in biological waste water
> treatment plants which are managed according to the state of the art.

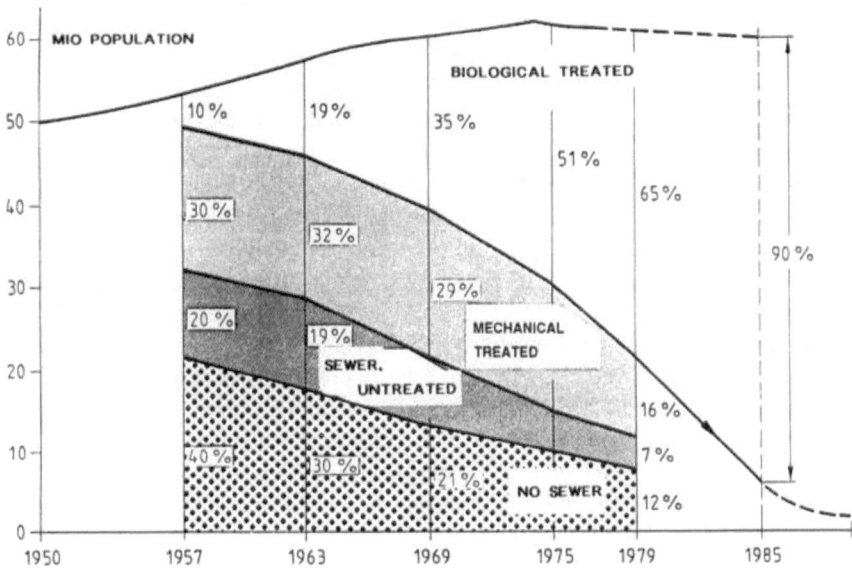

FIG. 11 Development of waste water treatment in municipal sewage treatment plants in Germany.

D. Effect: Conversion plus Construction of Sewage Treatment Plants (1964–1979)

Figure 11 shows the situation of waste water treatment in 1964 with about 20% biological sewage treatment plants for municipal waste water. In line with the conversion to biodegradable surfactants, the construction of biological sewage treatment plants started. Until the end of 1979, about 65% of the population was connected with biological sewage treatment plants. The surface water quality improved substantially (Fig. 5) and the surfactant concentration decreased significantly. The time between 1964 and 1979 is therefore especially important in view of the development of surfactant concentrations in surface waters.

The investigations of Fischer, Hellmann, Huber, Malz, and others show that the positive development observed in the conversion phase of 1964–1965 has continued as a result of the new biodegradable surfactants and the efficiency of the new biological sewage treatment plants. This trend continued with increasing degradation and elimination rates caused by improving the sewage treatment process and extending residence times in the biological treatment stage.

In 1964, the concentration of MBAS in the Neckar River, with its high waste water loading, was about 0.64 mg/liter MBAS at the mouth of the Rhine. In 1979 this concentration was about 0.06 mg/liter. This corresponds to a decrease of 90%. In the Main River, 0.56 mg/liter MBAS was measured in 1964; to 1979 this loading decreased by 86% to a MBAS concentration of 0.08 mg/liter. In the Ruhr River, the MBAS concentration decreased to 0.06 mg/liter between 1964 and 1979; this corresponds to a decrease of 85%.

In the area of the lower Rhine, the change in the situation has become particularly evident because of the combination of biodegradation and sewage treatment. Because of the installation of the sewage treatment plant Emschermündung in 1978 for about 5 million IE (inhabitant equivalents), the surfactant concentration decreased at the lower Rhine from 0.26 to 0.08 mg/liter MBAS.

However, note that according to Fischer considerable parts of the MBAS concentrations measured in 1979 have been proven false. The surfactant part (ABS) analyzed according to Waters and Kleiser and presented in 1979 was only 50% of the MBAS value in the Neckar, Main, and Ruhr rivers; it decreased to 15% in the Rhine River to the German-Dutch border; this means the human-generated anionic surfactant concentration was stabilized at 0.05 mg/liter in all rivers by 1979.

VI. PSEUDOSURFACTANTS

False MBAS values of around 0.1 mg/liter MBAS were observed by Fischer and Hellmann in the early 1960s when they investigated surface waters. In view of the high current surfactant concentration of about 0.8–1 mg/liter, the fictitious MBAS contributors to those false MBAS levels were not of the same importance as they are today.

Today, their occurrence can not longer be neglected when assessing surfactant concentration in surface waters. Surfactant concentrations are now in a range in which not only human-generated surfactant residues are detected by the classic analytical methods of sum parameter analyses, such as the methylene blue-active substances (MBAS), the disulfine blue-active substances (DSBAS), and the bismuth-active substances (BiAS), but also quite a number of biogenic organic compounds are measured that show surface-active properties and respond intensively to these analytical methods, for example, humic acids. Hellmann called these substances pseudosurfactants.

For anionic surfactants the analytical method of Waters reflects the real situation better than the MBAS method. However, only HPLC and thin-layer chromatography analyze synthetic surfactant residues (e.g., LAS) adequately, after appropriate sample preparation, which is necessary to enrich the sample from the complex mixture of biogenic substances that also react with methylene blue, for example. According to Gerike and Steber, analysis of Rhine River water showed that 0.05 mg/liter MBAS corresponds to about 0.025 mg/liter using the method by Waters (ABS) and <0.01 mg/liter using HPLC. The human-generated anionic residues are allocated to LAS.

Although the MBAS method according to DIN/DEV is still suitable for investigations of sewage treatment plants or to check biodegradability according to the regulation based on the Detergent Law, this method fails if applied to measurements in surface water, sludge, and soils. The results do not reflect reality: they are always too high.

Research by Hellmann using thin-layer chromatography confirmed that in the Rhine River at Bimmen residues of LAS occur only in concentrations below 0.01 mg/liter as human-generated anionic fraction at a MBAS background concentration of up to 0.17 mg/liter. Hellmann analyzed residual concentrations of nonionic surfactants in the Rhine of less than 0.005 mg/liter, but it is still unclear whether all biogenic substances had been separated to eliminate interference. He found concentrations of cationic surfactants of <0.001–0.005 mg/liter using thin-layer chromatography after a sample-specific cleaning step, but he suggests that these values are still influenced by interference from substances from the sample. Woltering, for example, found cationic surfactant concentrations between 0.004 and 0.016 mg/liter in the Rhine and other German rivers. Hellmann's results have confirmed again speculation about the impact of sample composition on the assessment of surfactants in surface water, sludge, and soil.

The MBAS method analyzes mainly biogenic pseudosurfactants if surface waters are investigated. This is confirmed by the research of Steber at the Rhine at Düsseldorf-Himmelgeist from 1990 to 1992 (Fig. 12):

The MBAS level showed a clear correlation with the water flow.

The LAS load is only a very little influenced by the water flow and, thus, varies less around the compensation line.

The BiAS load characteristic for nonionic surfactants is also influenced very little by the water flow.

This confirms that only the LAS load is of human origin, because surfactant discharge into surface waters has been relatively consistent during recent years.

If the MBAS load presents predominantly biogenic surface-active substances to more than 90%, Hellmann calls them pseudosurfactants and there must be a clear correlation with water flow.

VII. CONTRIBUTION OF SEWAGE TREATMENT PLANTS TO TODAY'S SURFACE WATER LOADING BY SURFACTANTS

The development of the residual surfactant concentration in surface waters is influenced by two factors:

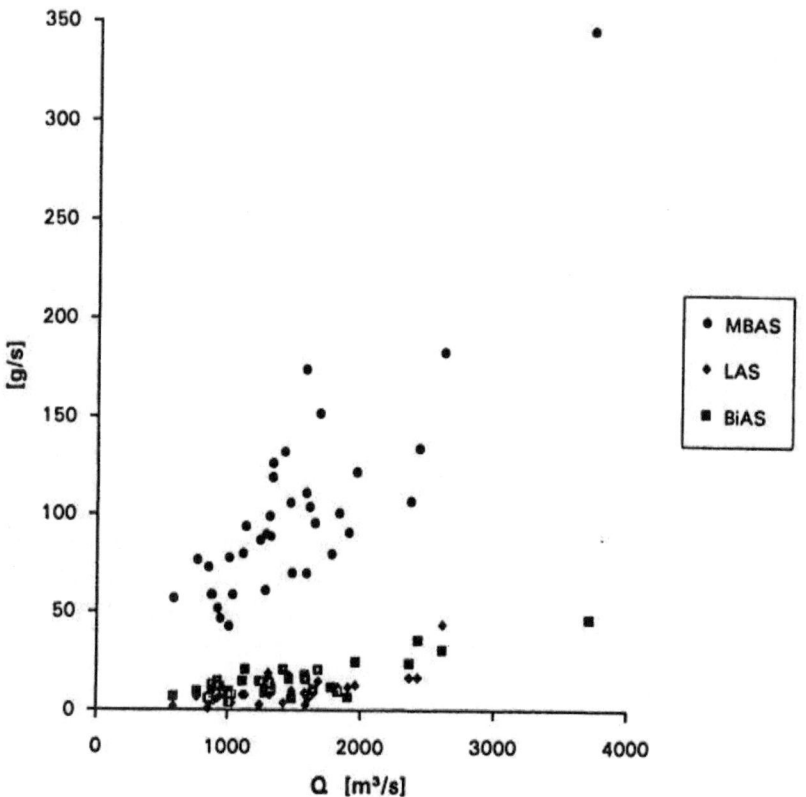

FIG. 12 Dependence of surfactant load (MBAS, LAS, and BiAS) on the water flow of the Rhine River 1990–1992 (according to Steber).

1. Improvement in the biodegradability of surfactants
2. Completion of municipal sewage treatment plants and improvement in their purification performance

Figure 11 shows the development of sewage treatment construction in Germany since 1966 (old Federal States). Today, the 95% level is being exceeded. In the regions of the water management association in Nordrhein-Westfalen, for example, almost 100% of all inhabitants for whom it makes sense and for whom it can be justified are connected to biological sewage treatment plants.

The retention of surfactants happens in municipal sewage treatment plants by biodegradation and by adsorption on the sewage sludge. About 30% of the anionic surfactants, 4% of the nonionics, and up to 60% of the cationics can be absorbed on the sludge. Elimination by degradation and adsorption in sewage treatment plants is more than 97%. Elimination starts on the way to the sewage

TABLE 3 Minimum Requirements for Municipal Sewage Treatment
Plants According to Attachment 1 "Communities of the
Administrative Ordinance for Waste Waters"
(*Rahmenabwasserverwaltungsvorschrift* as of January 1, 1992)

Inhabitants (1000s)	COD (mg/liter)	BOD$_5$ (mg/liter)	N inorganic (mg/liter)	P$_{tot}$ (mg/liter)
1	150	40	—	—
5	110	25	—	—
20	90	20	18	—
100	90	20	18	2
100	75	15	18	1

treatment plant, which adds to elimination in the sewage treatment plant so that
the total elimination between the point of use of the surfactants in household and
industry to discharge into surface waters accounts for more than 98%.

For the biological stage in sewage treatment plants, Malz could demonstrate
a very good correlation between surfactant biodegradation and the sludge load.
For B_{TS} values around 0.3 degradation rates <90% are measured (B_{TS} = sludge
load based on BOD$_5$, kg BOD$_5$/kg TS × days), which increase substantially
above 95% for B_{TS} values of 0.1. In general, there is a direct correlation between
BOD$_5$ degradation and the degradation of anionic and nonionic surfactants. Cationic surfactants are eliminated to 60% by adsorption but also biodegrade; this
results in a total elimination rate of about 95%. The new generation of cationic
surfactants is ready biodegradable. Therefore, they are no longer predominantly
eliminated by adsorption.

With the Administrative Ordinances for Waste Water (*Rahmenabwasserverwaltungsvorschrift*) dated September 8, 1989, regulatory minimum requirements
for the quality of sewage treatment effluents are prescribed for various types of
waste water. According to attachment 1, Municipals, of January 1, 1992, the
values listed in Table 3 belong to municipal sewage treatment plants. Nitrogen
and phosphorus elimination is mandatory. To achieve this purification performance, the dimensioning of the sewage treatment plants is reduced to $B_{TS} < 0.1$
for sludge loading and the retention time increases to 6–12 h.

This development is relevant to the discussion of surfactants and water protection because the official test procedure—the OECD confirmatory test—suits
the sewage treatment plant characteristics of 1962 with high sludge loading and
short retention times ($B_{TS} > 0.3$–0.7; 3 h aeration time), and thus degradation
rates of more than 80% were required. Under these stringent test conditions, the
biodegradation values de facto exceed very clearly the 90% reported by the Main
Committee on Detergents in their expert report. The task is now to adjust the
official test method to today's state-of-the-art waste water treatment.

The experimental work by an expert working group within the Main Committee on Detergents, chaired by Schöberl (1991), concentrated on the task of adopting the existing official test procedure established in 1962 to the generally accepted rules of today's technology for sewage treatment plants with extensive nitrification and denitrification (Fig. 13). The sludge loading was reduced to B_{TS} = 0.16, corresponding to DOC_{TS} = 0.1 kg/kg × days, (DOC_{TS} = sludge loading related to dissolved organic carbon, kg DOC/kg TS × days) and the retention time was increased to 6 h. Correlation of the volume between denitrification and nitrification is about 1–2.

The proposal for the modified test procedure is being referred to DIN to initiate a revision of the test procedure on a national and international level. Ring tests to validate the method have shown very consistent biodegradation rates for LAS, for example, between 96 and 99% expressed as MBAS, compared with 93–95% based on today's standard method. The effluent contained less than 0.02 mg/liter of the initial concentration of 10 mg/liter LAS. The DOC (dissolved organic carbon) degradation was about 92%.

Further development of the test method to an extensive nitrification and denitrification unit also occurred against the background of testing surfactant residues in a continuously working system at various trophic levels. After the simulated sewage treatment plant step with extensive N elimination, the coupled *Daphnia* test over three generations showed no measurable effects. Thus, a concrete indication for exposition considerations is obtained and a theoretical safety factor for the tested product is no longer needed. By using the step-by-step simulation, these indications can also be obtained for other trophic levels. This has only now become possible because biologically treated sewage test water that does not have a high ammonium nitrogen content, as in the current official test method, can be fed into this continuously working system.

The modified dynamic biodegradation test can be used as a standard. Continuous biotests at the various trophic levels, however, require specially trained experts, so that these investigations are restricted to few specialized laboratories.

What is today's situation?

The surfactant concentrations of sewage treatment plant influents as well as effluents undergo local deviations that depend on the infrastructure of the catchment area, the unknown water inflows, mixed or separated waste water drainage systems, and so on. The influent and effluent concentrations are in the following ranges:

Mg/liter		Influent		Effluent	Degradation (%)
MBAS	5–16	average 10	0.05–0.6	average 0.3	97
BiAS	3–8	average 5	0.05–0.4	average 0.15	97
DSBAS	2–4	average 3	0.01–0.2	average <0.1	<97

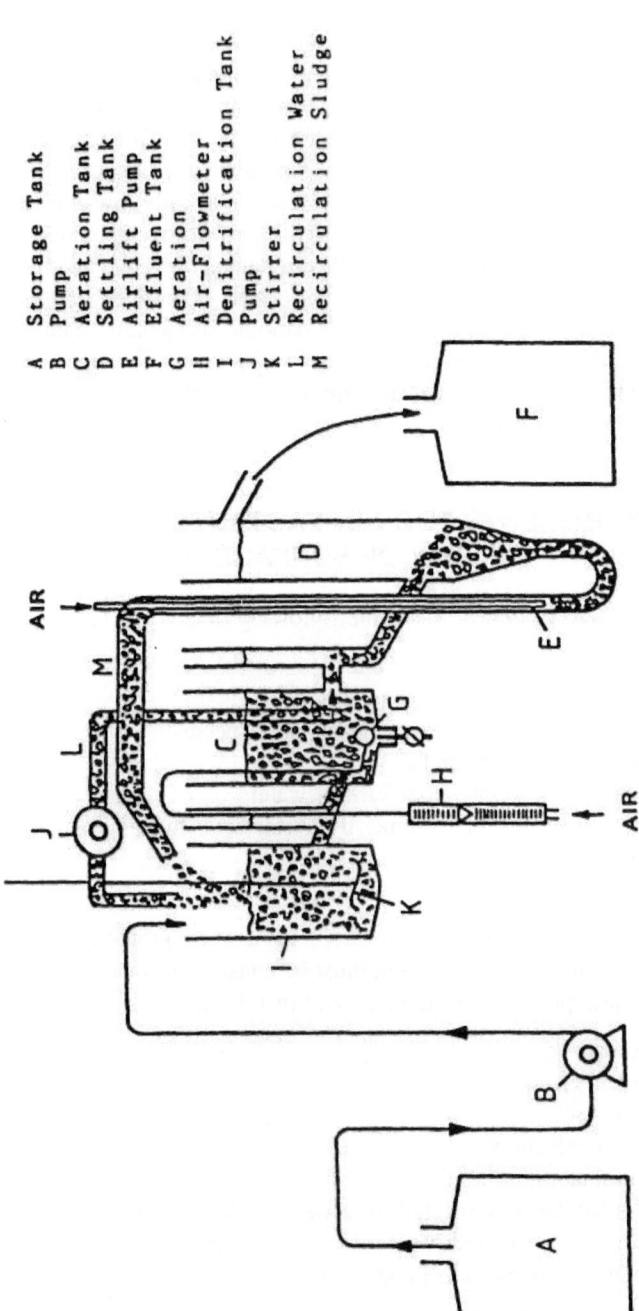

A Storage Tank
B Pump
C Aeration Tank
D Settling Tank
E Airlift Pump
F Effluent Tank
G Aeration
H Air-Flowmeter
I Denitrification Tank
J Pump
K Stirrer
L Recirculation Water
M Recirculation Sludge

FIG. 13 Modified test procedure for the official dynamic biodegradation test with nitrification and denitrification (according to the working group of Dr. Schöberl, Main Committee Detergent).

Parallel to the continuous increase in the numbers of sewage treatment plants with nitrification and denitrification as well as phosphate precipitation according to the minimum requirement (attachment 1, Communities), BOD_5 effluent values are decreasing to concentration of less than 5 mg/liter BOD_5, practically 0.

Because there is a direct correlation between the biodegradation of surfactants and the BOD_5 decrease, it can be expected that the already low surfactant concentrations will decrease further in the future. Because of the extended residence times, however, larger quantities of pseudosurfactants that have an effect on the "real" effluent concentration values will be formed in sewage treatment plants. The new generation of good biodegradable cationic surfactants, such as quaternary diesters (DEQ), are of special interest for degradation effects in sewage treatment plants. These substances are eliminated mainly by biodegradation under aerobic as well as anaerobic conditions.

VIII. REQUIREMENTS FOR SURFACTANTS FROM THE POINT OF VIEW OF SURFACE WATER

From a surface water point of view, the following objectives and requirements are postulated:

Achieve and stabilize surface water quality class II
Protect the food chain and the variety of aquatic species
Protect the estuaries and tidelands
Protect drinking water supply from surface waters
Identify residual loading, for example by surfactants, by test methods on the various trophic levels
Protect ground water

Regarding the correlation between surfactants and water protection, the following developments in sewage treatment must be considered for the assessment of surfactants despite product formulations and test designs:

Lower sludge loading
Extended resident times
Multistep processes
Nitrification-denitrification
Phosphorus elimination
High level of inhabitants connected to sewage treatment plants
Pretreatment in case of indirect waste water discharge
Extension of sewage sludge use in agriculture

Especially for this reason, it is very important that the current official test method, which is based on the state of wastewater treatment technology of 1969, be replaced by the modified test system described earlier, with nitrogen elimination and extensive DOC degradation. Even with the test method of 1962, however, surfactants show the ecorelevant properties presented in Table 4.

The expert report of the Main Committee on Detergents of 1986 confirms that the basic requirements of <80% surfactant degradation/elimination, <60% organic C degradation, and <1 mg/liter aquatic toxicity are substantially exceeded, and based on the experience of 1992, the objectives for the 1990s have been achieved: >95% surfactant degradation/elimination, >>80% organic C degradation, and >0.2 mg/liter of aquatic toxicity (based on "degraded" substances).

The individual surfactant raw materials are essential as test substances for biodegradability in the dynamic test or screening test. To assess the properties of surfactants in surface waters, however, a degraded substance or, in other words,

TABLE 4 Biodegradability and Aquatic Toxicity of Surfactants Used in Detergents and Cleaning Products (Consumption Volumes Above 1000 tons/a) (1986)

| | Biodegradation/elimination | | Aquatic toxicity | |
| | | | Fish | *Daphnia* |
Type of surfactant	Surfactant (primary degradation)	DOC (total mineralization)	toxicity (LC_{50}, mg/liter)	toxicity (EC_{50}, mg/liter)
Basic data	>80	>60	>1	>1
Anionic surfactants				
Alkyl benzene sulfonates	90–97	73–93	3.2–4.9	8.9–14
Alcohol ether sulfates	96	67–89	1.4–20	1–50
Alkylsulfonates	97–98	83–96	3–24	8.7–13.5
Alcoholsulphate	98–99	97–99	3–20	5–70
Sulfosuccenic acid ester	96	49	39	33
Soaps	—	90[a]	6.7–150	22–72
Nonionic surfactants[b]				
Alcohol ethoxylate	83–98	62–90	1–50	2–200
Alkyl phenol ethoxylates (APEO)[c]	85–97	50–94	1.5–1000	4–50
Cationic surfactants				
Alkyl ammonium Quarternary compounds	94	100	1–6[d]	0.1–1.0[d]

[a]Not DOC, but CO_2 analysis because the calcium and magnesium salts of soaps cannot be analyzed by DOC analysis.
[b]The range depends on the degree of ethoxylation.
[c]No longer used in detergents as of 1990. The range of data is mainly reasoned by the fact that biodegradation/elimination and ecotoxicity very much depend on the chemical structure. Primary degradation and toxicity are counted rotating.
[d]Detoxification by formation of neutral salts with anionic surfactants and by adsorption on minerals.

the toxicity "corrected by the degradation," is the essential criterion. Recall the correlation of surfactants and surface waters in Sec. II. Many wrong interpretations are simply caused by the fact that the acute toxicity has been measured with the original material. This is not scientifically correct, nor does it reflect reality. The expert report of the Main Committee on Detergents to the German government about the ecologically relevant properties of surfactants contains toxicological basic data for raw materials. It is now being proven by many measures that the toxicity in the effluent of biological sewage treatment plants is by a factor of 10 lower (based on the same amount of surfactants) than that of the starting material before biodegradation. Table 5, developed by Huber, allows us to decide which detergent ingredients are of special importance in which respect. Based on product composition, the ecological relevance of an ingredient can now be assessed.

IX. SUMMARY

In the correlation between surfactants and water protection, we can see one part of the technological history of modern times, which must be recognized by the public in its positive development. If we follow the most important historical events:

1954　Beginning of the surfactant boom
1958–1959　Foam on German rivers
1960　Establishing of the Main Committee on Detergents (Hauptausschuss Detergentien)
1961　First detergent law
1962　Decree on biodegradation of anionic surfactants
1964　Conversion from "hard" to "soft" surfactants
1975　Second detergent law
1977　Decree on biodegradation of anionic and nonionic surfactants
1981　Phosphate regulation, stage 1
1984　Phosphate regulation, stage 2
1987　Third law on detergent and cleaning products

and if we quantify the effects, the following can be stated with good reason:

1. The promises of 1962 to solve the problem were right,
2. The surfactant concentration in the Rhine River, for example, has decreased many times from 1958 to today, although surfactant consumption has increased in the same period by about a factor of 4.
3. The good results provide evidence of the effective cooperation between the detergent industry, waste water treatment, and government in which the Main Committee on Detergents played a major role.

TABLE 5 Ecological Criteria for Detergent Ingredients[a]

	Biodegradability/ elimination	Fish toxicity	Daphnia toxicity	Other aquatic toxicity	Ecologically unacceptable metabolite formation	Bioaccumulation	Complex formation/ mobilization	Ion exchange/ metal	Fertilization effect	Salt pollution
Anionic surfactants	++	++	+	++	0	0	—	—	—	—
Nonionic surfactants	++	++	+	++	++	0	—	—	—	—
Cationic surfactants	++	++	++	++	0	0	—	—	—	0
Phosphate	0	—	—	—	—	—	0	—	++	—
Na Al silicates	—	—	—	—	—	0	—	+	—	—
Polycarboxylates	0	—	—	—	—	—	0	0	—	—
Citric acid	++	—	—	—	—	—	0	—	0	—
EDTA	++	—	—	—	—	—	++	—	0	—
NTA	+	—	—	—	—	—	++	—	0	0
Phosphates	—	—	—	—	—	—	++	—	0	0
Perborate/borate	—	0	0	0	—	—	—	—	—	—
Sodium carbonate	+	—	—	—	—	—	—	—	—	—
TAED/DAED	+	—	—	—	—	—	—	—	—	—
CMC	—	—	—	—	—	—	—	—	—	—
Sodium sulfate	—	—	—	—	—	—	—	—	—	+
Silicates	—	—	—	—	—	—	—	—	0	—
Enzymes	+	—/0	—/0	—/0	—	—	—	—	—	—
Optical brighteners	++	—	—	—	—	0	—	—	—	—
Cumene sulfonate/ toluene sulfonate	+	—	—	—	—	—	—	—	—	—

[a]++ = very important; + = important; 0 = little importance; — = not important.

All experts who have contributed to these exciting achievements in the history of new technologies should be paid high recognition. All those who accept the challenge of necessary future work should feel encouraged that it is possible to practice active environmental protection.

BIBLIOGRAPHY

1. Book, K.J., Effect of the conversion of biologicaly degradable raw materials in large technical filter plant and and main drainage channel. Year-book Vom Wasser, (1966), in press.
2. Bucksteeg, W., Degradation of detergents in filter plants and main drainage channels and after conversion on soft laundering raw materials. Wasser, Luft und Betrieb, Issue 1 (1966).
3. Fischer, W.K., Detergents in surface waters. Fette, Seifen, Anstrichmittel 51–59 (1964).
4. Fischer, W.K., and Winkler, K. Investigations in the river basin of the Rhein 1958–1975, Vom Wasser pp. 81–129 (1976).
5. Fischer, W.K. Development of the concentration of surfactants in German waters 1960–1980. Tenside 5:250–251 (1980).
6. Gericke, P., The quality of water of the Rhein near Düsseldorf. Tenside 1 (1989).
7. Gericke, P., Investigations of waters in the river basin of the Rhein and Oncological consequences. Tenside 4 (1989).
8. Gericke, P., et al., LAS-residue in German rivers. Tenside 2:136 (1989).
9. Gericke, P., Winkler, K., Schneider, W., Jacob, W., and Steber, J., Quantitative balance of wash and cleaning agents—capacity with its effect on the waters, Tenside 2/1991, pp. 86–89.
10. Schöberl, P., Bock, K.J., and Huber, I., HA Status Report on Detergents: Oncological relevant data of tensides and non-tensidic contents in wash and cleaning agents Tenside 2:86–107 (1988).
11. Hellmann, H., Analysis of Surface Waters. Thieme, Stuttgart, 1986.
12. Hellmann, W., Tensides and pseudotensides in surface waters and effluents. Tenside 2:111–117 (1991).
13. Hannes, E.C., and Rapaport, R.A. Calculations and analytical proofs of LAS—concentrations in surface waters, sediments and residues. Tenside 2:141–147 (1989).
14. Husmann, W., Malz, F., Jendreyko, H., "Removal of detergents from effluents and waters." Progress Report of Nord-Rhein-Westfalen No. 1153, (1963).
15. Husmann, W., and Malz, F., The problem of detergency—development, present status prospects for the future. Report of the Federal Ministry of Health, 1967.
16. Huber, L., Investigations of the detergent content of important waters in Bavaria. GWF 1221–1225 (1964).
17. Huber, L., Oncological aspects of detergents. Vom Wasser B68 (1987).
18. Huber, L., Detergent and Waterhardness—oncological aspects. Detergency, water hardness, environment. Bensiker Collections 3, 31, 87.
19. Huber, W., Analysis of tensides in water, soil and mud. Tenside 2:106–110 (1991).

20. Jendreyko, H., The properties of novel detergents in anaerobic cleaning of effluents. Contributions from Munich 9:233–241 (1962).
21. International Commission for the protection of the Rhein. Tables of physico-chemical investigations, 1986.
22. Kaufmann, H.P., and Malz, F., Adsorptive precipitation and foam fractionation of detergents from aqueous solutions. Fette, Seifen und Ansrtichmittel 1024–1030 (1960).
23. Kordik-Kolb, E., LAS and complexing agent in Bavarian rivers. Contribution from Munich, 44:493–507 (1990).
24. LWA-NW, Report of Rhein Quality, NRW 1988.
25. Malz, F., The state of "new" detergents in aerobic effluent cleaning. Contribution from Munich 9:266–275 (1962).
26. Malz, F., Rebel, M., and te Hessen, D., The border-development of the biological water quality in the last 20 years. Audience City Hygiene 4:212–216 (1982).
27. Malz, F., Tenside-water-environment. Seifen-Fette-Ole-wax 18:591–594 (1985).
28. Malz, F., Chemical, physical and biological analysis of effluent investigations. Biological analytical methods Part 4/B. Effluent Technic Issue 3:14–31 (1987).
29. Malz, F., Tensides and protection of water. Tenside 6:354–360 (1987).
30. Malz, F., The principal commission for detergents. Structure, goals and projection of aims. Tenside 2:72–77 (1988).
31. Malz, F., The function of the wash and cleaning agent in the water charge. Contributions from Munich. 44:433–457 (1990).
32. Malz, F., New results from the activities. Tenside and water protection. Tenside 6:482–486 (1991).
33. Noll, L., Application of tensides in detergency and cleaning products. Tenside 2:90–92 (1991).
34. Iawr and Riwa, Development of the Rhein water quality. Annual Report 1991, Part A, p. 19.
35. Schöberl, P., Coupling of the OECD—confirmatory tests with continuous oncotoxicity tests. Tenside 2:97–105 (1991).
36. Schöberl, P., Further development oncological test systems. Tenside 2:93–96 (1991).
37. Statistical Federal Office for Environmental Techniques, File 2.1. Public water supply and effluent cleaning 1987. Stat BA Wiesbaden, 1990.
38. Waters, I., et al., New component for softener rinser with improved ecological compatibility. Tenside 6:460–468 (1991).
39. Wickboldt, R., Intermediate products in the biological degradation of a straight-chain alkylbenzene sulfonate. Presented at the International Congress for Surface-Active Substances, Brussels, 1964.

2

Physicochemical Interactions of Surfactants and Contaminants in Soil

ERWIN KLUMPP and MILAN JOHANN SCHWUGER Institute of
Applied Physical Chemistry, Research Center Jülich GmbH, Jülich, Germany

I. INTRODUCTION

Surfactants and contaminants pass into soil through products and waste flow. The input pathways of contaminants are largely known [1] and are not discussed here. Possible surfactant sources are as follows:

Agricultural application of pesticides and fertilizers
Agricultural sewage sludge
Sewage sprinkling on irrigation fields and irrigation with river water containing surfactants

Households not connected to sewage treatment plants
Defective sewers
Sewage plant overflows caused by heavy rain events
Soil remediation
Surfactant-loaded layer silicates as barrier layer materials in landfill engineering
    and as carriers for pesticides
Tertiary oil recovery

Although these input pathways have long been known, studies on the behavior of surfactants in soils, for example their mobility, degradability, and damaging effects, are available to only a limited extent. This might be attributed to the fact that the quantities passed into the soil are regarded as small and ecotoxicologically harmless and that the apparently satisfactory degradation rates in test systems and in the aquatic ecosystem are directly extrapolated or transferred to the conditions in soil. Several facts indicate, however, that the behavior of surfactants in soil demonstrates numerous ambiguities.

It is known, for example, that the degradation of surfactants in soil is retarded and that it is difficult to distinguish in test results between biodegradation and other elimination mechanisms, such as adsorption or precipitation [2]. On the other hand, we accept that unique statements about the influence of surfactants are hardly possible because of the complexity of the processes involved and the diversity of the soil.

Special groups are the so-called major controlled surfactant inputs, as in soil remediation and the application of organoclays [3,4]. The effect and fate of these substances are studied more intensively and elucidated better because of more comprehensive investigations and the larger amounts of surfactants used.

The interactions of surfactants and soil are varied. These so-called surface-active substances influence the soil-water balance, the surface properties of soil constituents, and thus the soil structure. For example, they change the activity and population density of microorganisms and the growth of plants. To describe these surfactant influences successfully, parameters of soil physics, soil chemistry, and soil biology are required.

The soil-water balance constitutes a central control variable of soil processes because it is closely coupled with transport processes, sorption processes, and biotic activity. Surfactants reduce the surface tension of soil water as a result of their surface-active properties, which occurs even at very low concentrations with nonionic surfactants. One of the consequences is modified water distribution in the different pore fractions of the soil [2].

To describe the soil structure, such parameters as pore size distribution and aggregate stability are crucial. These change as a result of a reduction in the surface tension of the soil water and surfactant adsorption on soil particles. A possible consequence is changed dispersion of the primary particles. This, again, has an effect on pore size distribution, and the particles present in the dispersed form (soil colloids) can be washed into deeper soil layers by percolating water [2].

In summary, it may be assumed that a comprehensive description and evaluation of the behavior and potential effects of surfactants in soils is not readily possible on the basis of current findings.

This overview is mainly restricted to the adsorption of various classes of surfactants on soil and sediment and their most relevant inorganic soil constituents, in particular surfactant (adsorbed surfactant)-contaminant interactions. It does not claim to be complete but may initiate future research approaches.

II. ADSORPTION OF SURFACTANTS IN SOIL AND ON INORGANIC SOIL COMPONENTS

Understanding of surfactant sorption onto soil and its components is needed to assess surfactant-facilitated transport of organic compounds in soil-aqueous systems.

Another reason for investigating surfactant adsorption onto soil is to understand the transport in soil of the surfactants themselves, as well as their microbial degradation products. The latter may be relatively toxic, as is the case for alkylphenolethoxylates [5].

In soil washing, surfactant adsorption is a side effect and takes place at elevated concentrations. In this case, surfactant adsorption is a threat to the success of surfactant-enhanced remediation because it reduces the active concentration of surfactant, and the adsorbed surfactant on the soil and/or sediment also poses a problem [3].

These three examples demonstrate the heterogeneity of approaches in the literature to the topic of surfactant adsorption in soil. Depending on the problems to be investigated, the concentrations studied range from nanomoles to millimoles. Moreover, a uniform systematic treatment of the results is also impaired by the fact that commercial surfactants are frequently used, which are not clean enough for physicochemical investigations.

Surfactant and organic contaminant adsorption is frequently described by the same partition processes, although the physicochemical behavior of surfactants differs strongly from that of NOCs (nonionic organic contaminants). This also indicates that an improved fundamental understanding of the sorption of surfactants on environmental surfaces is necessary. In most cases, however, this is possible only by investigating elevated concentrations on well-defined soil constituents. For this reason, systems are also presented here that, to a first approximation, do not appear to be environmentally relevant, but they may serve as a basis for understanding contaminant-surfactant sorption interactions.

A. Anionic Surfactants

Linear alkylbenzene sulfonate (LAS) has been the major anionic surfactant material used in detergent formulation for nearly 25 years. Consequently, its fate in the environment was investigated in numerous studies. Commercial LAS consists of a complex mixture of homologs and isomers. The mixture most frequently used in detergent formulations has an average chain length of 12–13 and a phenyl position between 2 and 6. This makes it difficult to evaluate and interpret the published results.

The sorption of LAS on river sediments has been investigated relatively intensively [6–9]. Matthijs and De Henau reported a correlation between the sorption of LAS and the organic carbon content of sediments in the ppm range, suggesting a hydrophobic sorption mechanism. Their desorption study under environmental conditions indicates that the sediment is a sink for LAS. If the chain length of LAS is increased, that is, the hydrophobicity of the surfactant, the sorption of LAS increases consistently. A correlation between octanol/water partitioning and sediment adsorption has been found, suggesting a hydrophobic sorption mechanism [7]. Hand et al. have found that at environmentally relevant concentrations (ppb range) there is no clear correlation between sorption and organic carbon content and desorption is nearly reversible [8]. These contradictory results demonstrate that the role of organic carbon associated with sediments is still unclear.

Marchesi et al. compared the sorption of linear alkyl sulfates and of linear alkylbenzene sulfonates on two sediments. They also varied the chain length and calculated the free adsorption enthalpy for a methylene group in the alkyl chain (G_C^o). Compared to the Henry's law adsorption constants, the results are interpreted in terms of a hydrophobic bonding mechanism [9].

Little is known about adsorption of LAS in soil. It is assumed that this so-called soft detergent is easily degraded, which limits its accumulation in soil and ground water compartments [10–12]. However, interactions between anionic surfactants and soil are also relevant at high concentrations because these are significant in soil washing. On natural solids, however, it is difficult to draw general conclusions from sorption experiments. To understand these processes it is necessary to discuss the sorption processes of surfactants with the relevant components of soil and sediment. We may refer here to the literature on flotation [13], detergency [14], and enhanced oil recovery [15]. In these application fields, the adsorption of ionic surfactants on charged solids is the subject of considerable research.

Fuerstenau reported on the adsorption of sodium dodecyl sulfonate on alumina [16]. Figure 1 shows the adsorption isotherms and the corresponding zeta potential values as a function of the equilibrium concentration of the surfactant at pH = 7.2. The course of the isotherms is typical of the adsorption of ionic

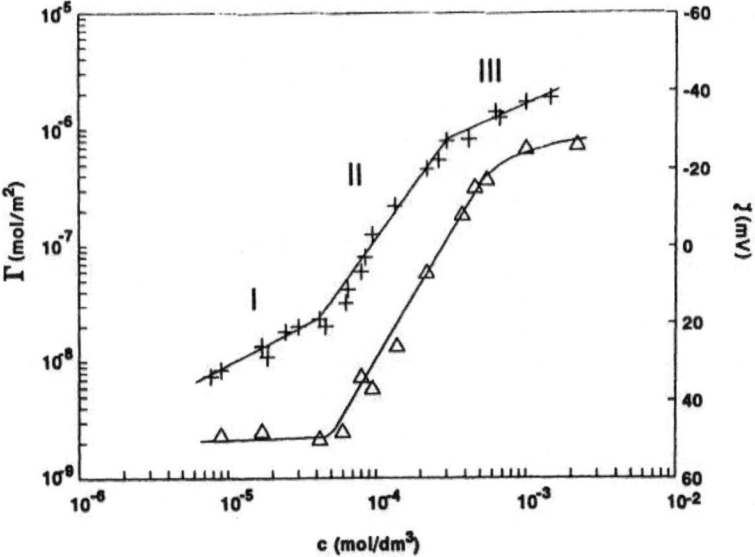

FIG. 1 Adsorption isotherm (+) for sodium dodecyl sulfonate on alumina and corresponding zeta potentials (Δ) of alumina particles as a function of the equilibrium concentration at pH 7.2 and 2×10^{-3} mol/dm^{-3} ionic strength. (From Ref. 16.)

surfactants on oppositely charged surfaces. At low concentrations the curve has a slope of approximately 1, and the zeta potential also changes proportionally (stage I). It decreases continuously: that is, the surface is less and less positively charged. It can be assumed that the adsorption in this region is electrostatically controlled. With increasing surfactant concentration (stage II), the slope of the isotherms changes distinctly, which shows that the rising degree of coverage increases the affinity of the surfactant to the surface even if the isoelectric point is exceeded, although the surface and the surfactant have the same polarity. The adsorption is mainly determined by interactions between the alkyl chains of the surfactants. Surfactant aggregates can form on the surface due to the analogy of micelle formation, so-called hemimicelles. The minimum concentration required for hemimicelle formation is the "hemimicelle concentration" (HMC). The HMC is strongly dependent on the alkyl chain length and the number of aromatic groups of the surfactant molecule, as well as on the manner in which surfactant molecules may be arranged at the surface. In stage III further adsorption takes place by supplementing the adsorbate layer to bilayers. Another sorption mechanism of anionic surfactants on inorganic soil constituents is precipitation [17]. This is discussed in Sec. IV.A using the example of SDS (sodium dodecyl sulfate)/Me^{n+}-clay mineral systems.

B. Cationic Surfactants

Cationic surfactants are used in a wide range of consumer and industrial products, such as detergents and fabric softeners. They are toxic, and their elimination from surface waters should therefore be observed particularly intensively. Cationic surfactants adsorb strongly onto soil and sediments because of favorable electrostatic interactions with the predominantly negatively charged surfaces of natural materials. Despite the importance of these compounds, there has not been any extensive systematic investigation of their adsorption onto natural materials, such as soils and sediments [18].

Brownawell et al. [19] studied the factors that control the sorption of dodecylpyridinium bromide (DP) on environmental and pristine surfaces. The adsorption isotherms are nonlinear, even at very low surface coverages of DP. This behavior can be described by assuming a heterogeneous mixture of adsorption sites. The sorption depends primarily on the CEC (cation exchange capacity) of the sorbent. The effects of electrolyte type and concentration on adsorption also indicate that the primary adsorption mechanism is cation exchange. A judgment about a correlation with organic carbon is hardly warranted. Similar conclusions were also drawn by other authors [20]. The adsorption of cationic surfactants on other sorbents, which are important constituents or potential analogs of environmental matrices, has been studied in more detail [21–23]. Of particular importance is cationic surfactant adsorption on layer silicates because cationic surfactant/clay mineral complexes are used for landfill sealing.

The cationic surfactant isotherm on clay mineral is of the high-affinity type [24–26]. The slope of the isotherm is very steep because of the electrostatic attraction between cationic surfactants and negatively charged clay minerals. Beyond the CEC, cationic surfactant is physically adsorbed by hydrophobic interactions between the alkyl chains of the surfactant molecules. A surfactant bilayer is formed on the surface of the clay mineral, and the isotherm exhibits a plateau. For the so-called organoclays, the quantitative exchange with the cationic surfactant corresponds to the CEC. If this takes place on a swelling layer silicate, such as montmorillonite, cationic surfactants are intercalated in the interlayers and the layer interspaces thus become organophilic. Consequently, the adsorption properties of the layer silicates change fundamentally. The basal spacings and thus the thickness of the surfactant layer formed can be determined by x-ray diffraction. Figure 2 demonstrates the change in the basal spacings d_L for alkylammonium smectites as a function of the alkyl chain length n_C and schematically shows the structures of intercalated surfactants calculated from basal spacings. If the length of the alkyl chain of the adsorbed surfactant is increased from C_{10} to C_{18}, the basal spacing changes from 1.7–1.8 to 2.1–2.2 nm [27]. The adsorption isotherms of the cationic surfactant HTAC (hexadecyl trimethylammonium chloride) on the oxide minerals SiO_2 and Al_2O_3 are shown in Fig. 3 [28].

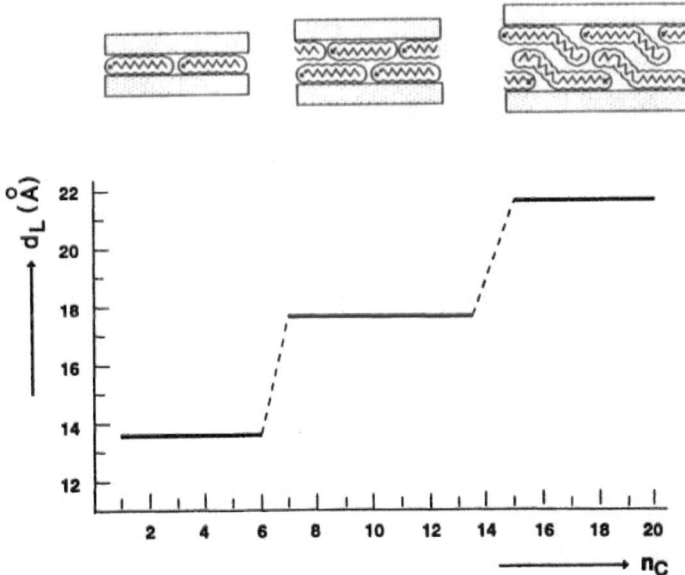

FIG. 2 Change in the basal spacings d_L for alkylammonium smectites as a function of the alkyl chain length n_C. (From Ref. 27.)

The shape of the isotherms can be partitioned into three segments, as is frequently the case for the adsorption of ionic surfactants on oppositely charged surfaces. The sorption mechanisms in the three regions then correspond to those shown in Sec. II.A for the anionic surfactant SDS on Al_2O_3.

C. Nonionic Surfactants

Recently, the entire detergent industry has been undergoing substantial change. The fraction of nonionic surfactants in total surfactant production has increased. Also, this surfactant class has been proposed for soil washing and enhanced subsurface remediation. However, few investigations have addressed the sorption of nonionic surfactants onto subsurface media [5,6,29–33].

Palmer et al. conducted a study to analyze the movement of nonionic surfactant in the subsurface, namely on two aquifer materials [32]. Their hypothesis was that the sorption of nonionic surfactants is similar to that of neutral organic chemicals and thus a function of their hydrophobicity. As expected, sorption decreased with decreasing hydrophobicity within the series of nonylphenolpoly-oxyethoxylates (as a result of increasing oxyethylene groups). However, the adsorbed amounts did not agree with predictions based on hydrophobic partitioning because of that strong sorption on the mineral surfaces.

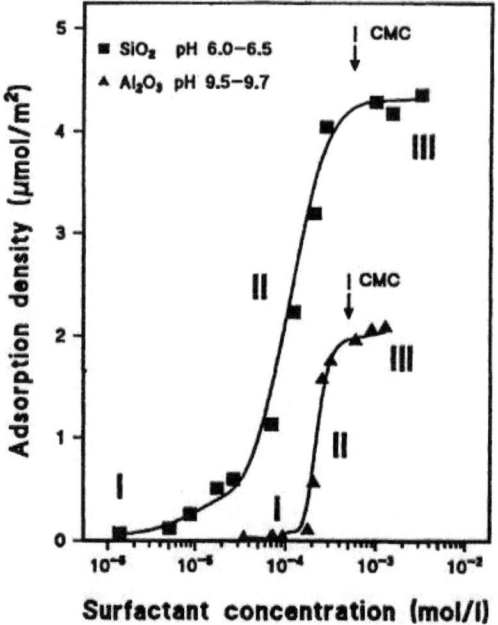

FIG. 3 Adsorption isotherms of hexadecyl trimethylammonium chloride (HTAC) on SiO$_2$ and Al$_2$O$_3$. (From Ref. 28.)

Liu et al. [31] evaluated the sorption of four nonionic surfactants with a soil containing an organic carbon content of 0.96%. Sorption was observed to follow the Freundlich isotherm at concentrations below the CMC. The isotherms showed a plateau above the CMC. However, little was reported on the sorption mechanism, as is frequently the case in studies with natural soil samples.

While the adsorption of cationic surfactants on montmorillonites has been frequently discussed, nonionic surfactant adsorption on layer silicates is a relatively new field [34,35]. Recently, Rheinländer compared the adsorption of dodecyl octaethylene glycol ether (C$_{12}$E$_8$) on four different clay minerals on the basis of physicochemical investigations [26]. Figure 4 shows the adsorption isotherms in log-log plotting. The isotherms are of the L$_2$ type (Langmuir type) and thus mostly based on physisorption. The adsorption of C$_n$E$_m$-type surfactants on clays probably takes place predominantly via the hydrated ethylene glycol groups due to ion-dipole interactions and hydrogen bridge bonding. The extent of adsorption strongly depends on the surface area of the respective layer silicate. Thus, the isotherm on Na$^+$ kaolinite is below that on Na$^+$ illite, and both isotherms are clearly below those on the montmorillonite and bentonite, respectively. The adsorption plateau of calcium bentonite (the bentonite consisted of 95% mont-

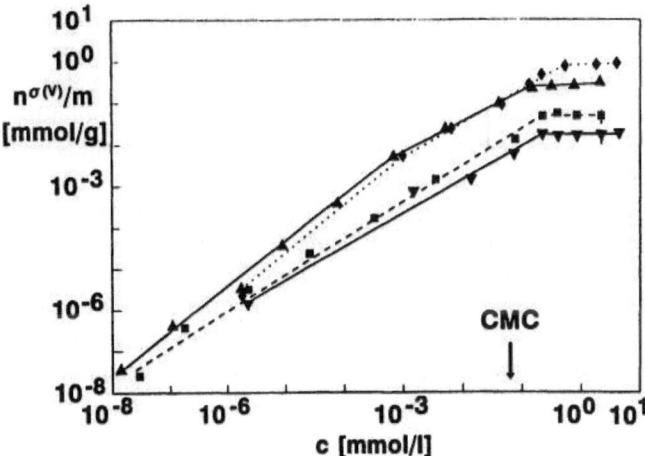

FIG. 4 Adsorption isotherms of the $C_{12}E_8$ nonionic surfactant on different layer silicates: (upside-down triangles) Na^+ kaolinite, (squares) Na^+ illite, (diamonds) Na^+ montmorillonite, (triangles) Ca^{2+} bentonite. (From Ref. 26.)

morillonite) is three times lower than that of sodium montmorillonite. This is attributed to the lower dispersion of calcium bentonite in water [26]. However, for both montmorillonites it is confirmed by x-ray measurements that the nonionic surfactant penetrates into the layer interspace, displacing water and forming bilayers. On the other hand, microcalorimetric measurements reveal that the differential molar exchange enthalpy $\Delta_{21}h$ (heat of adsorption) is clearly higher on calcium layer silicates than on the sodium forms. It is assumed that, analogously to cyclic crown ethers, the ethylene glycol chain of the surfactants forms stable complexes with alkaline-earth ions like Ca^{2+}.

III. PHYSICOCHEMICAL INTERACTIONS OF CONTAMINANTS AND SURFACTANTS IN WATER

It is well known that the aqueous solubilities of hydrophobic compounds can be modified by several factors, such as the presence of organic solvents, macromolecules, and surfactants. Thus, such cosolvents as alcohols and ketones increase the solubility of hydrophobics in aqueous solutions [36,37]. Recent studies with some extremely water-insoluble organic solutes have shown that their apparent water solubility can be significantly enhanced by low concentrations of some fractionated humic and fulvic acids. This enhancement effect was attributed to a partition-like interaction of solutes with the microscopic organic environment of dissolved organic macromolecules [38]. The study of the effects of surfactants upon solubilities of hydrophobics of environmental significance presented similar results. Surfactants can produce significant solubility enhancement even at concentrations below the CMC, which is attributed to a partition-like interaction of the otherwise extremely water-insoluble organic compounds with the nonpolar content of the surfactant [39,40].

At surfactant concentrations above the CMC, the organic interior of micelles acts as an organic pseudophase into which organic contaminants can be partitioned. This phenomenon is called solubilization [41]. The extent to which a solute will concentrate in a micelle can be related to the octanol-water partition coefficient K_{OW} of the solute [42,43]. In general, the larger the K_{OW} of a solute, the greater is its tendency to concentrate inside the micelle. The enhanced solubility of contaminants has been described using a two-phase separation model for solute behavior [39].

IV. PHYSICOCHEMICAL INTERACTIONS OF CONTAMINANTS AND SURFACTANTS IN SOIL

A. Ionic Contaminants

Little is known so far about the interactions of ionic contaminant, surfactant, and soil. As in the studies of surfactant adsorption, it is also assumed here that, for the most part, some organic and inorganic components of the soil determine these interactions. For example, clay minerals are among the most important soil components and are therefore used as model substances for the mineral soil horizon. Very little has been published on the interactions of surfactants with heavy metal ions on clay minerals. These studies show that cationic surfactants compete with heavy metals for adsorption sites on the mineral surface [44,45], whereas anionic surfactants can precipitate with metal ions on the clay mineral surface [46–48]. All publications only describe these effects qualitatively without addressing the mechanisms in more detail.

It can be seen from Sec. II that, from among the different surfactant classes, only the cationic surfactants adsorb on soil and sediments in a quantitative and almost irreversible manner. During their adsorption, the exchangable cations are mobilized according to a stoichiometric ion exchange. In this process, the structure of the hydrophobic portion of the surfactant molecule plays only a minor role [49]. Although alkali ions and alkaline-earth ions are almost completely mobilized, heavy metals can be only partially displaced in contaminated soils requiring remediation.

Figure 5 shows the mobilization of Cd^{2+} and Pb^{2+} by didodecyl dimethylammonium bromide (DDDMA) on the A_p and B_t horizons of an orthic luvisol. For

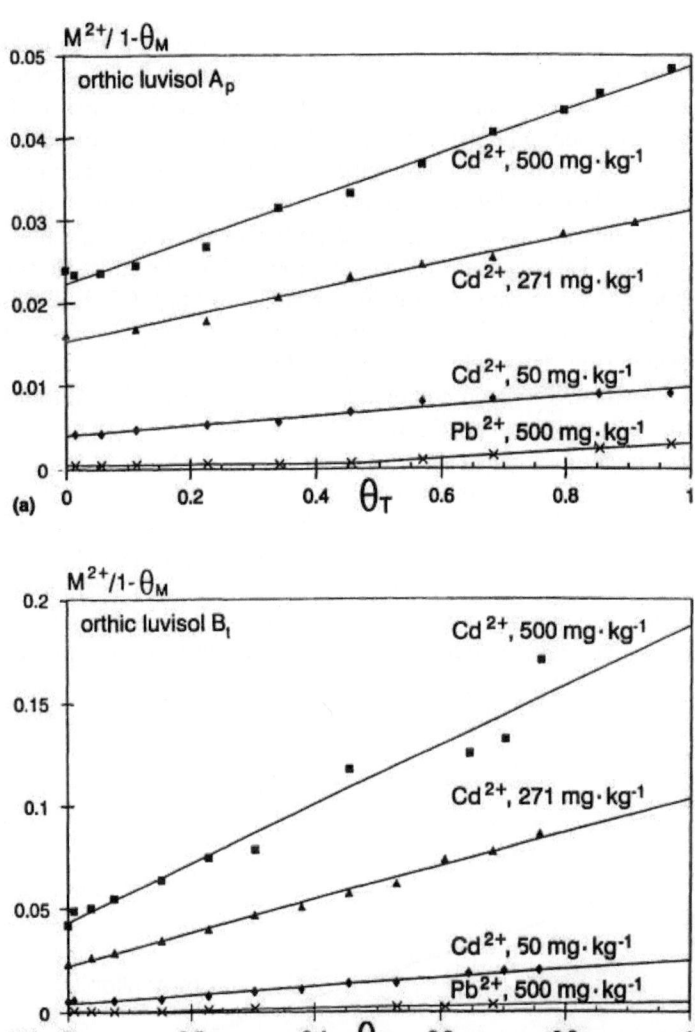

FIG. 5 Mobilization of Cd^{2+} and Pb^{2+} by didodecyl dimethylammonium bromide (DDDMA) on A_p (a) and B_t horizons (b) of an orthic luvisol (θ_T = bound DDDMA$^+$, related to CEC; $1 - \theta_M$ = mobilized heavy metal). (From Ref. 50.)

a direct comparison of the mobilization of the two cations, both soil horizons were contaminated with the same equimolar amount of heavy metal ion. A clearly stronger metal ion mobilization from the subsoil can be observed. This is attributed to the greater clay mineral fraction in comparison to the humic fraction in this soil horizon. In contrast to the mobilization of essential ions from orthic luvisol, mobilization is incomplete for both heavy metal ions. This is largely attributed to energetic reasons in the case of Cd^{2+}. The cause for the hardly mobilizable Pb is assumed to be the bonding of Pb as hydroxide on the surface [50]. It was also found during these investigations that mobilization experiments on homoionic layer silicates offer themselves to a basic understanding of these mechanisms. Precipitation is assumed for the adsorption of anionic surfactants on a clay mineral surface. This mechanism was also proposed for the interactions of dodecyl sulfate ions (DS^-) with Ca^{2+} kaolinite and Ca^{2+} illite [46,51]. If other (e.g., heavy metal) ions are adsorbed on the mineral surface, these processes are influenced depending on the solubility products. Figure 6 shows the specific surface excess $n_{DS}^{\sigma(v)}$ for DS ions on homoionic montmorillonites. Two different mechanisms can be recognized here for the course of $n_{DS}^{\sigma(v)}$.

Ca^{2+} and Pb^{2+} montmorillonites show an S-shaped curve. Surface precipitation takes place after exceeding the respective solubility product at a. X-ray diffraction measurements have shown that precipitation takes place only on the outer montmorillonite surfaces, which means that there is no intercalation in the interlayers. Space requirement calculations with the cross-sectional area of 28 A^2 for the sulfate group indicate that the outer surface is completely covered with DS^- in point b. The steeper slope in the a-b region for Pb^{2+} montmorillonite

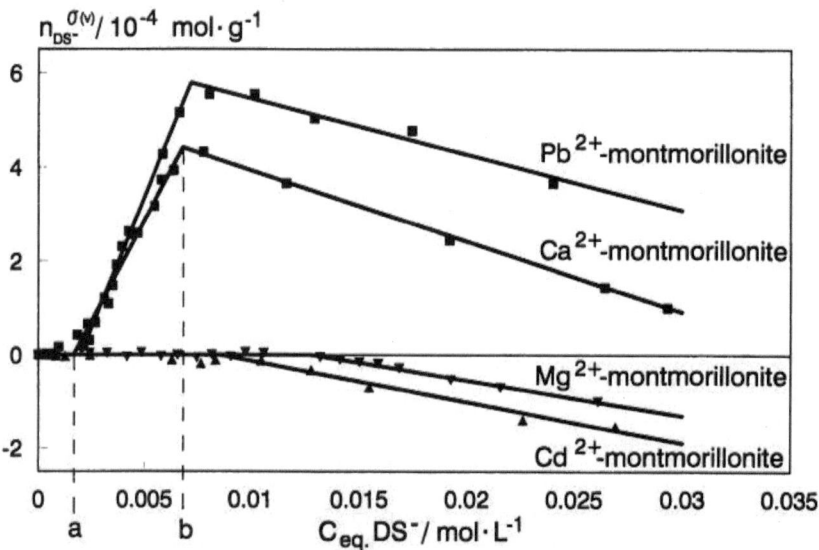

FIG. 6 Adsorption isotherms of dodecyl sulfate ions on homoionic montmorillonites. (From Ref. 50.)

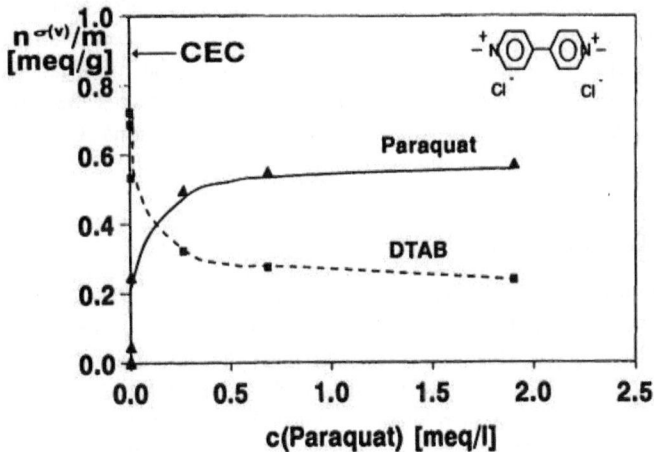

FIG. 7 Adsorption of paraquat on Ca^{2+} bentonite pretreated with cationic surfactant: (triangles) adsorption isotherm of paraquat; (squares) displacement of dodecyl trimethylammonium bromide (DTAB). (DTAB pretreated 0.73 meq/g.) (From Ref. 52.)

in comparison with Ca^{2+} montmorillonite is attributable to the smaller solubility product of $Pb(DS)_2$ on montmorillonite. The equilibrium concentration in point b corresponds to the CMC of the respective surfactant salt. Solubilization of $Ca(DS)_2$ and $Pb(DS)_2$ is assumed for the decline in $n_{DS}^{\sigma(v)}$ after exceeding the CMC [50].

The interaction of DS ions with Mg^{2+} and Cd^{2+} montmorillonite proceeds according to another mechanism. Initially, no reaction of DS^- with the montmorillonite surface takes place under the present experimental conditions, since the solubility products are not exceeded, $n_{DS}^{\sigma(v)} = 0$. A negative course of $n_{DS}^{\sigma(v)}$ occurs after the formation of the micellar phase, which can be explained by the preferential adsorption of water on the montmorillonite surfaces [50].

In the same way that surfactants can determine the mobility behavior of environmental chemicals in soil, surfactant transport can also be influenced by other ionic organic compounds. For example, surfactants adsorbed on layer silicates can be exchanged by ionic pesticides, as illustrated in Fig. 7 for a clay mineral precovered with cationic surfactant [52]. The preferentially adsorbed organic dication paraquat displaces the cationic surfactant DTAB (dodecyl trimethylammonium bromide) only partially (for better comparison, the amounts of substance are specified in equivalent amounts). For each equilibrium concentration, the associated adsorbed amounts of both components approximately correspond to the CEC. The competition of paraquat with DTAB is thus indicative of the presence of an exchange equilibrium.

Because nonionic surfactants, in contrast to the cationic surfactants, are mostly physisorbed on clay mineral surfaces, they can be displaced relatively easily by monocationic organic substances. If, for example, the monocationic pesticide cyperquat is present in the soil electrolyte, it also displaces the physisorbed nonionic surfactant $C_{12}E_8$ in addition to the essential cations (Ca^{2+} and Na^+) [26,53].

B. Nonionic Contaminants

The sorption of nonionic organic contaminants (NOCs) from water on the mineral soil components is slight because they preferentially adsorb water. The uptake of NOCs by soils is strongly correlated with the soil organic matter content and is a partition process. Correspondingly, these sorption processes can be described by linear isotherms at relatively high NOC concentrations, too. If the water solubility of an NOC is relatively higher, higher mobility is observed in soils. The bonding of surfactants on soil has the same consequences as the immobilization of humic matter. The hydrophobization of the surface fundamentally changes the adsorption properties of the soil components, so that the adsorption of NOCs is favored and these are immobilized and not available for further transport and degradation processes. In addition to the application of surfactants for soil protection, the accumulation on soils and sediments of surfactants passed into the environment in different ways can also lead to this situation.

Recently, Lee et al. used cationic surfactants to enhance the sorption capacity of soil with respect to NOCs. The sorption characteristics of soils have been modified by the addition of HDTMA (hexadecyl trimethylammonium) cations in an amount equivalent to the CEC of the soil [54,55]. These studies have shown that the sorption of NOCs by HDTMA-exchanged soils was greatly enhanced. Figure 8 represents the sorption uptake of aromatic hydrocarbons by untreated and HDTMA-treated soils. To quantify the dramatic improvement in sorptive capabilities by HDTMA soil, the sorption coefficients (equal to the slope of the linear isotherms) were calculated. The adsorption enhancement is 100-fold and independent of the clay content of the respective soil. Comparing the organic matter normalized sorption coefficients K_{OM}, it was found that the adsorbed surfactant is 10- to 30-fold more effective than natural soil organic matter and the log K_{OM} values of the HDTMA-treated soils agree closely with the corresponding log K_{OW} values.

Because of the complexity of the soil, it is not readily possible to speculate about the mechanism of sorption. The findings acquired in the past few decades with organoclays (or quite recent studies on oxides), however, may provide further details on the mechanisms typical of these processes. As already discussed, the originally hydrophilic layer silicate surface becomes organophilic by

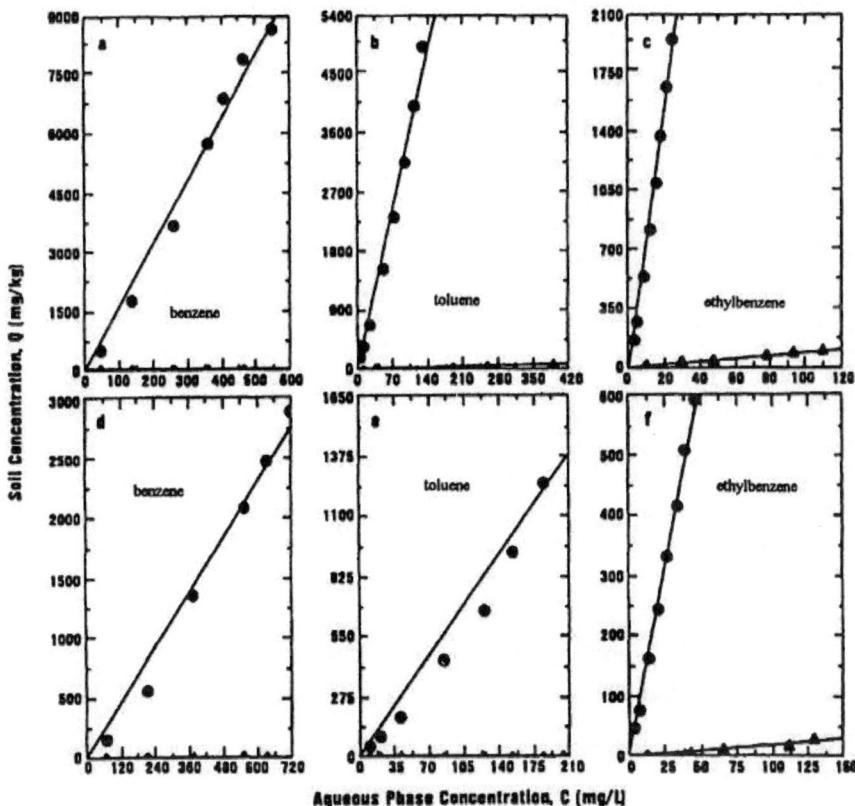

FIG. 8 Sorption of NOCs on untreated (triangles) and hexadecyl trimethylammonium-treated (circles) St. Clair (a–c) and Oshtemo (d–f) B_t horizon soils. (From Ref. 54.)

the exchange of hydrated cations, such as Ca^{2+} or Na^+, with cationic surfactants. This leads to an increased adsorption of NOCs. The extent of enhanced adsorption depends on the degree of hydophobization of the layer silicate surface. It therefore increases with growing surfactant load, with increasing surfactant chain length, and with the number of alkyl chains per surfactant molecule [56–60].

Boyd et al. have shown that the NOC adsorption on HDTMA smectite complexes depends on the exchanged amount of surfactant (Fig. 9) [56]. Whereas the HDTMA smectite complexes are equally effective for an exchange of 70 and 100% of the CEC (13 and 17.3% OC, organic carbon), the "35% CEC" HDTMA smectite (7.1% OC) exhibits a comparatively smaller enhancement for benzene adsorption. The latter is represented as a dual adsorbent, which in addition to the organic phase also exhibits noncovered hydrophilic surface areas.

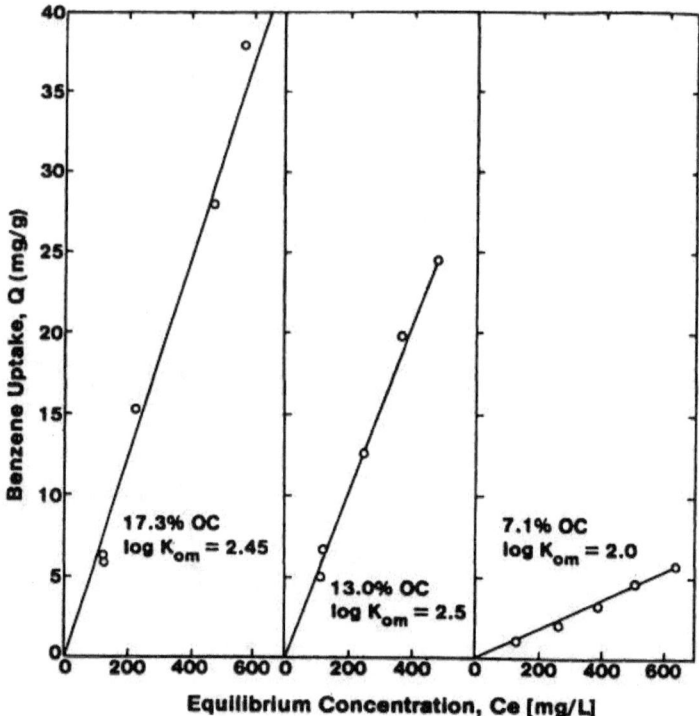

FIG. 9 Sorption of benzene from water by hexadecyl trimethylammonium smectites. (From Ref. 57.)

Because for high coverages the log K_{OM} values (2.45–2.5) coincide with the corresponding partition coefficient in heptane-water (log K_{HW} = 2.3), it is assumed that the conglomeration of the alkyl chains of the HDTMA ions forms a partition medium for the benzene molecules to be adsorbed.

Heitmann carried out a relevant systematic study with nitrophenol on alkyl and dialkyl ammonium/bentonite complexes (the bentonite consisted of 95% montmorillonite) [61]. The equilibrium studies reveal that the enhanced adsorption of relatively readily water-soluble 4-nitrophenol only takes place from an exchanged amount of 32% of the CEC (32% CEC) upward in the case of alkyltrimethylammonium (C_n) derivatives. If the hydrophilic layer silicate surface is covered by a dialkyldimethylammonium ($2C_n$) derivative, this effect "consequently" already occurs at 16% CEC.

The kinetic studies carried out with these systems have also confirmed this finding. Figure 10a shows the normalized adsorption kinetics for 4-nitrophenol on noncovered bentonite in comparison to 32 and 100% CEC C_{12} bentonite. On

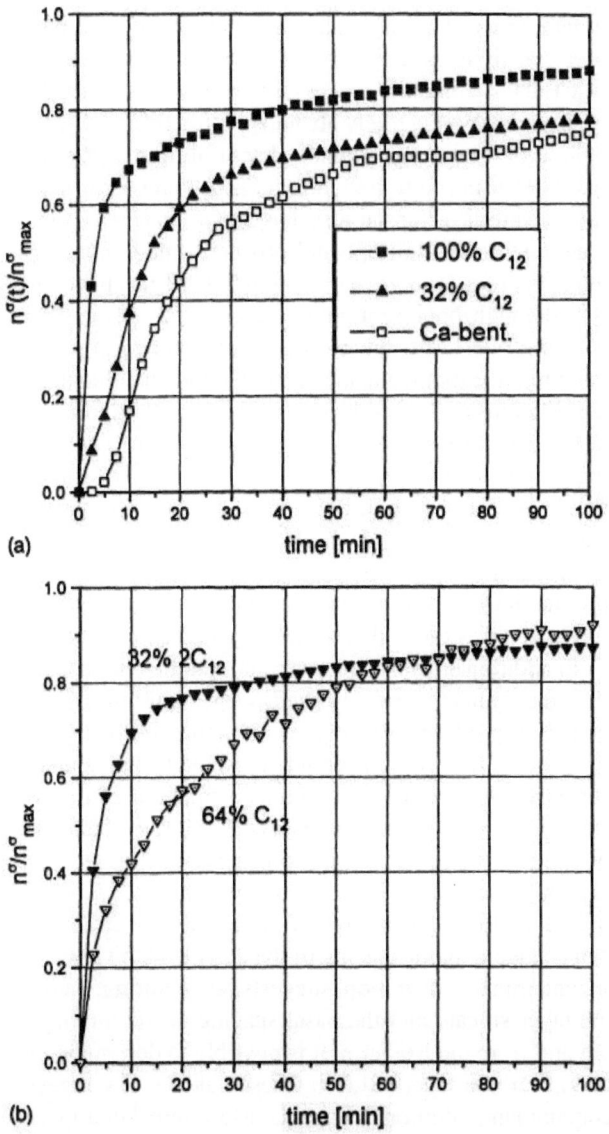

FIG. 10 (a) Normalized sorption kinetics for 4-nitrophenol on native and dodecyl trimethylammonium bentonite with increasing surface coverage (related to the CEC). (b) Normalized sorption kinetics for 4-nitrophenol on dodecyl trimethylammonium (C_{12}) and didodecyl dimethylammonium ($2C_{12}$) bentonite covered by the same amount of alkyl chain. (From Ref. 61.)

native layer silicate the kinetics curve has an S shape and the rate of adsorption is initially very slow. After 100 minutes only 70% of the equilibrium value is reached. Whereas on 32% CEC C_{12} bentonite a clear acceleration is observed in the initial phase and the S-shaped curve is largely abandoned, for 100% CEC C_{12} bentonite the very rapid phase of adsorption extends almost up to the plateau.

It can be seen from Fig. 10b that in addition to the exchanged amount of alkyl chains their orientation in the interlayer space also has an influence. On 32% CEC $2C_{12}$ bentonite, the adsorption of 4-nitrophenol proceeds much more rapidly (half-life = 2.94 minutes) than for the corresponding 64% CEC C_{12} bentonite (half-life = 8.57 minutes), although both bentonites are covered by the same amount of alkyl chain. On the one hand, the closeness of neighboring alkyl chains seems to be important for adsorption enhancement. On the other hand, the sorption process is an intercalation in the interlayers; that is, in addition to the structure of the surfactant layer, other parameters also play a role, such as basal spacing and the intercalated amount of water.

It has already long been known that x-ray diffraction is an effective method for elucidating the structure of the intercalated cationic surfactant molecules in the interlayers of swelling clay minerals [62,63]. Heitmann has shown that such measurements can also provide important information about the mechanisms of contaminant intercalation. Thus, it was found that the adsorption of 4-nitrophenol on organobentonites leads to additional layer expansion and that this not depend only on the volume of the contaminant to be adsorbed. Depending on the surfactant, nitrophenol intercalation produces a change in the structure of the adsorbed layer [61]. Figure 11 shows from which surfactant or alkyl chain amount onward an enhanced layer expansion can be observed as a function of the adsorbed nitrophenol amount. It is clearly seen that at least 64% of the exchangable ions must be replaced by C_{12} ions in C_{12} bentonite to observe additional layer expansion as a result of 4-nitrophenol adsorption. For double-chained $2C_{12}$ only half of this amount is required, that is, 32% of the CEC, to identify an effect that is even greater. The sudden change in basal spacing of the $2C_{12}$ bentonite, which is overproportional to the contaminant adsorption, suggests restructuring. Knowing the surface area of the layer silicate and the basal spacing or the interlayer space calculated therefrom and from the isotherm, it is possible to determine the precise composition of the adsorbate layer [61,64]. These calculations suggest that, in addition to the contaminant, additional water is also intercalated in the interlayers, which, however, is displaced again in the case of larger amounts of contaminant adsorbed. If the surfactant/nitrophenol/water mixtures are prepared in a simple test tube according to the calculated compositions of the adsorbate layer, it becomes apparent that a multiphase system is involved here depending on the concentration conditions [61]. It must therefore be accepted that a unique interpretation of the sorption mechanisms is hardly possible despite different investigation methods (kinetic, equilibrium, and x-ray studies).

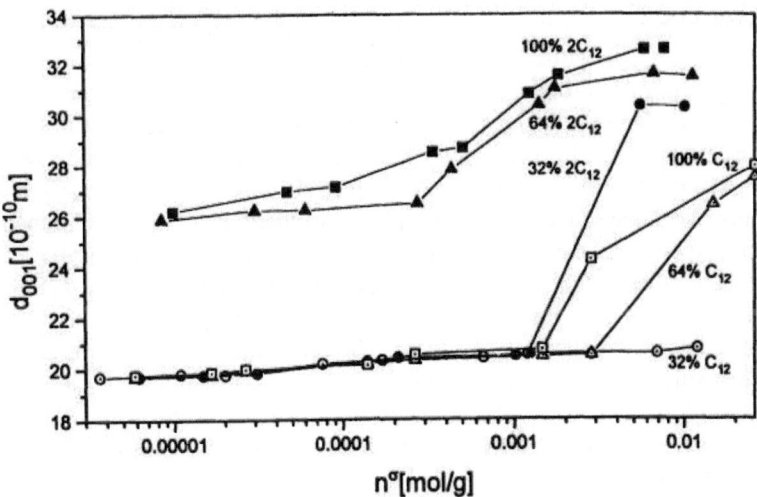

FIG. 11 Basal spacings of C_{12} and $2C_{12}$ bentonite complexes with different amount of exchanged cationic surfactant as a function of amount of adsorbed 4-nitrophenol. (From Ref. 61.)

A similar conclusion can be drawn from adsorption studies with different types of layer silicate (all 100% CEC covered). To relate mineralogy to sorbent efficiency, organoclays were prepared from vermiculite, illite, smectite, and kaolinite clay minerals using HDTMA [65]. For nonsubstituted aromatic compounds like benzene, naphthalene, and biphenyl, the degree of uptake depends primarily on the amount of exchanged HDTMA, which is a function of the CEC of the clay. The log K_{OM} values show no significant difference.

Surprisingly, illite and kaolinite proved to be as effective as the swelling clay minerals with correspondingly similar charge densities. However, for the alkyl benzenes like toluene and ethyl benzene, the higher charge density vermiculite and smectite clays were more effective sorbents than the lower charge smectites, and the magnitude of difference increased with the alkyl chain length. This can be attributed to steric reasons. The closer packing of exchanged HDTMA in these high-charge clays resulted in greater basal spacings, which appeared to enhance sorption. Consequently, the description of these sorption processes by K_{OM} values also functions to only a limited extent.

In addition to the adsorption of the cationic surfactant DTAB on four different layer silicates, Rheinländer also studied their hydrophobization by the nonionic surfactant $C_{12}E_8$ and the influence of the two surfactants on the sorption of biphenyl [26]. Although the adsorption mechanism of the two surfactants is completely different at small surface coverages, the plateau values of the adsorption isotherms are of the same order of magnitude and the same layer expansion can be observed for the swelling layer silicates.

The isotherm of hydrophobic biphenyl on the noncovered clay mineral (see Fig. 12, bottom curve) is linear (of the Henry type). If the surface is only very weakly hydrophobized by the nonionic surfactant $C_{12}E_8$ ($c_{surfactant} \ll$ CMC), biphenyl adsorption increases significantly (central curve). With growing hydrophobization the slope of the still linear biphenyl isotherm increases (0.35 mmol/g $C_{12}E_8$ corresponds to the plateau value of the surfactant isotherm). Table 1 compares the effect of the two surfactants on biphenyl adsorption at different layer silicates. Although the organic carbon adsorption coefficient K_{OC} (the Henry coefficient is related to the organic carbon content of the adsorbent) for DTA^+ layer silicates is hardly influenced by the type of layer silicate, $C_{12}E_8$ layer silicate complexes universely show higher but also different K_{OC} values.

The reason is found in the different surfactant layer structure, which, however, is known in some detail only for Ca^{2+} and Na^+ montmorillonite. Nonionic surfactants form a complex with Ca^{2+} (see Sec. II.C). The different adsorption behavior of the various Na^+ clay minerals covered by nonionic surfactants, however, remains to be clarified. It can be assumed that no classic monolayer formation takes place here and a formation of surfactant aggregates is possible. As can be seen from the preceding examples, knowledge of surfactant adsorption plays a crucial role in understanding the possible interactions between the adsorbates in the surfactant/nonionic contaminant/soil (soil component) system. The mechanisms of surfactant adsorption seem to be clarified best for oxides (see Sec. II).

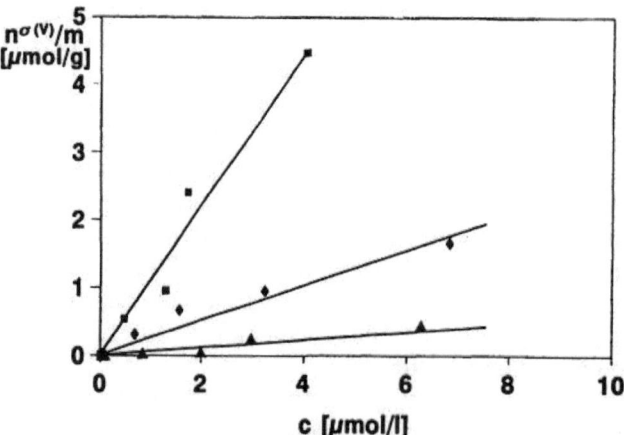

FIG. 12 Adsorption isotherms of biphenyl on sodium montmorillonite for different adsorbed $C_{12}E_8$ amounts: (triangles) native Na^+ montmorillonite, (diamonds) 0.01 mmol/g $C_{12}E_8$, (squares) 0.35 mmol/g $C_{12}E_8$. (From Ref. 26.)

TABLE 1 Adsorption Constants of Biphenyl on Different Layer Silicates Covered by Dodecyl Trimethylammonium Bromide (DTAB) and Dodecyl Octaethylene Glycol Ether $(C_{12}E_8)$[a]

| | DTAB | | $C_{12}E_8$ | |
Adsorbent	K_S (1/mmol)	K_{OC} (1/g)	K_S (1/mmol)	K_{OC} (1/g)
Na^+ kaolinite	3.3 ± 0.22	18 ± 1	2.5 ± 0.57	7.4 ± 2
Na^+ illite	3.2 ± 0.68	18 ± 4	5.9 ± 0.64	18 ± 2
Na^+ montmorillonite	2.8 ± 0.26	16 ± 1	7.3 ± 0.54	22 ± 2
Ca^{2+} bentonite	2.7 ± 0.14	15 ± 1	10.3 ± 0.26	31 ± 1

[a]K_S, adsorption constant (K is related to the adsorbed surfactant amount in moles); K_{OC}, organic carbon adsorption constant.
Source: From Ref. 26.

The adsorption behavior of mixtures of cationic surfactant HTAC and phenol on SiO_2 and α-Al_2O_3 is shown in Fig. 13 [28]. In these experiments the initial phenol concentration was kept constant. The general shapes of surfactant isotherms (see Sec. II.B) are not affected by the addition of phenol, but the isotherms are shifted to lower concentrations and the CMC is lowered.

The course of the isotherms is very similar in both systems (Fig. 13). At low surfactant concentrations, only a weak increase in surfactant adsorption was observed with increasing equilibrium concentration. The influence of HTAC on phenol adsorption is not significant, either. The phenol molecules can be adsorbed on either the mineral surface or the surfactant anchors. It must be mentioned here that from alcohol solutions a clearly enhanced alcohol adsorption was also found in this concentration range [66].

With growing surfactant concentration (region II, $c_T >$ HMC) the large increase in surfactant adsorption leads to a strongly enhanced phenol adsorption. Phenols are coadsorbed in the surfactant surface aggregates, which can be present both as monolayers and as bilayers. This process can be interpreted as adsolubilization [28].

Similar results were also obtained by Valsaraj with readily volatile hydrophobic substances, such as chloroform, benzene, and dichlorobenzene on anionic surfactant/alumina complexes. The partition constants correlated well with traditional hydrophobicity indicators, such as octanol-water partition constants and molar solubilities of these hydrophobic compounds in the aqueous phase. However, a simple estimation of adsolubilization on the basis of these constants is not proposed because of specific surfactant/(ad)solubilizate interactions in micelles and hemi- or admicelles [67].

In region III (Fig. 13), above the CMC, the surfactant adsorption density

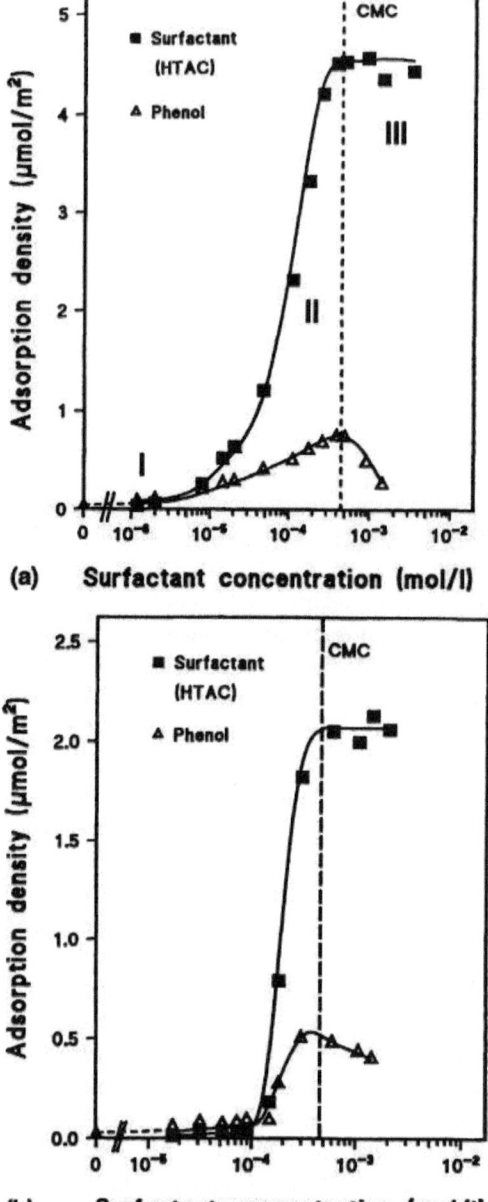

FIG. 13 (a) Mixed adsorption of hexadecyl trimethylammonium chloride (HTAC) and phenol on SiO_2 (pH = 6.3). (b) Mixed adsorption of HTAC and phenol on Al_2O_3 (pH = 9.6). (From Ref. 28.)

TABLE 2 Partition Coefficients and Free Energies of Solubilization of
Phenols in Hexadecyl Trimethylammonium Chloride (HTAC) Aggregates
on SiO_2 Surface and in the HTAC Micelles (pH = 6.0–6.7)

Compound	$K(a)$	$K(m)$	$\Delta G°$ (a) (kJ/mol)	$\Delta G°$ (m) (kJ/mol)
Phenol	10,119	3,279	−22.8	−20.1
4-Chlorophenol	70,510	21,110	−27.7	−24.7

Source: From Ref. 28.

attains a plateau and there is a decrease in phenol adsorption. The growth of the
number of surfactant micelles leads to an increased solubilization of phenol in the
micelles. This implies that the decrease in phenol adsorption may be considered
as a result of the competition between the solubilization of phenol in the adsorbed
surfactant layer and in the micelles [28,59,60].

To obtain insight into the solubilization properties of organic pollutants in the
surfactant aggregates on the mineral surface compared with those in the micelles,
it is important to know the partition coefficients of NOCs between the surface
aggregates and the bulk phase on the one hand and between the micelle phase and
the bulk phase on the other hand. The partition coefficient $K(m)$ for a NOC
between the micelles and water can be calculated using an approach that con-
siders the micelles as a kind of pseudophase [68]. Similarly, the surfactant
surface aggregates formed on the mineral surface may also be treated as some
kind of pseudophase [28].

Above the CMC, where the surfactant isotherm reaches a plateau value, the
partition coefficient of NOC between adsorbed surfactant layer and bulk phase
$K(a)$ remains constant. For this case, Table 2 presents the partition coefficients
$K(a)$ and $K(m)$ and the standard Gibbs free energies $\Delta G°$ (a) and $\Delta G°$ (m) of
phenol and 4-chlorophenol in the HTAC adsorbed layer on the SiO_2 surface and
in the HTAC micelles.

All the values for 4-chlorophenol are significantly higher than for phenol. This
effect may be explained in terms of the higher hydrophobicity and lower water
solubility of the chloro-substituted species.

The values of $\Delta G°$ (a) and $\Delta G°$ (m) indicate that the solubilization of both
phenol and 4-chlorophenol in the HTAC-adsorbed layer is more favored than
their solubilization in micelles. The interpretation of the difference observed
could be related to the structure of the adsorbed and micellar surfactant aggregates.

Similar conclusions were drawn by Monticone et al. from investigations carried out in the 2-naphthol/cationic surfactant/silica system [69]. It must be mentioned here, however, that these findings cannot be generalized for NOCs. Park

et al. found that the distribution of NOCs in the NOC/anionic surfactant/alumina system can be different depending on the hydrophobicity of the NOCs. Whereas $K(a)$ and $K(m)$ are about the same for CCl_4, naphthalene and phenanthrene display a greater $K(m)$ than $K(a)$. Possible explanations may be the molecule size of these NOCs or structural differences between micelles and surfactant surface aggregates [70]. From these latter examples, it becomes apparent that the mechanism of the adsolubilization of NOCs in surfactant aggregates can be more complicated than the description with partition coefficients seems to indicate.

REFERENCES

1. O. Fränzle, *Contaminants in Terrestrial Environments*, Springer-Verlag, Berlin, 1993.
2. G. Kuhnt and K. Knief, Ökologisches Verhalten von Tensiden in Böden, UBA Report, Kiel, 1990.
3. C. C. West and J. H. Harwell, Environ. Sci. Technol. 26(12):2324 (1992).
4. S. A. Boyd, W. F. Jaynes, and B. S. Ross, in *Organic Substances and Sediments in Water*, Vol. 1 (R. A. Baker, ed.), Lewis, Chelsea, 1991.
5. P. H. Brunner, S. Capri, A. Marcomini, and W. Giger, Water Res. 22:101 (1988).
6. K. Urano, M. Saito, and C. Murata, Chemosphere 13(2):293 (1984).
7. E. Matthijs and H. De Henau, Tenside Detergents 22(6):299 (1985).
8. V. C. Hand and G. K. Williams, Environ. Sci. Technol. 21(4):370 (1987).
9. J. R. Marchesi, W. A. House, G. F. White, N. J. Russel, and L. S. Farr, Colloids Surfaces 53:63 (1991).
10. M. S. Holt and S. L. Bernstein, Water Res. 26(5):613 (1992).
11. R. J. Larson, T. W. Federle, R. J. Shimp, and R. M. Ventullo, Tenside Surfactants Detergents 26(2):116 (1989).
12. E. M. Thurman, L. B. Barber, Jr., and D. LeBlanc, J. Contam. Hydrol. 1:143 (1986).
13. B. M. Moudgil, H. Soto, and P. Somasundaran, in *Reagents in Mineral Technology* (P. Somasundaran and B. M. Moudgil, eds.), Marcel Dekker, New York, 1987.
14. M. J. Schwuger, in *Anionic Surfactant: Physical Chemistry of Surfactant Action* (E. H. Lucassen-Reynders, ed.), Marcel Dekker, New York, 1981.
15. P. Somasundaran and P. Chandar, in *Solid-Liquid Interactions in Porous Media* (J. M. Cases, ed.), Edition Technip, Paris, 1985.
16. D. W. Fuerstenau, in *The Chemistry of Biosurfaces*, Vol. 1 (M. L. Hair, ed.), Marcel Dekker, New York, 1971.
17. K. L. Stellner and J. F. Scamehorn, Langmuir 5:70 (1989).
18. R. J. Larson and R. D. Vashon, Dev. Ind. Microbiol. 24:425 (1983).
19. B. J. Brownawell, H. Chen, J. M. Collier, and J. C. Westall, Environ. Sci. Technol. 24(8):1234 (1990).
20. J. Gerke and W. Ziechmann, Chem. Erde 50:247 (1990).
21. J. P. Law and G. W. Kunze, Soil Sci. Soc. Am. Proc. 30:321 (1966).
22. R. Schwarz, K. Heckmann, and J. Strand, J. Colloid Interface Sci. 124:50 (1988).
23. B. H. Bijsterbosch, J. Colloid Interface Sci. 47:186 (1974).

24. A. de Keizer, Progr. Colloid Polym. Sci. *83*:118 (1990).
25. T. Mehrian, University of Wageningen, Ph.D. Dissertation, 1992.
26. T. Rheinländer, University of Düsseldorf, Ph.D. Dissertation, 1993.
27. K. Jasmund and G. Lagaly, *Tonminerale und Tone*, Steinkopff, Darmstadt, 1993.
28. D. Schieder, B. Dobias, E. Klumpp, and M. J. Schwuger, Colloids Surfaces *88*:103 (1994).
29. B. W. Vigon and A. J. Rubin, J. Water Pollut. Control Fed. *61*:1233 (1989).
30. S. A. Abdul and T. L. Gibson, Environ. Sci. Technol. *25*:665 (1991).
31. Z. Liu, D. A. Edwards, and R. G. Luthy, Water Res. *26*(10):1337 (1992).
32. C. Palmer, D. A. Sabatini, and J. H. Harwell, in *Transport and Remediation of Subsurface Contaminants* (D. A. Sabatini, R. C. Knox, eds.), ACS Symposium Series 491, American Chemical Society, Washington, D.C., 1992.
33. D. A. Edwards, Z. Adeel, and R. G. Luthy, Environ. Sci. Technol. *28*(8):1550 (1994).
34. J. Rouquerol, S. Partyka, and F. Rouquerol, in *Adsorption at Gas-Solid and Liquid Solid Interface* (J. Rouquerol and S. W. Sing, eds.), Elsevier, Amsterdam, 1982.
35. W. Röhl, W. von Rybinski, and M. J. Schwuger, Colloid Polym. Sci. *272*:324 (1994).
36. R. Pinal, P. S. C. Rao, L. S. Lee, P. V. Cline, and S. H. Yalkowsky, Environ. Sci. Technol. *24*(5):639 (1990).
37. P. S. C. Rao, L. S. Lee, and R. Pinal, Environ. Sci. Technol. *24*(5):647 (1990).
38. C. T. Chiou, R. L. Malcolm, T. I. Brinton, and D. E. Kile, Environ. Sci. Technol. *20*(5):502 (1986).
39. D. E. Kile and C. T. Chiou, Environ. Sci. Technol. *23*(7):832 (1989).
40. D. E. Kile, C. T. Chiou, and R. S. Helburn, Environ. Sci. Technol. *24*(2):205 (1990).
41. M. J. Rosen, *Surfactants and Interfacial Phenomena*, Wiley, New York, 1989.
42. K. T. Valsaraj and L. J. Thibodeaux, Water Res. *23*(2):183 (1989).
43. D. A. Edwards, R. G. Luthy, and Z. Liu, Environ. Sci. Technol. *25*:127 (1991).
44. A. Beveridge and W. Pickering, Water Res. *17*:215 (1983).
45. S. Srivastava, R. Tyagi, N. Pant, and N. Pal, Environ. Technol. Lett. *10*:275 (1989).
46. H. Hanna and P. Somasundaran, J. Colloid Interface Sci. *70*:181 (1979).
47. J. Zundel and B. Siffert, in *Interactions solide-liquide dans le milieux poreux* (J. Cases, ed.), Editions Technip, Nancy, 1984
48. B. Siffert and J. Zundel, Clay Min. *20*:189 (1985).
49. J. Gonzalez, A. Pohlmeier, H. D. Narres, and M. J. Schwuger, Mitteil. Deutsch. Bodenkundl. Ges. *68*:235 (1992).
50. J. Gonzalez, University of Düsseldorf, Ph.D. Dissertation, 1995.
51. B. Siffert, A. Jada, and E. Wersinger, Colloids Surfaces, *69*:45 (1992).
52. T. Rheinländer, E. Klumpp, M. Rossbach, and M. J. Schwuger, Progr. Colloid Polym. Sci. *89*:190 (1992).
53. H. W. Dürbeck, E. Klumpp, J. D. Schladot, and M. J. Schwuger, Progr. Colloid Polym. Sci. *95*:48 (1994).
54. J. F. Lee, J. R. Crum, and S. A. Boyd, Environ. Sci. Technol. *23*:1365 (1989).
55. S. A. Boyd, J. F. Lee, and M. M. Mortland, Nature *333*:345 (1988).
56. S. A. Boyd, M. M. Mortland, and C. T. Chiou, Soil Sci. Soc. Am. J. *52*:652 (1988).
57. S. A. Boyd, S. Shaobai, J. F. Lee, and M. M. Mortland, Clays Clay Minerals *36*(2):125 (1988).

58. J. A. Smith, P. R. Jaffè, and C. T. Chiou, Environ. Sci. Technol. 24(8):1167 (1990).
59. E. Klumpp, H. Heitmann, H. Lewandowski, and M. J. Schwuger, Progr. Colloid Polym. Sci. 89:181 (1992).
60. E. Klumpp, H. Heitmann, and M. J. Schwuger, Colloids Surfaces 78:93 (1993).
61. H. Heitmann, University of Dortmund, Ph.D. Dissertation (1996).
62. I. Dekany, F. Szanto, A. Weiss, and G. Lagaly, Ber. Bunsenges. Phys. Chem. 89:62 (1985).
63. I. Dekany, F. Szanto, A. Weiss, and G. Lagaly, Ber. Bunsenges. Phys. Chem. 90:427 (1986).
64. I. Regdon, Z. Kiraly, I. Dekany, and G. Lagaly, Colloid Polym. Sci. 272:1129 (1994).
65. W. F. Jaynes and S. A. Boyd, Soil Sci. Soc. Am. J. 55:43 (1991).
66. V. Monticone and C. Treiner, J. Colloid Interface Sci. 166:394 (1994).
67. K. T. Valsaraj, Sep. Sci. Technol. 24(14):1191 (1989).
68. K. Shinoda and E. Hutchinson, J. Phys. Chem. 66:577 (1962).
69. V. Monticone, M. H. Mannebach, and C. Treiner, Langmuir 10:2395 (1994).
70. J. W. Park and P. R. Jaffè, Environ. Sci. Technol. 27(12):2559 (1993).

3

Analytical Methods for Surfactants and Complexing Agents at Concentrations Relevant to Environmental Occurrence

GERD KLOSTER[†] Institute of Applied Physical Chemistry, Research Center Jülich GmbH, Jülich, Germany

---

[†]Deceased.

I. INTRODUCTION

From ancient times onward soaps have been used as surfactants in washing and personal hygiene. Starting in 1946 with the introduction of tetrapropyleneben-zene sulfonate (TPS) as a synthetic anionic surfactant in the United States, these synthetic surfactants rapidly displaced soaps as the main ingredient in detergents owing to their excellent performance and their low cost. In contrast to soap, however, which is easily biodegraded, TPS was rather resistant to biodegrada-tion. As a consequence, concentrations of TPS in the environment increased in parallel to increasing consumption of detergents. Finally, a point was reached at which, during the drought years 1959 and 1960, visible foams were formed at weirs on major rivers in Germany (as well as in different parts of the world), making the problem of rising surfactant concentrations evident to the general public [1].

In reaction, detergent laws were passed in most industrialized countries re-quiring detergents to be biodegradable. As a consequence, TPS was phased out before 1965 in many industrialized countries and substituted by linear alkylben-zene sulfonates (LAS) as the main biodegradable anionic surfactant. Conse-quently, surface water concentrations of anionic surfactants, which had exceeded a level of 1 mg/liter in some German rivers in 1964, started to decline soon afterward. In 1987, concentrations of anionic surfactants in large rivers in Ger-many were consistently below 200 µg/liter [2–4], the limit value stated in the German drinking water statutory order for anionic surfactants [5].

Apart from the change from TPS to LAS, detergent formulations have become increasingly sophisticated in recent years by the addition of nonionic and cationic surfactants and different builders. Thus, increasingly complex analytical proce-dures must be developed to monitor surfactants and complexing agents (phos-phate substitutes), compounds expected to have the largest impact on the envi-ronment, in environmental matrices. These procedures must be designed to meet stringent criteria: analysis of very low (µg/liter) concentrations of surfactants and complexing agents in fairly clean surface waters, and high (g/kg) concentrations of surfactants in such difficult matrices as sewage sludges must be equally possible.

In contrast to the analytical procedures used for product analysis by manu-facturers [6], environmental analysis must deal with limited sample amounts containing analytes at trace concentrations, requiring a strongly matrix-depen-dent preconcentration and isolation procedure and leading to extracts in which a large number of homologs and isomers must be determined as selectively and specifically as possible. The possible presence of metabolites further complicates this task. Nonspecific methods (summary parameters) are often used for screen-ing purposes, whereas for specific analyses chromatographic methods are widely used.

Methods developed for the analysis of anionic, nonionic, or cationic surfactants in environmental matrices (including preconcentration procedures), as well as methods for analysis of the major phosphate-substituting complexing agents, such as aminopolycarboxylic acids or organic phosphonates, are reviewed in this chapter.

II. SURFACTANTS

A. General Remarks

Surfactants are amphiphilic molecules with hydrophilic and hydrophobic parts (Fig. 1). Whereas the hydrophobic part is usually a hydrocarbon chain, either aromatic or aliphatic, the hydrophilic residue can bear a negative charge (anionic surfactant) or a positive charge (cationic surfactant), both a positive and a neg-

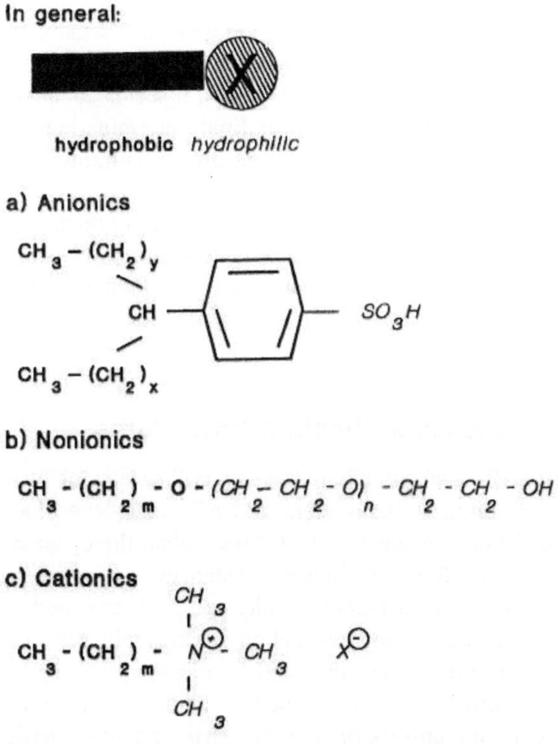

In general:

hydrophobic *hydrophilic*

a) Anionics

$CH_3 - (CH_2)_y$
$CH_3 - (CH_2)_x$
CH — ⟨benzene ring⟩ — SO_3H

b) Nonionics

$CH_3 - (CH_2)_m - O - (CH_2 - CH_2 - O)_n - CH_2 - CH_2 - OH$

c) Cationics

$CH_3 - (CH_2)_m - N^{\oplus}(CH_3)_3 \quad X^{\ominus}$
(with CH_3 groups on the nitrogen)

FIG. 1 Chemical structures of typical surfactants, with the hydrophobic part in roman type and the hydrophilic part in italics.

ative charge (amphoteric surfactant), or a polymeric polyoxyethylene chain (nonionic surfactant; for details see, e.g., Ref. 7). As the name *surfactant* (surface-active agent) implies, these compounds have a high affinity toward phase boundaries of surfaces. This affinity is highly desired during detergent application, because it increases the efficacy of a detergent formulation. When analyzing for surfactants in environmental matrices, however, this affinity is highly troublesome because it seriously complicates extraction, preconcentration, and separation procedures.

In environmental matrices like surface waters, sediments, or sewage sludges, surfactants are usually present in complex mixtures of the different surfactant classes. Furthermore, their metabolites may be present as well, some of which have retained surfactant properties but others have lost them. Thus elaborate and sophisticated procedures for extraction and enrichment of analytes as well as removal of interfering matrix components are mandatory for specific and sensitive analysis. The successful accomplishment of this task is decisive for the outcome of the intended analytical determination method, especially at trace concentration levels.

The exact requirements for the necessary preconcentration and separation procedures, however, are highly dependent on the analytical method employed for the final determination of the separated surfactants. Thus, it seems adequate first to delineate the analytical methods employed for final determinations to learn about possible interference. In succession, different preconcentration and separation procedures are reviewed here.

The presentation of analytical methods is further subdivided into procedures that employ off-line methods for identification and/or quantification after isolation of a suitable fraction containing the surfactants and those employing on-line methods for identification and quantification (usually chromatographic methods using specific detectors).

B. Methods Using Off-Line Quantification Procedures

Historically, the use of nonspecific analysis methods using colorimetry or titrimetry for the determination of summary parameters (MBAS, methylene blue-active substances; BiAS, bismuth-active substances; CTAS, cobalt thiocyanate-active substances; or DSBAS, disulfine blue-active substances) as surrogate parameters for the various classes of surfactants (anionic, nonionic, or cationic) was the earliest attempt to analyze these compounds at environmentally relevant concentrations. Similar nonspecific data were obtained by electrochemical detection methods. Furthermore, metal complexes formed by anionic or nonionic surfactants were analyzed by atomic absorption spectrometric or voltammetric detection of the metal ion involved.

Other off-line methods allow more specific determinations of individual compounds or groups of compounds within the various classes of surfactants. These methods involve the spectroscopic analysis of fractions isolated by a suitable isolation and separation procedure. Spectroscopic techniques employed are infrared (IR), nuclear magnetic resonance (NMR), or mass spectrometry (MS). More emphasis must be placed on isolating a fraction of sufficient purity for these determination methods than is necessary for the colorimetric methods.

1. Colorimetry and Titrimetry

(a) Anionic Surfactants. Anionic surfactants form complexes with suitable cationic dye compounds; these complexes are soluble in organic solvents, in contrast to both the anionic surfactant and the cationic dye, both of which are soluble in water when present alone. The cationic dye used most often is methylene blue (MB), originally proposed by Epton in 1948 [8]. It can be employed in either the original titrimetric method [8] or in a colorimetric determination method measuring the extinction of the extracted complex at 650 nm [9–11].

In the original procedure [8], an acidic solution of MB is added to the sample containing the anionic surfactant; chloroform is added to this mixture. After shaking, the complex of anionic surfactant and methylene blue is in the chloroform phase and a small excess MB is in the water phase. By titration with a suitable cationic surfactant (which must have a higher complexing ability than MB), MB is displaced from the complex and back-extracted into the aqueous phase. The end point is reached when both phases have identical color intensities. The method must be calibrated against a suitable primary standard anionic surfactant. Originally, sodium tetradecan-2-sulfate was used as calibration standard [8]. Later, sodium dioctylsulfosuccinate [9,10], sodium dodecylsulfate [12], or the methyl ester of dodecylbenzenesulfonic acid (after alkaline hydrolysis) [11] was employed as calibration standard. It was shown that many anionic surfactants can be used as calibration standards; of prime importance, however, is their availability in a chemically pure state and their stability on storage [12].

Extractable complexes between anionic surfactants and MB are obtained only for alkyl sulfates or alkylsulfonates with a hydrocarbon chain of eight carbon atoms or more, as well as for alkylbenzenesulfonates. Benzenesulfonates, xylenesulfonates, or naphthalenesulfonates (which are used as hydrotropes) do not form extractable complexes with MB.

Cationic surfactants, when present in the same solution as the anionic surfactant, compete with MB and form colorless complexes that can also be extracted into organic solvents. Thus, in the presence of cationic surfactants, low results are obtained. Therefore, cationic surfactants must be separated before using the MB method [11].

Determination of the end point in the original two-phase titration method, however, is difficult; long times are needed for equilibration near the end point. To avoid these difficulties, either the use of a colorimetric method [9–11] is recommended. On the other hand, a more complicated two-phase titration method using a mixed indicator system (disulfine blue as an anionic dye and dimidium bromide as a cationic dye) was proposed [12]. Using this procedure (for details see the original paper), the color observed in the chloroform layer changes from pink in the presence of an excess of anionic sample to blue in the presence of excess cationic titrant. At the end point a gray-blue color is observed in transition from the pink to the blue solution. It is claimed that the clarity of this end point is superior to that of a number of other indicator systems investigated [12].

All anionic surfactants having a minimum chain length of eight carbons form complexes with MB. Because both the titrimetric method and the colorimetric method yield results that are proportional to molar concentrations, not to mass concentration, some conversion factor must be used to obtain mass concentration. Complexes between anionic surfactants and MB have been shown to be 1:1 complexes [8]. Because of the changing molecular structure and chain length [linear alkyl-benzenesulfonate (LAS), alkylsulfonates, alkyl sulfates, and α-sulfofatty acid esters, for example], some form of averaging is employed in calculating the results. Results are usually expressed as mg or μg MBAS (methylene blue-active substances), calculated as (for instance) sodium dodecyl-benzenesulfonate [11]. Thus, actual values are dependent on both the particular composition of the anionic surfactants in the sample and the standard used for quantification.

Apart from anionic surfactants, other anionic organic compounds present in environmental samples may also form extractable complexes with MB and thus introduce a systematic error. This is especially true of samples from waste water, sludges, soil, or sediment, but even for surface waters, interferences have been reported [13–16] that result in high values for MBAS. The chemical nature of the interferences involved is unknown; only speculation about their structure is available. Interfering compounds, however, can be eliminated by using more specific methods of identification and quantitation, such as infrared (IR) spectroscopy, high-performance liquid chromatography (HPLC), and gas chromatography (GC) after desulfonation (see later). Thus, the portion of MBAS that originates from interactions of MB with anionic surfactants can be assessed. It ranges from up to 93% for untreated waste water to a minimum of 2% in activated sludges or river sediments [17]. Consistently, higher values are obtained when determining MBAS compared with more specific methods.

Despite these drawbacks, MBAS is still used widely for the determination of anionic surfactants and has found application in the German Standard (DIN-Norm) [11] for these compounds. The advantage of the colorimetric method is that it can be performed by using relatively simple apparatus in a fairly short time; it yields concentration values that are systematically high (thus, a conservative procedure with a significant margin of safety is available for monitoring campaigns), and it obtains data that are directly comparable to monitoring results that have been available for more than 30 years employing this methodology [2–4]. For environmental monitoring, colorimetric MBAS is preferred as a screening method, and for product analysis the two-phase titrimetric method has found wider application for the determination of anionic surfactants.

Cationic dyes other than MB have also been investigated as possible counterions for anionic surfactants in colorimetric procedures [18–21]. The purpose of these investigations was to find improvements in reaction conditions that avoid some of the difficulties still encountered in the MBAS procedure. In particular, cationic dyes were desired that have higher molar extinction coefficients than MB and that form complexes with anionic surfactants with a higher extraction coefficient into organic solvents like chloroform at the same time. Both effects would improve detection limits for samples containing low concentrations of anionic surfactants. However, none of the competing cationic dyes reported [18–21] has found application instead of MB until now, although some of the intended improvements were achieved.

Finally, methods have been reported in which color is formed only in the presence of anionic surfactants [22] or that utilize a color change of cationic dyes in the presence of anionic surfactants [23]. The method using the reoxidation of reduced leuco bases of triphenylmethane dyes rests on the exclusion of a reducing agent (sulfite ion) from micelles of anionic surfactants [22]. It is thus applicable only at concentrations above the critical micelle concentration, which is not normally reached in environmental samples. The method employing a cationic dye, namely 1-(10-bromodecyl)-4-(4-aminonaphthylazo)pyridinium bromide, which suffers a spectral shift in its absorption spectrum in the presence of anionic surfactants, has the principal advantage that no extraction of the colored complex is necessary for quantitation of the complex. Reduction of the absorption at 595 nm (absorption maximum of the dye in water) was found to be proportional to the concentration of anionic surfactants. A change in the microenvironment of the dye in the ion associated with anionic surfactants was invoked as the possible cause; different absorption maxima were recorded for sodium dodecylsulfate and sodium dodecylbenzene-sulfonate, respectively [23]. The synthesis of the dye is tedious, however; thus, this method has not yet been widely applied in competition with the MBAS method.

(b) Cationic Surfactants. Cationic surfactants form complexes with suitable anionic dyes that are extractable into organic solvents. The situation is completely analogous to the anionic surfactant/cationic dye complexes described in detail previously.

The anionic dye most widely used for the nonspecific determination of cationic surfactants is disulfine blue (DSB), a triphenylmethane dye [24–27]. Because the extractable dye/cationic surfactant complex must be extracted for colorimetry, the presence of anionic surfactants (usually present in excess over cationic surfactants in environmental samples) results in serious interference; they must therefore be separated before addition of DSB.

In the original procedure [24], hexadecyltrimethylammonium chloride was used as the calibration standard cationic surfactant. However, any cationic surfactant that is stable on storage and can be purchased in sufficient purity may be employed. Quarternary ammonium compounds having at least one hydrocarbon chain with 12 carbon atoms can be completely extracted by DSB. A 1:1 stoichiometry between hexadecyltrimethylammonium chloride and DSB in complex formation was claimed [24]. Doubts have been expressed, however, that this is true for dioctadecyldimethyl-ammonium chloride (DSDMAC) as well [28].

In comparison with MB and anionic surfactants, the determination of cationic surfactants with DSB is hampered by a number of problems not encountered with MBAS. In particular, cationic surfactants are strongly adsorbed to almost any surface; therefore, measures must be taken to reduce these adsorption processes as completely as possible to obtain valid results [24,25]. Extraction efficiencies for the complex are reported to be low; furthermore, they seem to be strongly dependent on the exact procedure used for stirring or shaking [28]. The stability of the color of the extracted complex over time was also lower than one would wish [24,26]. Thus, the determination of DSBAS is not as straightforward as that of MBAS.

Naturally occurring compounds containing long-chain alkylamine or quarternary ammonium residues that are not surfactants also form extractable complexes with DSB under the weakly acidic reaction conditions necessary for the determination [24,25,26,28,29]. As a consequence, high values for DSBAS will be recorded for samples containing significant concentrations of these nonsurfactant cationic compounds, like activated sludges, sediments, or polluted waters. In complete analogy to the situation with anionic surfactants, more specific methods of identification and quantification, such as IR spectroscopy [28], thin-layer chromatography (TLC) [26], and HPLC [29], are necessary to eliminate these interferences.

The DSBAS method is used widely for screening purposes in monitoring programs for cationic surfactants despite these drawbacks described previously. It has found application in the German Standard (DIN-Norm) [27] for these compounds. The reasons for this frequent application are largely the same as detailed for the MBAS method: inexpensive equipment, fairly easy procedure, inherent margin of safety because of systematically high results, and comparability of data with previous monitoring time series.

Cationic surfactants can also be analyzed by a variation in the two-phase titration method [12] described previously for anionic surfactants, because cationic surfactants are used to titrate anionic surfactants in the sample. By adding a known excess of anionic surfactant to the sample (from which anionic surfactants had been removed by, for instance, ion exchange), the amount of cationic surfactant present can be calculated as the difference between anionic surfactant added to the sample and cationic surfactant needed to titrate the excess. This method is used mainly for product analysis.

Other anionic dyes have also been investigated in competition with DSB to achieve better extractability and higher extinction coefficients, among others. Methyl orange is employed in food residue analysis for quarternary ammonium compounds [30], and bromophenol blue was studied as a potential substitute for DSB [31], to name only two examples. None of them, however, has been able to reduce the use of DSB.

Anionic dyes that suffer fluorescence quenching [32] or a spectral shift in the absorption band [33,34] in the presence of cationic surfactants have also been investigated. The inherently higher sensitivity of fluorescence emission compared with absorption in the visible region is favorable. Because concentration-dependent quenching of the fluorescence of the dye 8-octadecyloxypyrene-1,3,6-trisulfonate results in only a small reduction in a large fluorescence signal for lower concentrations [32], this method does not seem to be applicable to trace analysis. The use of reagents undergoing a spectral shift in the presence of cationic surfactants avoids the extraction step necessary for the DSBAS method. A method using propyl orange as the anionic dye was recently proposed [33,34]. Only a few results are available now, so that no comparison with the DSBAS method is possible at the moment.

(c) *Nonionic Surfactants.* By definition, nonionic surfactants do not embody ionic charges, so that no ion associate complexes can be utilized for analytical purposes. The characteristic molecular building block of nonionic surfactants is an oligomeric polyoxyethylene chain of variable chain length. Consequently, donor-acceptor complex formation with the ether oxygen atoms (Lewis bases) has been instrumentalized in determination methods for this class of compounds. Hard Lewis acids (alkali or alkaline earth metal ions) are preferentially used to form donor-acceptor complexes with nonionic surfactants. A minimum number of oxygen atoms is mandatory for formation of stable complexes (in much the same way as crown ethers are used to complex these cations). The nonionic surfactants are thus transformed into cationic coordination complexes that form ion pairs with such complex anions as tetraiodobismuthate or tetrathiocyanato-cobaltate. Properties of these anions are then utilized for the final determination step. Cationic surfactants severely interfere with these methods as they compete with the cationic coordination complexes for the complex anions; they must therefore be eliminated from the sample before the final determination.

Barium tetraiodobismuthate (modified Dragendorff reagent) forms a red precipitate with nonionic surfactants in moderately to strongly acidic solutions [12,35–40]. A minimum chain length of five ethylene oxide (EO) groups in the hydrophilic chain is necessary for precipitate formation [12,35]. Longer EO chain length compounds complex progressively more tetraiodobismuthate per µg nonionic surfactants [39]. Thus, standards of a fixed EO chain length must be used for calibration of this procedure; mean response factors are then applied for quantification of BiAS by this procedure (actual numbers in the results depend on the particular standard used) [12]. After collecting and washing the precipitate, it is redissolved using a suitable solution, such as ammonium tartrate. The bismuth ions released are then converted to their pyrrolidinedithiocarbamate [38,40] or ethylenediaminetetraacetate [39] complexes and determined colorimetrically [38–40], by potentiometric titration [12,36,37], or by x-ray fluorescence spectroscopy.

Polyoxyethylene glycol also reacts under the conditions reported [35]. It must be separated during preconcentration of the sample to avoid interference. Nevertheless, the BiAS method is used in the standardized (DIN-Norm) procedure in Germany [12], as well as in other countries. In contrast to the situation with anionic and cationic surfactants, in which MBAS and DSBAS values were consistently high as a result of interference from natural compounds, BiAS values may be systematically low if nonionic surfactants with short polyoxyethylene chains (<5 EO) are present. This situation may prevail in samples that have been largely biodegraded, resulting in a concomitant shortening of the EO chain. Thus, large amounts of nonylphenol (NP), nonylphenolmonoethoxilate (NP1EO), and nonylphenoldiethoxilate (NP2EO) were detected in environmental samples from biodegraded nonylphenolpolyethoxilates using more specific HPLC methods [41,42]. These residues went undetected using the nonspecific BiAS procedure.

In contrast to Europe, where the BiAS procedure is used frequently, the CTAS procedure has found wide application in the United States. Using this procedure, ammonium tetrathiocyanatocobaltate is added to a nonionic surfactant solution; a blue water-insoluble complex is formed that can be extracted into organic solvents [43–48]. Again, a minimum of 6 EO residues is necessary for complex formation [44]; the moral response of the complex formed varies with EO chain length, having a maximum around 10 EO [43,45–47]. Cationic surfactants lead to interference, resulting in high values; they must therefore be separated before complex formation. The CTAS method involves fewer steps in sample preparation than the BiAS method and is thus simpler to perform. The results of both methods, although both variable in themselves depending on EO chain length, compare favorably within experimental error [48]. The reservations concerning the nonspecific nature of complex formation and the systematically low values for biodegraded samples apply to both methods equally.

Other complex salts have been used for donor-acceptor complex formation [37], but none of them has reached a stage of application comparable to BiAS or CTAS.

2. Electrochemical Methods

Most of the electrochemical methods (polarography and tensammetry) are even less specific than the colorimetric/titrimetric procedures reviewed earlier; only ion-selective electrodes (surfactant-specific electrodes) offer a potential of specific surfactant analysis.

In polarography and tensammetry, the indirect effects of surfactants on polarographic currents are instrumentalized for determination (for a review see Ref. 49), because surfactants do not have any direct electrochemical activity. Surfactants adsorb to the surface of electrodes and thus either modify the capacitance of the mercury electrode, decreasing the capacitive current, or alter the diffusion rate or convection rate of electroactive species to the electrode surface, thus decreasing the diffusion current or the polarographic maximum, respectively [50–59]. The exact mechanism by which surfactant concentrations at the electrode surface are transformed into readable current or voltage values is beyond the scope of this review; the interested reader is referred to the original literature.

Polarographic or tensammetric methods have a number of potential advantages: they tolerate large concentrations of electrolytes, which most other methods do not, sample preparation is often very simple, measurements can be performed at very low surfactant concentration, and equipment needed is usually available in most laboratories. However, polarographic methods have not found wider application for the determination of surfactants in environmental samples because of a number of serious disadvantages. Any surfactant—anionic, nonionic, or cationic—can adsorb to the electrode surface and generate a signal [52,54,55,58]. In mixtures of surfactants, the signal is not necessarily proportional to the total concentration; in some cases only the most strongly adsorbed surfactant generates a signal and all others are silent [50]. Natural compounds, such as humic acid or fatty acids, may adsorb to the electrode surface and generate signals analogous to surfactants [58]. Therefore, application of these methods in trace environmental analysis can be envisaged only for general screening purposes (to demonstrate that surfactant concentrations are below certain specified numbers). For applications in other fields, such as physicochemical investigations using single compounds, they are well suited.

Ion-selective electrodes may be constructed that generate signals in the presence of specific surfactant molecules. In favorable cases, all other surfactants do not result in measurable signals, thus incorporating a high degree of specificity into the procedure. Ion-selective electrodes for quarternary ammonium halides have been reviewed [60]; applications were reported for anionic as well as cationic surfactants. The price for specificity in analysis is a rather high sensitivity of these electrodes toward potential interferences, which limits their applicability in environmental analysis. In other fields involving pure compounds in clean solvents, their applicability is better [61,62].

3. Spectroscopic Methods

(a) Atomic Spectroscopy. Atomic spectroscopy, both atomic absorption (AAS) and atomic emission (AES) spectroscopy, can be utilized as indirect methods for the determination of surfactants. Metal ions or complex metal-containing anions (such as tetraiodobismuthate or tetrathiocyanatocobaltate) can form either precipitates or extractable complexes with various surfactants. As in colorimetry (see Sec. II.B.1), metal ions extracted or precipitated are proportional to the concentrations of the surfactants coextracted or coprecipitated with them. After digestion of the extract or precipitate, metals can be analyzed by AAS or EAS with a high sensitivity.

The main field of application seems to be in the determination of nonionic surfactants in which metal complex anions are widely used in colorimetric or titrimetric procedures (see Sec. II.B.1.c). In one procedure, the excess of the reagent used for precipitation of the nonionic surfactant (barium phosphomolybdate) was analyzed by AAS for Mo content to determine the nonionic surfactant concentration by difference [63]. On the other hand, the extract prepared for the CTAS method was analyzed for Co by AAS instead of using a colorimetric finish [64–67]. The same disadvantages, however, are expected to affect the AAS method as the colorimetric method mentioned previously: nonspecific complex formation, different stoichiometry of metal ion incorporation for varying EO chain length, and systematically low values for biodegraded samples. Furthermore, more often than not sample preparation for AAS is more tedious than for colorimetry or titrimetry. Consequently, AAS methods are only rarely applied for environmental samples compared with colorimetric methods.

For anionic surfactants, however, the Australian Standard Procedure uses the AAS determination of Cu ions extracted from an aqueous diaquobis(ethylenediamine)copper(II) solution by anionic surfactants into chloroform as the bis-(surfactant)bis(ethylenediamine)copper(II) complex. After destruction of the complex by dilute acid, Cu is determined by AAS in proportion to surfactant concentration [68]. Alternatively, Cu can be quantified by voltammetry [69].

AAS has also been reported as an on-line detector for anionic surfactants separated by HPLC. LAS, alkylsulfonates, and alkyl sulfates were separated by HPLC; the column effluent was connected to the nebulizer of an inductively coupled plasma AAS used as a sulfur-specific detector [70]. Compared with other on-line detection systems for HPLC, however, this method seems rather tedious, although it is very specific (Sec. II.C.1.a).

(b) Ultraviolet (UV) Spectroscopy. Molecular spectroscopy is widely used for structure identification in organic chemistry. Thus, different kinds of molecular spectroscopy may be utilized for quantification of known compounds in isolated fractions. Some spectroscopic methods enable the analyst to analyze single ho-

mologs or isomers qualitatively or quantitatively; others give a more general summary parameter monitoring functional groups.

UV spectroscopy is dependent on the presence of a suitable chromophore in the molecule to be analyzed. Although the major application of UV is in on-line detection following HPLC separation, some off-line applications have been reported as well [28,71,72]. Use is made of the absorption of UV radiation by the aromatic nucleus of LAS. Extensive precleaning of samples is necessary before analysis, because most other compounds comprising an aromatic nucleus will interfere and must be removed for that reason. Because the extinction coefficient at the 254 nm line is not always sufficient for quantitative determination, use is made of derivative spectroscopy at shorter wavelengths (see Fig. 2). Using this technique, increased contrast is generated on a rather unstructured absorption band (Fig. 2A). The increasing contrast of the second derivative (Fig. 2C) is claimed to improve qualitative and quantitative analysis. Results comparable to other methods are claimed for surface water, waste water, and sludge samples [71,72]. Much experience, however, seems to be necessary to obtain valid results.

(c) Infrared (IR) Spectroscopy. IR spectroscopy is widely used in surfactant analysis. A classic textbook contains a large number of IR spectra characterizing pure chemical compounds belonging to the different classes of anionic, cationic, and nonionic surfactants [73].

In environmental analysis, IR has mainly been employed for the differentiation of substances determined by the nonspecific colorimetric procedures, such as MBAS, BiAS, or DSBAS, into surfactant and nonsurfactant compounds [16,28,74–78]. To this end, a separation of the MB-anionic compound complex, for example, into different fractions is performed by use of thin-layer chromatography (TLC). Isolated fractions are then characterized by IR. A typical example is shown in Fig. 3, demonstrating the IR spectrum of about 30 mg MBAS/kg river sediment (Fig. 3B). For comparison, the spectrum of the LAS-MB complex is shown in Fig. 3A. Characteristic bands for the LAS-MB complex at 1212/1197 cm^{-1} (SO$_3$ residue) and fingerprint region bands at 1032, 1009, and 580 cm^{-1} are typical of LAS [16]; these bands are absent in the isolated fraction of the environmental sample (at a detection limit of 1 mg MBAS/kg sediment for IR), leading to the conclusion that substances other than LAS have formed colored complexes with MB, leading to high values in MBAS [16]. Similar cases can be made for nonionic or cationic surfactants as well [16,74,77,73]. During biodegradation experiments, IR was used to monitor metabolites that still contain major parts of the original molecule intact, even though its surfactant properties (and its reaction with MB or DSB) were lost after primary biodegradation [75,76]. As can be seen in Fig. 3, the IR method relies heavily on the availability of suitable standard compounds or mixtures to identify these substances in isolated fractions from environmental samples. In samples

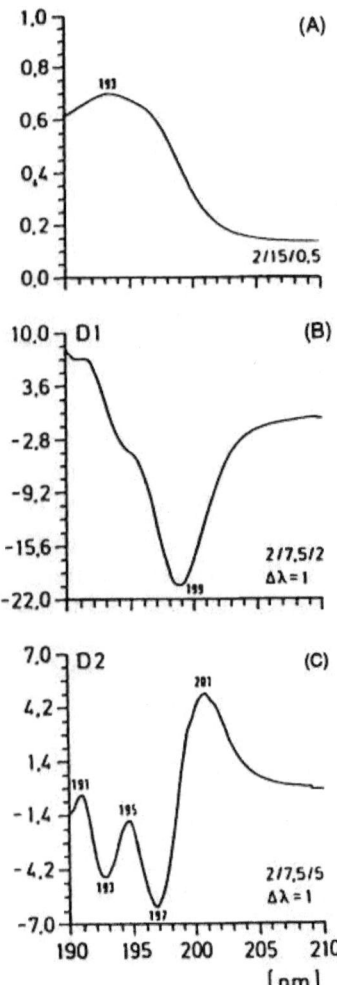

FIG. 2 UV absorption spectrum of LAS; spectral region 190–210 nm: (A) original spectrum; (B) first derivative; (C) second derivative. (Reprinted from Ref. 71 by permission of VCH Verlagsgesellschaft.)

containing only trace amounts of surfactant-dye complexes, considerable experience is needed for interpretation and evaluation of IR spectra recorded. Consequently, only a few laboratories regularly use IR for environmental analysis.

(d) Nuclear Magnetic Resonance (NMR) Spectroscopy. NMR spectra regularly contain far more information on the molecular structure of the particular compound investigated than either IR or UV spectra. The special strength of NMR is

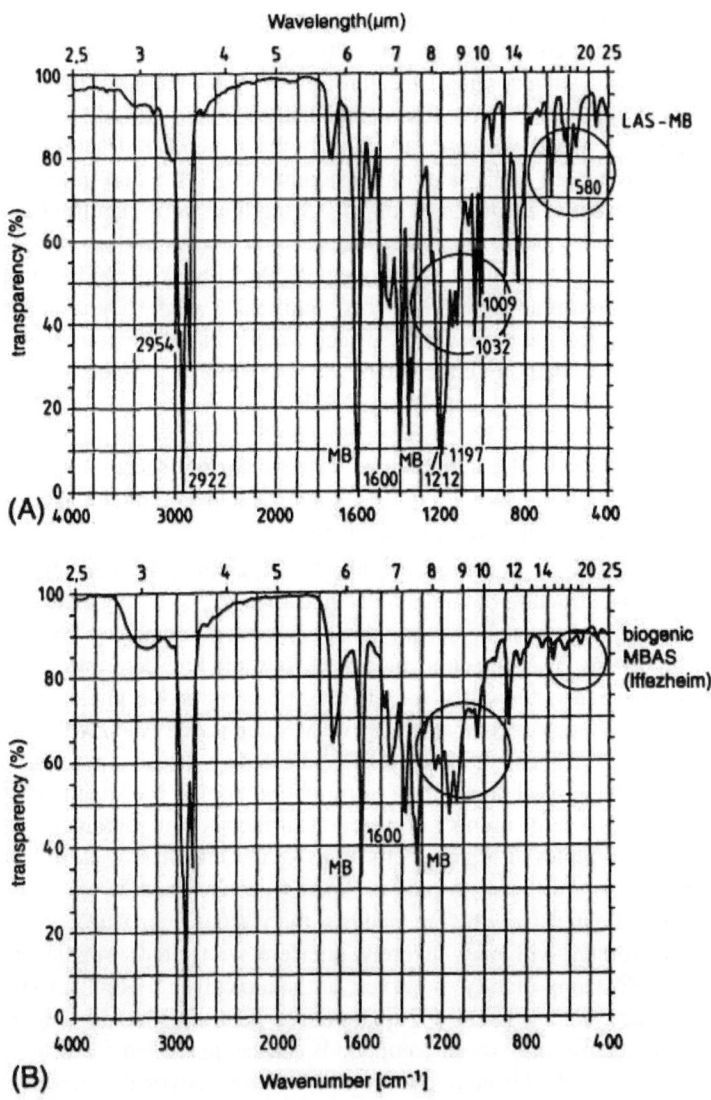

FIG. 3 A (top) IR spectrum of standard LAS-MB complex. B (bottom) IR spectrum of pseudo anionic surfactants from river Rhine sediment. (Reprinted from Ref. 16 by permission of Carl Hanser Verlag.)

the simultaneous qualitative analysis of isomeric compounds [79]. Relative quantitative determination of different isomers in mixtures is usually possible, whereas absolute quantification must rely on standard addition methods. Proton NMR and ^{13}C NMR are used most often; the latter spectra contain more structural information, but longer recording times are involved in obtaining them.

Only two cases of application in environmental analysis are reviewed here: in the first study [80], ^{13}C NMR was used to identify the molecular structure of anionic surfactants from deep wells sampling groundwater. At the 500 m level, anionic surfactants present at a concentration of 0.3 mg/liter were identified as LAS, whereas those at the 3000 m level (present at a concentration of 2.3 mg/liter) were identified as the older, nonbiodegradable TPS. A second study [81] demonstrated by proton and ^{13}C NMR that significantly higher concentrations of polyethylene glycols are present in surface waters than is obvious from the results of the CTAS method (undetectable to 145 µg/liter, respectively). Structural information was also obtained from NMR.

Again, as was the case with IR, extensive sample preparation is necessary and the availability of pure standard compounds is decisive for the success of this analysis method. Furthermore, expensive and sophisticated equipment is involved, thus limiting the use of NMR to special cases.

(e) Mass Spectrometry (MS). Mass spectra, like NMR, contain detailed information on the molecular structure of compounds investigated. MS is most widely used in environmental analysis as an on-line detection system coupled to GC or HPLC to yield highly specific detection modes. A few off-line determination methods, however, have also been used for surfactant determination in environmental samples. The different technical modes of MS applicable to nonvolatile surfactant compounds have been compared [82,83]. Fast atom bombardment MS was found to yield the largest amount of information useful for qualitative analysis (identification). A large number of different surfactant compounds could be identified in water samples at high sensitivity [84,85]. Quantitative information, however, is far more difficult to obtain. Apart from the need to preclean environmental samples extensively before analysis, the viscous matrix necessary in fast atom bombardment MS may severely interfere with either sample or standard response. Most importantly, however, an isotopically (^{2}H or ^{13}C) labeled internal standard is mandatory for quantitative determination of single surfactants. In a strict sense only those compounds can be quantified for which an isotopically labeled standard is carried through the entire analytical procedure [86]. When these requirements are met, quantitative analyses can be performed at high sensitivity and specificity, as demonstrated for selected cationic surfactants [86,87], anionic surfactants [88], or metabolites of nonionic surfactants [89]. The stringent requirements for isotopically labeled standards have restricted its use in environmental analysis. It is widely used, however, as a detector with

GC or HPLC as well as for structure assignment in the analysis of pure compounds.

C. Methods Using On-Line Quantification Procedures (Chromatographic Methods)

The ultimate goal in environmental analysis is the quantification of individual compounds, separately from all their isomers and/or homologs. This end cannot be reached by the application of nonspecific methods of analysis; on the contrary, sophisticated separation methods must be employed toward this end. Chromatographic procedures can be applied to achieve separations into classes of compounds in nearly all cases; frequently, further chromatographic separation is possible finally to realize qualitative and quantitative determination of individual surfactant compounds. Because of the amphiphilic nature of surfactants and their minimal volatility, high-performance liquid chromatography (HPLC) or supercritical fluid chromatography (SFC) is used far more often than gas chromatography (GC).

Because of the large number of isomers and homologs in commercial surfactants from detergent formulations (especially within the classes of anionic and nonionic surfactants), highly sensitive and efficient detection methods are mandatory to obtain low limits of detection for single surfactant compounds. Thus, requirements for detection methods are necessarily far more stringent than for the nonspecific methods reviewed earlier.

As a consequence, qualitative and quantitative determination of surfactants containing aromatic residues is significantly more advanced in HPLC (because of the possibility of using a UV or fluorescence detector that is very sensitive and versatile) than that of surfactants containing only aliphatic residues. In GC or SFC this difference is not observed because both can use the universal, sensitive, and versatile flame ionization detector.

1. High-Performance Liquid Chromatography (HPLC)

(a) Anionic Surfactants. The majority of HPLC applications in the determination of anionic surfactants is concerned only with the analysis of linear alkylbenzene sulfonates (LAS), the surfactant present in largest quantities in today's detergent formulations. Reasons for this were detailed earlier (Sec. II.B.3.b): a high extinction coefficient of the aromatic nucleus for the absorption of UV radiation as well as reasonable emission intensities in fluorescence spectroscopy. Thus, highly sensitive and versatile detectors, such as UV absorption or fluorescence detectors, can be used to quantify HPLC effluents; all UV-transparent solvents may possibly be used to effect chromatographic separations with almost no restrictions with respect to gradient elution programming [17,68,90–111].

Commercial LAS are complex mixtures of homologs and isomers. Usually, the alkyl chain length varies between 10 and 14 carbon atoms, with 11, 12, and

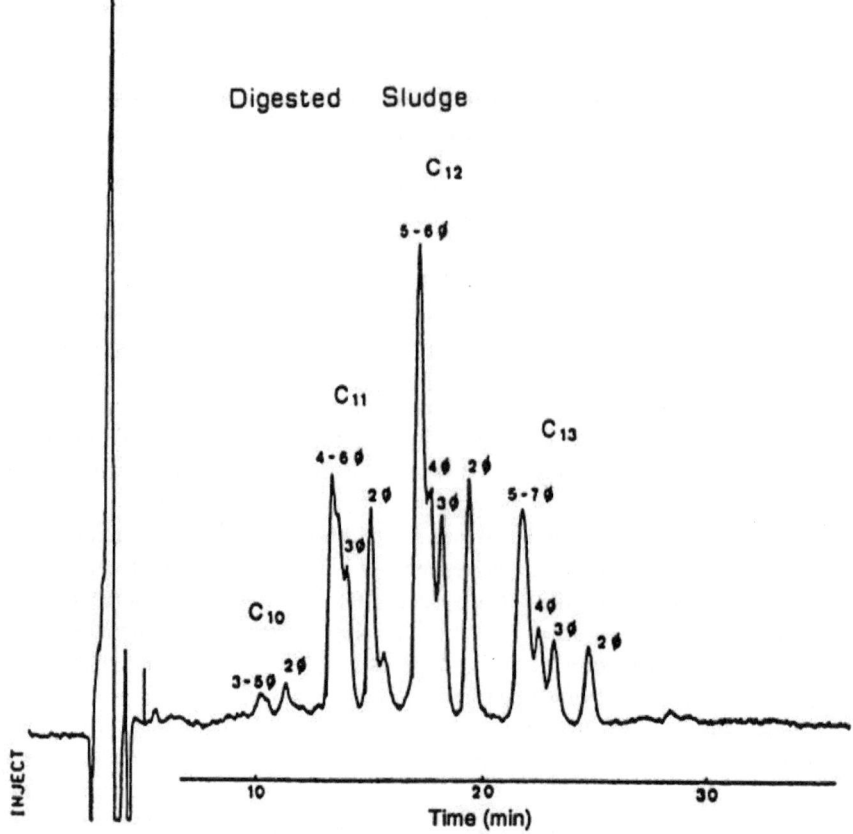

FIG. 4A Typical HPLC chromatogram of LAS with separated homologs and partially separated isomers, on a C-18 RP column. (Reprinted from Ref. 94 by permission of Gordon and Breach Science Publishers.)

13 carbon atoms as the dominant homologs. Apart from different chain lengths, the position of the 4-benzenesulfonate residue along the alkyl chain may be at any carbon atom except carbon position 1 (for dodecylbenzene-sulfonate a total of five isomers, for example). Thus, highly sophisticated separation procedures are mandatory to solve the analytical problem of precise qualitative and quantitative determination of LAS in environmental matrices.

In HPLC, separation is achieved by utilizing the different solid-liquid partition behaviors of the various analytes partitioning between the column material (solid) and the liquid mobile phase. Because the solubility of LAS in aqueous solution is much larger than in organic solvents, chromatographic separations on reversed-phase (RP) columns are almost exclusively used for the separation of

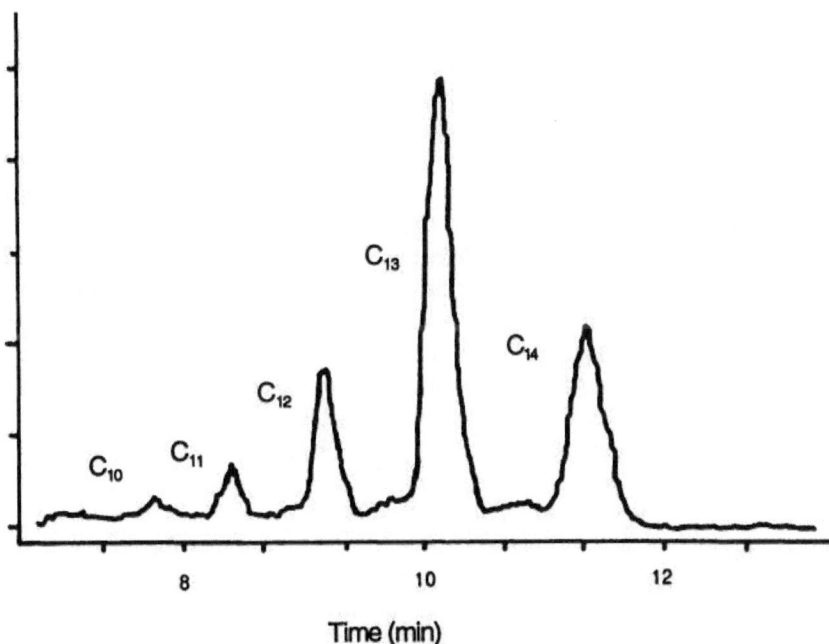

FIG. 4B HPLC chromatogram of a standard mixture of LAS with homologs only separated on a C-1 RP column. (Reprinted with permission from Ref. 103. Copyright 1989 American Chemical Society.)

LAS, with gradients of aqueous salt solutions and polar organic solvents as the mobile phase. Two different approaches are predominant: a RP column with a long-chain alkyl bonded phase (usually octadecyl) can be used to separate homologs and partly to separate isomers [90,91,93,94,96–98,102–104,108]. As can be seen in Fig. 4A, the isomers bearing the aromatic ring at position 2 (short-hand notation 2-ϕ) is separated best from the other isomers; as the aromatic ring moves farther to the interior of the alkyl chain, separation between neighboring isomers (4-ϕ, 5-ϕ, and 6-ϕ for dodecylbenzenesulfonate, for example) becomes increasingly difficult. The larger number of individual peaks for separated isomers and homologs results in the need to preconcentrate larger mass amounts of LAS for analysis to exceed the detection limit for individual isomers.

Alternatively, a RP column with a short-chain alkyl bonded phase (shorter than octyl) may be used. The shorter alkyl chain length leads to the result that the separation between homologs is retained and the separation between different isomers of the same homologs is completely lost [95,99,100,103,107,110,111]. As can be seen in Fig. 4B, the number of resolved peaks is strongly reduced compared with Fig. 4A. Consequently, more mass is accumulated in individual peaks for an identical sample size in Fig. 4B compared with individual peaks in Fig. 4A, leading to improved limits of quantification. A logical conclusion from these differences is to use the short-chain RP column for samples containing LAS in concentrations near the detection limit. For samples of LAS containing higher concentrations, such as biodegradation test mixtures, the long-chain RP column yields more information concerning the preferential degradation of individual isomers, albeit at a higher limit of detection.

Absolute detection limits vary from 1.5 ng for individual homologs using fluorescence detection to 50 ng for individual homologs using UV detection [93,94,96,98,103,110,111]. Although a method has been described that analyzes river water for LAS without pretreatment, except that filtration [90], preconcentration, and separation steps to concentrate and preclean the sample for HPLC are commonplace (see Sec. II.D) Using a suitable preconcentration and separation procedure for the different environmental matrices, method detection limits of 0.1–2 µg/liter for water [93,103] using fluorescence detection, 10 µg/liter for water using UV detection [98], 30 µg/kg for bottom sediment using fluorescence [93] and 100 µg/kg for solid samples using UV detection [98] have been reported. LAS was determined in matrices as different as waters [90,95,98,101,103,105,110–112], sludges, sediments and soils [93,94,96,98–100,104,112–114], marine waters [93,102,115,116], and even sludge-only landfills [109]. Compared with data obtained from identical samples using such nonspecific methods as MBAS, HPLC determinations of LAS consistently result in values significantly lower because of the separation of potential interference during the chromatographic process.

The situation for aliphatic anionic surfactants, such as alkylsulfonates and alkyl sulfates, is entirely different. Although these compounds are considered more environmentally acceptable alternatives to LAS (because they are more rapidly biodegradable and produced from renewable resources) and may be expected to increase in consumption, only a very few specific determination methods are available. A number of experimental methods will be reported instead, in strict contrast to LAS, for which reliable methods for environmental monitoring exist.

For charged aliphatic compounds, two general detection principles are available, refractive index and conductivity detection [117–119]. Both principles, although largely independent of molecular structure, have a number of serious disadvantages for application in environmental analysis: their inherent detection limits are rather high (samples must be precleaned and concentrated to a significantly higher extent than LAS), gradient elution is not normally possible because of baseline problems (leading to increasing loss of resolution and detectability for later eluting peaks), and compounds other than surfactants yield signals of comparable size, especially when using a refractive index detector.

Consequently, the search was directed at finding a detector that is more sensitive and, if possible, more specific than refractive index or conductivity and can be realized using not too expensive equipment.

Because all commercial anionic surfactants contain sulfur, an atomic emission detector (AES) has been used to quantify sulfur in anionic surfactants separated by HPLC into individual fractions [70]. Alkyl sulfates, alkylsulfonates, and alkyl ether sulfates were separated on a styrene-divinylbenzene column. Absolute detection limits were 15 ng sulfur (corresponding to about 100 ng anionic surfactant). The method is tedious and time consuming, however, because fractions must be analyzed off-line, and it requires expensive instrumentation.

Derivatization of aliphatic anionic surfactants to form esters is another possible method of increasing sensitivity and specificity. Esters containing aromatic residues in the alcohol portion allow UV or fluorescence detection of the derivatized anionic surfactants after HPLC separation [120]. The use of HPLC for esters containing aromatic residues, however, has found far less application than the formation of methyl esters for GC in conjunction with a flame ionization detector.

Indirect photometric or conductimetric detection relies on monitoring changes in physical properties of the bulk mobile phase caused by the analytes to be determined. Either compounds absorbing in the UV or visible spectral region or compounds with a high specific conductivity are added as indicators to the solvent used for chromatographic separation. Thus, a constant absorption or conductivity is recorded as a baseline. Nonabsorbing aliphatic surfactants with low inherent conductivity usually dilute the concentration of the indicator compound in the mobile phase, generating a negative signal [121–125]. Gradient elution is principally possible if both components contain the indicator in identical concentration and no spectral change is induced by the different solvents. The major setback of these indirect methods, however, is that they result in a signal for every compound, diluting the indicator in the mobile phase. Thus, indirect detection methods are not specific for structural elements of individual molecules.

Yet another approach is the utilization of a nonspecific colorimetric procedure (see Sec. II.B.1.a) as an on-line derivatization method for the determination of anionic surfactants [126–130]. More specifically, the anionic surfactants are separated into individual compounds using a suitable HPLC method; to the eluent cationic dyes (either absorbing in the UV or visible region or fluorescent) are added in a solvent immiscible with the eluent. Ion pairs between the anionic surfactants and the cationic dye are formed that are soluble in apolar solvents. The technical problem to be solved is the separation of the two phases (polar phase containing the cationic dye in excess and apolar phase containing the ion pair) on-line without excessive loss in resolution caused by dead volume of the phase separator. Only compounds having anionic charge and structural elements (like alkyl chains) that favor extraction of the ion pair are detected; the concentration of the ion pair extracted is proportional to the amount of anionic surfactant in the eluent in much the same way as described for MBAS. The results are more specific than those of the MBAS method, however, because ion pair formation takes place after separating individual compounds. Gradient elution is restricted by the necessity to keep extraction of the non–ion-paired cationic dye low as well as to maintain two immiscible phases after ion pair formation. Originally, solvent-segmented phase separators (adapted from Auto-Analyzers) were used to effect phase separation [126,127,129]. Alternatively, a porous polytetrafluoroethene membrane separator was used to accomplish phase separation [128]. Recently, a sandwich phase separator originally introduced for the analysis of cationic surfactants [130] has been employed for analysis of anionic surfactants using the fluorescent cation 1-cyano-[2-(2-trimethylammonio)ethyl]benz[f]isoindole for ion pair formation [131]. Alkyl sulfates and alkylsulfonates were separated according to alkyl chain length on a RP column by a suitable gradient. Absolute detection limits are below 30 ng for individual homologs of alkyl sulfates or alkylsulfonates. A drawback is that detector response is linear over only two orders of magnitude in concentration of anionic surfactant [131]. Although the introduction of on-line ion pair extraction systems as HPLC detector is more demanding than using a UV detector, automation of this system is possible. Once working routinely, these systems may prove a versatile and sensitive detection principle for anionic surfactants [126–131]. None of them has yet reached the stage of routine use in environmental analysis of surfactants, however.

The most universal and selective detector for HPLC effluents is the mass spectrometer: it can generate information on molecular identity for all classes of compounds without derivatization. Coupling the MS to the HPLC, however, is not a trivial problem. The various methods of coupling HPLC to MS were recently reviewed [132–134] and are not covered here. Applications of LC-MS to surfactant analysis have been described for various classes of surfactants. Flow injection analysis of drinking and waste water samples has been performed using tandem MS for analysis mainly of nonionic surfactants [135,136]. Alkyl sulfates [137] and alkyl ether sulfates [138] have been analyzed semiquantitatively using standard solutions. Nonionic surfactants (APEOs, AEOs, and APEO metabolites) have been quantified from standard solutions [139] and water samples [140]. Although quantification of single compounds using LC-MS was attempted, it seems fair to state that the strength of LC-MS at the moment is in qualitative analysis. Characterization of unknown peaks in HPLC chromatograms can be performed by recording a full spectrum for that particular peak; alternatively, qualitative identification can be corroborated by tandem MS. In quantitative analysis, however, problems are expected because of insufficient stability of interface and ionization conditions during the time needed for HPLC separation. Thus, isotopically labeled internal standards are needed for every compound to be quantified; these are not normally available. Furthermore, equipment is expensive and demand on operator time is high. Consequently, HPLC-MS has not yet found as much application as GC-MS, although it is offering great perspectives for future developments.

(b) Cationic Surfactants. Cationic surfactants are used as fabric softeners and antistatic agents. Their annual consumption is significantly less than that of anionic or nonionic surfactants, respectively. The major cationic surfactant in use in Europe today is distearyldimethylammonium chloride (DSDMAC), a compound containing only aliphatic residues. Cationic surfactants containing benzyl or pyridinium residues (which implies a significant UV absorption useful for quantitative determination [141]) have only an insignificant share of the market. Thus, the situation for HPLC analysis of cationic surfactants [108] resembles largely that for aliphatic anionic surfactants: the presence of only aliphatic residues in the molecules to be analyzed necessitates special research efforts to develop sensitive and specific detection methods.

Conductimetric detection has been most widely used for the determination of the positively charged cationic surfactants [142–147]. Because cationic surfactants interact very strongly with polar surfaces, normal phase HPLC separations are commonplace. In this separation mode, a polar stationary phase (usually a cyanopropyl bonded phase) is used to effect separation with an organic mobile phase (usually mixtures of chloroform and methanol) to elute individual compounds.

As stated earlier, with conductimetric detection an isocratic elution mode is mandatory. Thus, early eluting peaks are expected to be sharp and later eluting peaks become increasingly broad, lowering resolution and rendering quantification more difficult (see Fig. 5).

An analytical detection limit of 0.2 µg was reported for DSDMAC in HPLC using conductimetric detection [148]. To be useful in environmental analysis, a highly efficient preconcentration and isolation procedure is mandatory for the different environmental matrices to accumulate enough cationic surfactant in the final extract before HPLC and thus to exceed the detection limit. Nevertheless, analytical procedures have been reported for river water [142,143,145,147], waste water [142–146], sludges [144–146], and sediments or soils [144]. DSDMAC concentrations found were as low as 13 µg/liter for river water [143], 175 µg/liter for influent sewage [143], 28 µg/liter for sewage effluent [143], 3 mg/kg for activated sludge [144], <50 mg/kg for river sediment [144], and <1 mg/kg for agricultural soils without sludge application [144]. In comparison with data obtained with the nonspecific DSBAS method (see Sec. II.B.1.b) from identical samples, values from HPLC with conductimetric detection were significantly lower, indicating successful separation of nonsurfactant cationic compounds by HPLC [144].

Indirect photometric detection methods have been employed for HPLC anal-

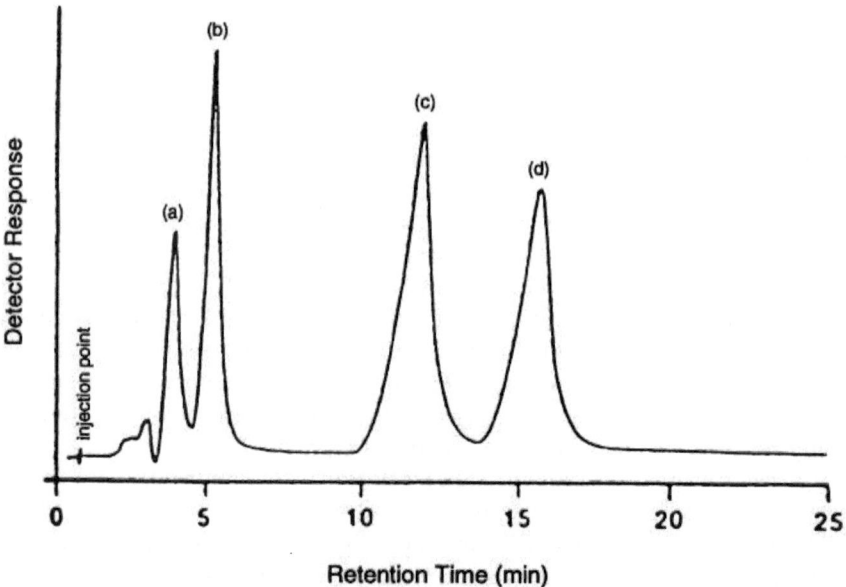

FIG. 5 HPLC chromatogram of four different cationic surfactants on a normal phase column using isocratic elution and conductivity detection: (a) DSDMAC, (b) hexadecyldimethylbenzylammonium, (c) 1-hexadecylpyridinium, and (d) dodecyltrimethylammonium. (Reprinted with permission from Ref. 142. Copyright 1982 American Chemical Society.)

ysis of cationic surfactants as well, albeit mainly for product analysis in detergent formulations [148,149]. As stated earlier for aliphatic anionic surfactants, the monitoring of physical properties of the bulk mobile phase (as is pertinent for indirect photometric detection) has a number of serious setbacks. For this reason, application of indirect detection modes in environmental analysis appears to be rather improbable. An evaporative light scattering detector has also been used for analysis of cationic surfactants in commercial formulations [150]. Specific conditions for successful quantification, however, seem too stringent for application to trace analysis.

Combination of HPLC separation of individual cationic surfactants with nonspecific colorimetric ion pair formation (see Sec. II.B.1.b) as an on-line postcolumn detection mode has great potential as an alternative method for qualitative and quantitative analysis of cationic surfactants. Analogous to the situation detailed earlier for aliphatic anionic surfactants (Sec. II.C.1.a), a suitable fluorescent or UV-absorbing anionic dye is added to the HPLC eluent in an immiscible solvent. Ion pairs are formed between the cationic surfactant and the anionic dye;

these are extracted into the nonpolar, organic phase. This organic phase is separated on-line from the aqueous-rich phase using an efficient phase separator and analyzed by optical spectroscopy in a UV or fluorescence detector [151–155]. Signals are generated that are proportional to the concentration of individual cationic surfactants in the eluent. Although separations on RP columns have been reported [152], normal phase HPLC seems to be better suited for the analysis of cationic surfactants [153–155]. A typical chromatogram is shown in Fig. 6. Using 9,10-dimethoxianthracene-2-sulfonate as the anionic dye in conjunction with fluorescence detection, an absolute detection limit of 2 ng can be realized using standard cationic surfactants [154]. Calibration curves are linear over two orders of magnitude in concentration. Gradient elution, in contrast to conductimetric detection, is possible; it is restricted, however, by the necessity to keep extraction of the non–ion-paired anionic dye low while maintaining two immiscible phases. The main difference between the approaches is in the construction of different types of phase separators. Porous polytetrafluoroethene (PTFE) membrane separators make use of the fact that aqueous solutions are repelled from PTFE surfaces; consequently, only organic solvents can penetrate this membrane [151,152]. Alternatively, the different wetting characteristics of PTFE and steel surfaces are exploited in a sandwich phase separator [130,153–155]. The aqueous phase is mainly attracted to the steel surface and the organic phase is concentrated on the PTFE surface from which it is guided to the detector. In the analysis of cationic surfactants, the use of on-line ion pair extraction systems

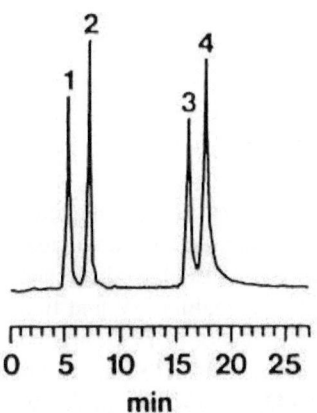

FIG. 6 HPLC chromatogram of four different cationic surfactants on a normal phase column using gradient elution in conjunction with postcolumn derivatization, a phase separator and fluorescence detection: (1) dioctadecyldimethylammonium, (2) didodecyldimethylammonium, (3) octadecyltrimethylammonium, and (4) dodecyltrimethylammonium. (Reprinted from Ref. 154 by permission of VCH Verlagsgesellschaft.)

as HPLC detectors (although they are technically more demanding) seems to be competitive with the standard conductivity detector because of its significantly higher sensitivity. Absolute detection limits of 2 ng are reported for DSDMAC from standard solutions [154]. Automation of this system is possible. After initial assembly of the detection system, its operation is fairly straightforward and easy. As a further advantage, aliphatic anionic surfactants can be analyzed by the detector assembly with only minor modifications.

(c) Nonionic Surfactants. The two groups of nonionic surfactants mainly used in Europe are alcohol ethoxylates (AEO) and alkylphenol ethoxylates (APEO), which have a combined market share of 80%. Typical alcohols used in the hydrophobic portion of nonionic surfactants span the range from octanol to octadecanol, typical alkylphenols are branched-chain octyl- or nonylphenol, and the typical degree of polymerization of the hydrophilic polyoxyethylene chain varies from 4 to 20 (EO) units. Nonionic surfactants are thus complex mixtures in both the hydrophobic (alkanol or alkylphenol) as well as in the hydrophilic (polyoxyethylene) parts of the molecule.

Consequently, HPLC separation of nonionic surfactants into individual surfactant molecules is a two-dimensional problem, which is inherently more complex than the separation of either anionic or cationic surfactants [156]. Using different HPLC stationary phases, one can separate nonionic surfactants either by interaction of the hydrophobic part of the molecules with RP columns (irrespective of the degree of polymerization of the polyoxyethylene chain) or by interaction of the polyoxyethylene chain with normal phase columns (irrespective of the nature of the hydrophobic residue; see Fig. 7) [96]. Both these sets of analytical data are incomplete, but complimentary. Attempts have been made to achieve complete separation of individual molecules of AEOs on a single column in a single run [157–159]. Although successful separation was achieved for standard compounds, chromatographic conditions are so sophisticated that application of this powerful, but a time-consuming procedure for environmental analysis seems to be out of reach at the present time.

Chromatographic separation and quantification of APEOs can be achieved at high sensitivity because the aromatic ring absorbs UV radiation and can be used in fluorescence emission spectroscopy. Thus, unchanged APEO have been analyzed in various environmental matrices using HPLC and mainly UV and fluorescence detection [42,95,96,99,102,109,156,160–169]. Octyl- (OPEO) and nonylphenol ethoxylates (NPEO), the majority of commercial APEOs, can be separated clearly into two peaks using RP columns [42,96,156], one peak containing OPEO and the other containing NPEO. Isomeric composition of the alkyl side chain of APEOs leads to broad peaks; no separation of different isomers was attempted, however. A minimum number of peaks is obtained under these conditions, leading to the lowest possible limits of quantitation (65 ng for NPEO)

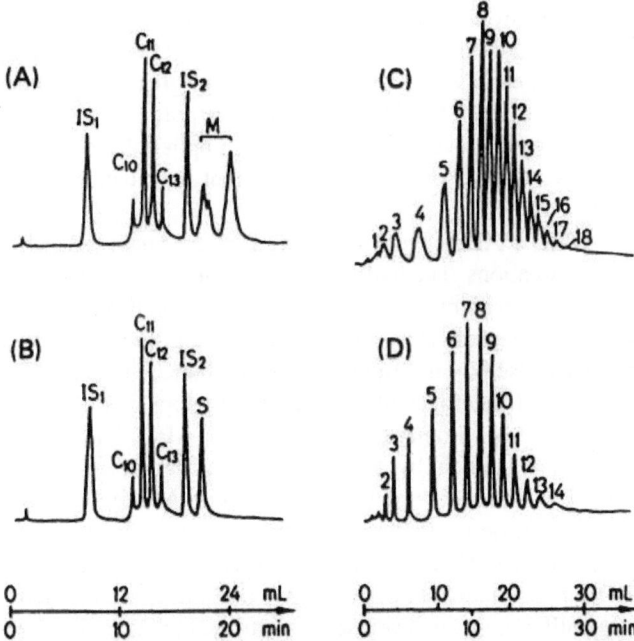

FIG. 7 Reversed phase (A, B) and normal phase (C, D) HPLC separation of LAS and APEO (A), LAS and OPEO (B), APEO (C), and OPEO (D) polyoxyethylene polymer homologs; for details see original paper. (Reprinted with permission from Ref. 96. Copyright 1987 American Chemical Society.)

[42,95,96,156]. The detector response, however, is proportional to molar concentration, because only the hydrophobic part of the molecule contributes to the recorded signal. To convert molar concentrations to mass concentrations, information on the mean length of the polyoxyethylene chain is needed. This information can be obtained by HPLC separation using a normal phase system; detection limits, however, are significantly higher using this approach because of the larger number of individual peaks (1 µg/liter for NP3EO and 3 µg/liter for NP18EO, respectively [156]). Molecules with a short polyoxyethylene chain (0–6) can be separated as efficiently as those comprising longer, more hydrophilic EO chains. This is in clear contrast to the determination of summary parameters, such as BiAS or CTAS (see Sec. II.B.1.c), for which a minimum of 5 EO units is necessary to generate a signal. Thus, short-chain, hydrophobic compounds, such as nonylphenol, NP1EO, and NP2EO, were quantified using normal phase HPLC from water, waste water, and sewage sludges [42,95,96,99,102,109,156,161,163,170]. These compounds, which cannot be quantified by BiAS or CTAS, are metabolites of NPEOs with longer polyoxy-ethylene chains. Because they are both more toxic and more lipophilic than the parent compounds, their presence in waters and sludges was of great importance for their ecotoxicological evaluation [41,42] because of their potential bioaccu-mulation. As a consequence of this evaluation, detergent suppliers have volun-tarily reduced or discontinued the use of APEOs in detergent formulations over most of Western Europe.

As replacement for APEOs, AEOs were introduced as the main nonionic surfactants in detergent formulations. The hydrophobic part of AEOs consists of n-alkanols with a chain length ranging from 8 to 20; only a small number of branched-chain alkanols are expected. Thus, the heterogeneity in the hydropho-bic part of AEOs is mainly caused by different homologs, whereas for APEOs, isomers were the main cause of heterogeneity. Because homologs can be sepa-rated more easily than isomers by chromatographic methods, AEOs lend them-selves to more specific determination by HPLC. Sensitive determination of AEOs, however, is a more difficult problem in comparison with APEOs, because AEOs contain only aliphatic residues that are difficult to detect by sensitive HPLC detectors. An evaporative light scattering detector has been used to quan-tify underivatized AEOs in commercial formulations; the specific conditions necessary for efficient quantification, however, seem so stringent that application for environmental samples does not appear viable [157,158]. Attempts have also been made to use UV detection at short wavelength (220 nm) to analyze under-ivatized AEOs in conjunction with normal phase HPLC. Successful separation was demonstrated using standard compounds. Addition of traces of anthracene was necessary to improve baseline stability [171]. Again, it seems improbable that it will be possible to apply this method to environmental samples, because UV absorption at 220 nm is rather nonspecific. Because of the large number of potential interfering compounds to be expected in environmental samples, very extensive precleaning operations would be the only chance to maintain speci-ficity for nonionic surfactants.

The method of choice to increase sensitivity in the analysis of aliphatic com-pounds, however, is the introduction of UV-absorbing or fluorescent residues by suitable derivatization reactions. For hydroxylic compounds like AEOs, the use of acid chlorides or isocyanates to form esters or urethans, respectively, is a rational choice. Sensitive detection is possible, for example, for esters derived from 3,5-dinitrobenzoyl chloride [172–174] or for urethans derived from phe-nylisocyanate [17,108,175,176]. In both cases, aromatic residues are introduced that enable UV or fluorescence detection. Furthermore, the amphiphilic nature of AEO is lost during the derivatization reaction at the hydrophilic end. The result-ing esters or urethans are far less prone to adsorption at surfaces or concentration at interfaces; thus, handling of small amounts of AEOs in environmental analysis is facilitated after derivatization.

Employing 3,5-dinitrobenzoyl chloride for derivatization requires the use of a number of auxiliary chemicals, such as pyridine, which results in a tedious reaction sequence [172,173]. Phenylisocyanate, however, can be added neat to the fraction containing AEOs in a nonpolar solvent; the resulting solution of urethans can be analyzed by HPLC after filtration or centrifugation [175,176].

In much the same way as described previously for APEOs, phenyl urethans derived from AEOs can be separated into homologous series of different ethoxylated n-alkanols on a RP column (irrespective of the degree of polymerization of the polyoxyethylene chain; see Fig. 8A) or into a polymer homologous series of polyoxyethylene oligomers on normal phase columns (irrespective of the nature of the hydrophobic alkanol; see Fig. 8B) [176]. As was true for APEOs, the detector response to phenyl urethans derived from AEOs is proportional to molar concentration. In conjunction with a suitable isolation and preconcentration procedure, it is possible to quantify AEOs at environmentally relevant concentrations. A round-robin study has shown good reproducibility of data obtained for waste water influent and effluent streams. The limit of quantification stated was 100 μg/liter [176]. This is higher than desired.

One reason for this fairly high limit of quantification is the coexistence of APEOs and AEOs in environmental samples. APEOs also form derivatives during the derivatization procedure; these APEO derivatives have a higher extinction coefficient than corresponding AEO derivatives. Furthermore, peak overlap between APEO phenyl urethans and AEO phenyl urethanes was observed in both chromatographic modes, making quantification especially difficult in samples containing low concentrations of both, such as effluent streams or surface waters [176]. Thus, further improvements are clearly desirable.

Experiments have been performed to separate complexes of APEO or AEO with metal cations (K$^+$ in this special case; see Sec. II.B.1.c) on a cation exchange resin column. Standards were separated according to EO chain length, and detection was achieved by UV for APEO and by refractive index or conductivity for AEO. Temperature programming was used to increase chromatographic resolution [177,178]. No attempts have been made to analyze environmental samples.

Finally, a method commonly used in the analysis of the hydrophobic part of AEOs in product analysis, namely the cleavage of AEO by HI to yield alkyl iodides, has been used to derivatize AEOs. HPLC separation of the alkyl iodides formed was achieved using standards [179]. Again, no attempts have been made to analyze environmental samples. Furthermore, GC separation of alkyl iodides should be a more favorable finish than HPLC.

2. Gas Chromatography (GC)

As a separation technique, GC is inherently more powerful than HPLC. This is especially true of capillary column GC, which has a very high resolving power.

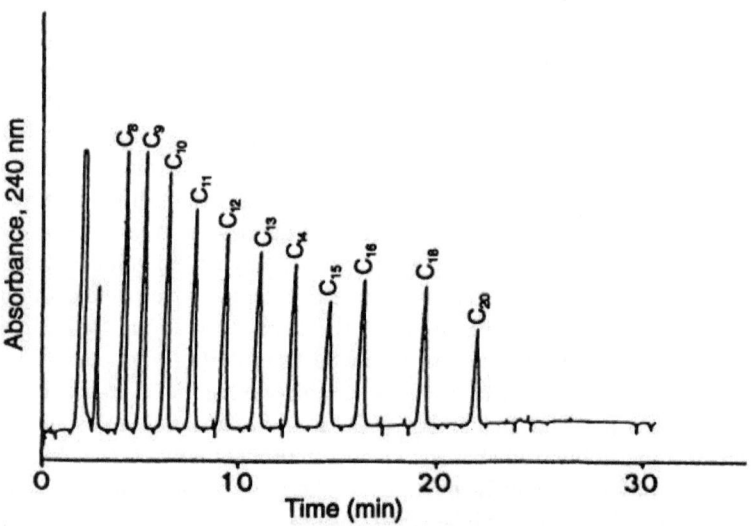

FIG. 8A HPLC separation of AEOs (derivatized to form phenyl urethanes) on a RP column showing separation according to hydrophobic chain length. (Reprinted from Ref. 176 by permission of the American Oil Chemist's Society.)

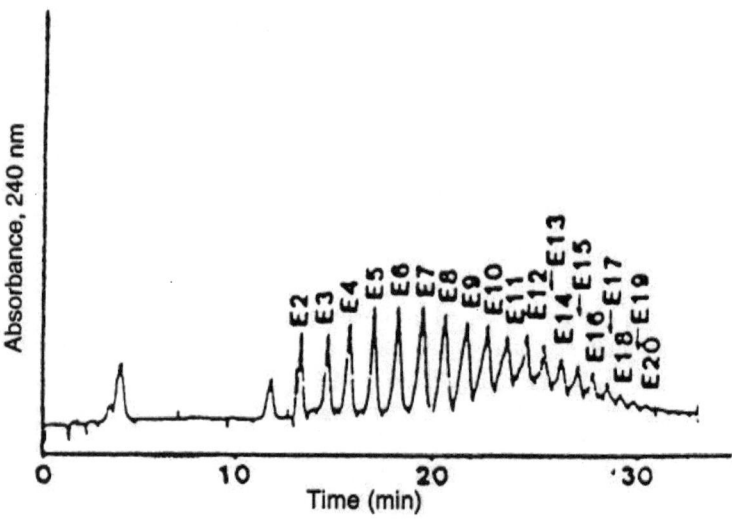

FIG. 8B HPLC separation of AEOs (derivatized form phenylurethanes) on a normal phase column showing separation according to EO number. (Reprinted from Ref. 176 by permission of the American Oil Chemist's Society.)

Furthermore, detection using a flame ionization detector (FID) is universal for carbon-containing compounds; FID has a linear dynamic range covering five orders of magnitude, which facilitates the determination of mixtures containing substances of greatly different concentrations.

Compounds that can be analyzed by GC must have a significant volatility at ambient pressure and at a maximum temperature of about 300°C. Furthermore, they must be thermally stable under these conditions. Among the surfactants, only the nonionic surfactants with a short polyoxyethylene chain are amenable to direct determination using GC [41,180–182]. Nonionic surfactants with longer polyoxyethylene chains, as well as both anionic and cationic surfactants, must be derivatized using suitable reagents to achieve sufficient volatility of the derivatives for GC analysis.

(a) Nonionic Surfactants.   APEOs can be analyzed by GC without derivatization as long as the polyoxyethylene chain is shorter than 8 EO units. This task was accomplished long ago using packed columns [180]. On a packed column, structural isomers of APEOs are not resolved; only one peak is recorded for every polymer homolog. NP, NP1E0, and NP2EO, for example, yield different peaks, and the different alkyl chain isomers of NP1EO coincide into the same peak on packed columns [180]. Using capillary columns increases the resolving power of the method: apart from separating the polymer homologs NP, NP1EO, and NP2EO, a complex pattern is obvious for every homolog, indicating separation of individual alkyl chain structural isomers (see Fig. 9) [181]. Identification of individual isomers was attempted using GC-MS and GC-Fourier transform IR [183]. Only a few applications of the GC method for environmental samples have been reported [41,181]. In these sewage sludge samples, the polyoxyethylene side chain was largely degraded; the main compounds observed were the more lipophilic NP, NP1EO, and NP2EO. Using the GC method it is possible to obtain further information on the molecular structure of compounds analyzed by employing a mass spectrometer (MS) instead of an FID as the detector. By GC-MS, identification of individual molecules may be possible in the presence of significant matrix interference.

AEOs can also be separated into single compounds by GC using a short packed column. Because there is no heterogeneity in the alkanol residues, separation into individual molecules is effected even on a low-resolution packed column without derivatization. Alkyl chain lengths from octanol to hexadecanol were investigated using standard substances; the oxyethylene side chain varied from zero to three EO units. Baseline separation was obtained for all individual compounds on a 50 cm packed column [182]. Only standards were investigated, and no attempts were reported to analyze environmental samples.

Derivatization reactions to transform the alkohols into esters or ethers normally increase the volatility while reducing the polarity of the parent compounds.

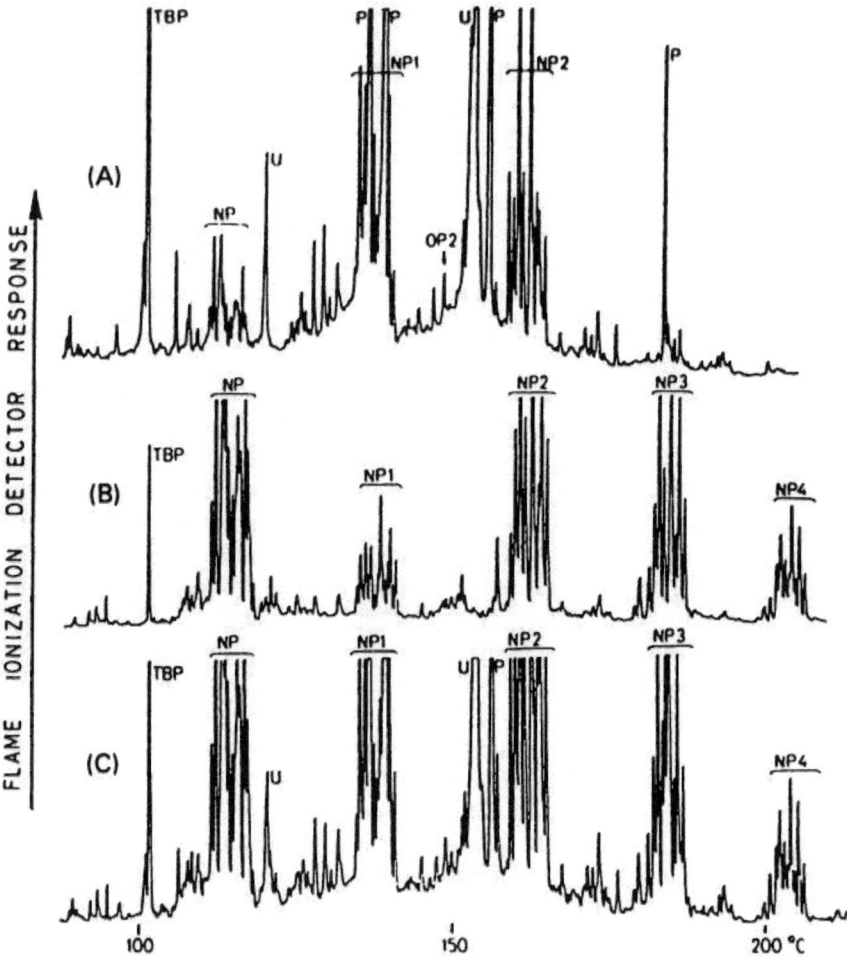

FIG. 9 Glass capillary GC separation of APEOs: (A) extract of secondary sewage effluent; (B) reference mixture of NP and NPEO; (C) coinjection of A and B. TBP: 2,4,6-tribromophenol, U: unknown, P: phthalate. (Reprinted with permission from Ref. 181. Copyright 1982 American Chemical Society.)

By intelligent choice of the derivatization reagents, derivatives can be synthesized that can be detected by more specific and/or sensitive detectors. Thus, using heptafluorobutyric anhydride or pentafluorobenzoyl chloride to derivatize NPEOs, the resulting perfluoro esters can be specifically and sensitively detected with an electron capture detector (ECD) after GC separation. On a 25 m BP-1 fused silica column, polymer homologs were separated from NP to NP7EO for the heptafluorobutyric ester and from NP to NP6EO for the pentafluorobenzoic ester, respectively. The complex pattern of alkyl chain structural isomers was also clearly evident on every single peak [184]. Analysis of environmental samples, such as water and sewage sludges, again clearly indicated that NP was the most abundant single compound, with NP1EO and NP2EO as minor components only traces of NP3EO were present [184].

Metabolites of NPEOs, namely nonylphenoxy carboxylic acids (formed by oxidation of the terminal hydroxyl group to a carboxylic acid), can also be analyzed by GC-MS after derivatization. Methyl esters were formed that were chromatographed on a glass capillary GC column coated with SE-54. MS was employed for molecular identification. With a detection limit of 1 µg/liter, these compounds were easily detected in sewage effluents [185].

(b) Ionic Surfactants.   Neither cationic nor anionic surfactants can be separated by GC without derivatization because their volatility is too low and their polarity too high.

Cationic surfactants have been degraded by a chemical reaction called Hofmann degradation to yield tertiary amines as products. This reaction has only rarely been applied with standard solutions [186] because the reaction more often than not yields multiple products from the same cationic surfactant. No applications to environmental analysis are known to the author.

For anionic surfactants, a small number of different reaction pathways can be utilized with the aim of forming volatile compounds closely related to the original anionic surfactants.

LAS are produced by sulfonation of a technical mixture of linear alkyl benzenes (LAB). This reaction is one of the few reversible electrophilic aromatic substitution reactions known in organic chemistry. Consequently, strong acids (like concentrated phosphoric acid) can be utilized to effect the reverse reaction, namely the regeneration of LAB from LAS. This desulfonation reaction was employed relatively early after the introduction of LAS to obtain specific analytical data on LAS (in contrast to the nonspecific MBAS data) [14,15,187–189]. The facilitated extraction of LAB (in comparison to LAS) into organic solvents and their relatively high volatility made this degradation derivatization the method of choice before the introduction of HPLC. Most of the individual isomers of LAB were effectively separated from one another even in the early days [188,189]. It is not entirely clear, however, whether the isomer distribution of LAS is conserved in LAB during the desulfonation reaction, because Friedel-Crafts alkylations are also reversible, resulting in equilibrium mixtures. Method detection limits of 50 µg/liter for environmental samples were reported [189]. Although improvements have been made that have reduced the method detection limit to 10 µg/liter of LAS and less than 1 µg/liter for individual isomers [14], the desulfonation method (even when adapted to micro amounts) is time consuming. HPLC methods that avoid derivatization have become competitive in recent years for LAS.

Alternative derivatization methods use synthetic methods for derivative formation, thus changing only the nature of the functional group. LAS were converted to their sulfonyl chloride derivative by PCl_5, $COCl_2$, or $SOCl_2$. The resulting sulfonyl chlorides can be directly analyzed by GC and GC-MS because they are very volatile [190,191]. Individual LAS isomers were clearly separated on a 19 m glass capillary column coated with PS 255 or SE54. LAS from sewage sludges was qualitatively and quantitatively determined using this derivatization process; an FID was used as a detector in routine analysis and MS was utilized in confirmatory qualitative analysis [190,191].

Sulfonyl chlorides are reactive and sensitive derivatives, however, that are fairly unstable in the presence of moisture. Further synthetic reactions to yield esters (from alcohols and sulfonyl chlorides) or amides (from amines) produce less reactive and sensitive derivatives that are easier to handle. Methyl esters [192], trifluoroethyl esters [193], and N-methylanilide derivatives [194] have all been synthesized from LAS via their sulfonyl chlorides. MS and ECD have been used as detectors, and environmental matrices were successfully analyzed using these derivatives [192–194].

For LAS, however, GC methods have not reached the stage of routine application in environmental analysis because HPLC methods produce comparable results with less sample preparation and operator maintenance. For confirmatory compound identification, in contrast, GC-MS methods are superior to all methods because a maximum of structural information can be obtained from a single chromatogram.

Methods of derivatization for aliphatic anionic surfactants have mainly been described for product analysis; environmental applications were seldom claimed. Methyl esters of aliphatic sulfonates [195], as well as dibutyl sulfonamides [196], were synthesized via their respective sulfonyl chlorides. Direct derivatization of sulfonic acids to yield *tert*-butyldimethylsilyl derivatives has also been described [197]. Alkyl sulfates were derivatized to form trimethylsilyl esters that were quantified using an FID. Alkyl sulfates were detected in influent waste water but they were below detection limit (5 µg/liter) for receiving waters [198]. Finally, the hydrolysis of alkyl sulfates and alkyl ether sulfates with HI leads to the formation of alkyl iodides, conserving only the information on the alkyl chain [199]. Alkyl iodides can easily be separated by GC.

(c) Surrogate Parameters for Ionic Surfactants. Anionic and cationic surfactants, as used in detergent formulations, are technical mixtures of various compounds that furthermore contain percentage amounts of their synthetic precursors. Thus, LAS generally contain significant amounts of LAB, and quarternary ammonium (cationic) surfactants are accompanied by tertiary amines or alkyl

nitrile precursors or by-products. In contrast to ionic surfactants, their precursors are not amphiphilic compounds. Consequently, they can be easier concentrated from environmental samples and are better suited to GC analysis.

Alkylbenzenes, precursors for both tetrapropylenebenzene sulfonates (TPS) and LAS, have been extracted and quantified by GC (with both FID and MS as detectors) from waters and marine sediments [200,201], river sediments [202], and sludge or sludge-amended soils [203]. To be useful as surrogate parameters for TPS or LAS, however, the corresponding alkylbenzenes should have a similar distribution pattern and similar biodegradability to LAS or TPS. Because of their different physical and chemical properties, this cannot be taken for granted. In contrast, LAB are not well biodegraded compared with LAS, which is readily biodegraded. Thus, the LAS/LAB ratio changes from 60:1 in anaerobic sludges (no LAS biodegradation) to 1:1 in aerobic sludges (LAS largely biodegraded) [203]. Further research is clearly needed to elucidate the potential use of LAB as surrogate parameters for LAS.

The same argument is applicable to precursors or by-products of cationic surfactants, namely alkyl nitriles or tertiary amines. These compounds were clearly identified in the marine environment near the sewage outfall of Barcelona [204]. Further investigations are necessary to assess the significance of these compounds as surrogate parameters for cationic surfactants.

3. Supercritical Fluid Chromatography (SFC)

In SFC, gases above their critical temperature and pressure are used as mobile phases. Under these conditions, the supercritical fluids have densities like liquids while retaining diffusion coefficients of typical gases. Simplifying, one may say that SFC allows combining the advantages of HPLC and GC in a single method. More specifically, analytes must be soluble only in the supercritical fluid for SFC to be applicable; no volatility of analytes is required (comparable to HPLC). In addition, universal and sensitive detectors like FID can be applied that require only the presence of carbon in the analytes (comparable to GC). Consequently, no derivatization of analytes is required, neither to increase volatility nor to increase detectability.

Among the surfactants, nonionic surfactants have been separated by SFC [17,205–209]. Polymer homologs with varying polyoxyethylene chain lengths are clearly separated; homologous series of two different alkanol (tetradecanol and pentadecanol) are also clearly separated in a single chromatogram (see Fig. 10) [208]. APEOs of different degrees of ethoxylation can also be clearly separated [207]. Even homologs with long polyoxyethylene chains do not pose specific problems because no volatility is required. SFC and high-temperature GC have been compared with respect to their ability to separate AEOs, leading to the conclusion that both techniques are complementary rather than competitive [210].

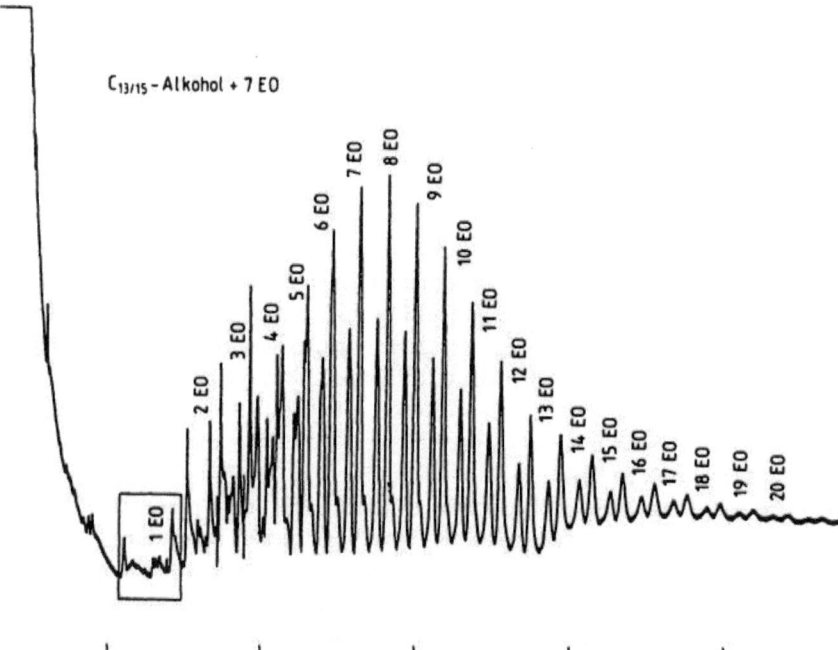

FIG. 10 SFC separation of underivatized AEO standard substances showing separation by both hydrophobic chain length (C_{13} and C_{15}) as well as by EO number. (Reprinted from Ref. 208 by permission of Carl Hanser Verlag.)

No reports on the determination of surfactants in environmental matrices using SFC is known to the author. Only applications in product analysis have been published. This may be because SFC equipment is not yet as common as HPLC or GC equipment. For the future, however, the potential advantages of SFC should make this method a valuable tool in environmental analysis of surfactants, especially of the nonionic type.

4. Electrophoretic Methods

Capillary electrophoresis (CE) or isotachophoresis are separation techniques that have a very high resolving power for charged compounds. Because they use empty capillaries to effect separation by electrophoretic movement, they are not chromatographic methods in the strict sense. They have, however, similar performance for ionic surfactants compared with HPLC. No analyses of environmental samples have been reported; only separations of technical or standard mixtures have been described for both anionic [211,212] and cationic [212,213] surfactants. The prime advantage of CE seems to be that the absolute sample size can be minimized because injection volumes are very low. The analyte concentration in the sample, however, must be at least as high as in HPLC. Because, normally, analyte concentration is the major problem in environmental analysis (not absolute sample size), only a few special cases may be amenable to CE analysis.

D. Isolation and Preconcentration Procedures

Because all the methods for qualitative and quantitative determination of surfactants described previously suffer various positive and negative interference from matrix components or other compounds similar to the analytes, separation methods are mandatory for almost all analytical procedures to isolate and preconcentrate surfactants from environmental matrices. The aim of these separation methods is to enrich the surfactants or surfactant classes to be analyzed while removing as much as possible of the potentially interfering compounds. Depending on the detection limits of the final determination method employed, concentration factors can vary by orders of magnitude for similar compounds; for example, mg amounts of cationic surfactants must be collected if conductivity or refractive index detection in conjunction with HPLC is employed, whereas μg amounts are sufficient for postcolumn ion pair fluorescence detection in conjunction with HPLC (see Sec. II.C.1.b).

Because of the amphiphilic nature of surfactants, they tend to concentrate at all phase boundaries. Losses to surfaces from aqueous solutions are therefore commonplace. Especially for the low concentrations normally encountered in environmental analysis of surfactants, quantitative recovery of analytes becomes a major problem, especially for such matrices as sewage, sludge, sediment, or soil. It is common practice in organic analysis to add suitable internal standards to correct for nonquantitative recovery during isolation and quantification of individual compounds. This approach can only be used with highly selective (usually chromatographic) methods of determination, however; it is useless for nonspecific methods because they cannot discriminate initially present surfactant from added internal standard. Furthermore, it is difficult in surfactant analysis to find suitable internal standards, because (1) many potentially useful candidates are used in various detergent preparations and therefore are suspected to be present in the environment; and (2) other candidates not present in detergent formulations may not be optimally suited because of altered partition behavior (as a consequence, for example, of their hydrophobic chain length). Despite these difficulties, internal standards have been selected and used in a variety of procedures, especially in chromatographic methods incorporating a (potentially nonquantitative) derivatization reaction [14,42,95,96,99,103,111,161,167,176,181,184,189–191, 193,214,215]. Alternatively, the analysis of spiked samples was used to correct for recovery, in nonspecific analytical procedures (for which it is the only possible correction method) as well as chromatographic methods [93,144,145,153,187, 188,216].

Most methods described in the literature have the aim of concentrating only one class of surfactants (anionic, cationic, or nonionic) while discarding the other classes; thus, methods can be optimized for a particular class of surfactants. The price to be paid for this approach in environmental monitoring, however, is the correspondingly increased number of samples to be handled. Because separation methods tend to be tedious and labor intensive, this may be a serious drawback in some circumstances.

Eliminating matrix components and nonsurfactant interferences as completely as possible is especially important when using nonspecific determination methods, such as the widely used MBAS or DSBAS methods (Secs. II.B.1.a and b), which respond to a large number of organic or inorganic anions or cations, respectively, whether these are surfactants or not. Matrix effects are therefore a constant challenge when using a nonspecific method. Using chromatographic, selective, and specific methods, matrix effects are generally less important because of the chromatographic process, although they are never completely absent, especially in highly polluted samples.

In general, isolation and preconcentration procedures must be optimized for every combination of environmental matrix and quantitative analytical method. A number of building blocks, however, are available as tools that can usually be assembled in a modular way to provide an efficient isolation and preconcentration procedure. Possible building blocks are reviewed concisely as single entities; finally, a few examples of complete isolation and preconcentration procedures as sample preparation schemes for specific chromatographic analyses are presented.

1. Extraction from Matrix

At the beginning of an isolation and preconcentration procedure, a homogeneous solution of surfactant and interfering soluble matrix components is prepared for further processing. Samples consisting largely of solid substances, such as soil, sediment, sewage sludges, and biota, are extracted to remove surfactants from the solids as well as from interstitial water and transfer them into homogeneous solution. (For surface water samples this step can often be omitted.) Fairly drastic conditions are often employed to desorb surfactants from the solids; at the same time, care is taken not to degrade the more labile surfactants to be determined.

Polar organic solvents have been employed widely for the desorption of surfactants from solids. Methanol alone [15,86,93,94,98,104,106,108,203,215], methanol in combination with NaOH [96, 109, 161], and methanolic HCl [28,108,144,145,153,217] have been the most popular extraction solvents. Their miscibility with water and the addition of ionic auxiliaries have facilitated the desorption and dissolution of both hydrophobically and ionically bound surfactants. These extractants have been used in both a batchwise fashion or continuously using Soxhlet extraction [96,106,109,161,203]. Addition of surfactants other than those to be quantified with the analytical method has been utilized to minimize adsorption (for example, Ref. 96). Other organic solvents used to extract surfactants from solid matrices are n-hexane for nonionic surfactants with a low degree of ethoxilation [109,161] or dichloromethane [204].

Extraction of cationic surfactants from soil or solid matrices is an especially difficult problem, because ionic interactions predominate. A large number of organic solvents blended with various acids have been tested for optimum recovery [76,218]. Even digestion with HF in conjunction with evaporation of SiF_4 was tested and found to result in optimum recovery for DSDMAC from clays [218].

Finally, supercritical fluid extraction has been investigated for the extraction of LAS from sewage sludge [215,219]. Although supercritical CO_2 or N_2O alone did not effect significant recovery of LAS, either the addition of modifiers, like methanol [219], or the addition of quarternary ammonium salts as ion pairing reagents [215] resulted in nearly quantitative recovery of LAS.

Extracts prepared by the methods just covered must be further purified before quantitative analysis is possible; their main effect is removal of the major part of the inorganic matrix.

2. Solvent Sublation

Aqueous samples are usually filtered to remove particulates. Particulates are then treated as described for solid samples. Concentration of surfactants from the aqueous filtrate can be effected by solvent sublation. Using this technique, a solvent-saturated gas stream (N_2) is passed through the aqueous sample. Surfactants are enriched at the gas-liquid phase boundary and carried by the gas stream into an immiscible organic solvent (usually ethyl acetate). Surfactants are then dissolved and concentrated in the organic phase [11,25,27,48,77,108,144,156, 220]. Partition of the surfactant between the aqueous and organic phases under these conditions is a kinetically controlled nonequilibrium process; care must therefore be taken to obtain reproducible results. The solvent sublation technique is widely used as a preconcentration for aqueous samples; its application outnumbers all other techniques described in the literature.

3. Liquid-Liquid Extraction

For hydrophobic organic compounds, liquid-liquid extraction is the method of choice to concentrate them from aqueous solutions. Surfactants, however, do not have such a distinct preference for the organic phase. On the contrary, they are concentrated at phase boundaries. Thus, the attempt to extract surfactants directly from aqueous solutions into organic solvents without auxiliary measures is usually futile [181,184,221]. Formation of persistent emulsions occurs frequently under these circumstances, rendering phase separation difficult.

Formation of hydrophobic ion pairs, however, circumvents these problems. Not only do these ion pairs partition more completely into the organic phase, but emulsion formation is also suppressed under these conditions because the ion pair complexes formed are no longer surfactants. Cationic surfactants have been used to form complexes with anionic surfactants, and vice versa. These complexes between cationics and anionics are highly lipophilic and can easily be extracted from aqueous solutions [24,142,145, 153]. Methylene blue can also be employed as a cation auxiliary to extract anionic surfactants in much the same way as in the determination of MBAS (see Sec. II.B.1.a) [28,188,190–192]. The ion pairs extracted must be cleaved by chromatographic methods; ion exchange methods are employed most frequently.

For the concentration of nonionic surfactants, continuous liquid-liquid extraction (percolation) has been used successfully [181]. Centrifugal partition chromatography, also a kind of continuous liquid-liquid extraction, has been employed to concentrate nonionic surfactants from waste water into a minimum amount of hexane or octanol [222]. Analogously, steam distillation/solvent extraction was successfully employed to concentrate nonylphenol, NP1EO, and NP2EO from water [42]. All procedures seem to be rather special cases and have not been widely used.

4.   Solid Phase Extraction

In solid phase extraction, a hydrophobic solid material is utilized to adsorb hydrophobic compounds from aqueous solutions or polar mixtures. This separation method is also useful for surfactants, because they effectively interact with these solid materials via their hydrophobic end. Various solid phases have been used to concentrate surfactants, namely modified silica (reversed-phase) materials, such as octadecylsilica [95,104], octylsilica [94,98,104,106,193], or ethylsilica [103,214]. Carbopak B (a graphitized carbon black) has also been employed [89,111], as have styrene-divinylbenze copolymer resins, such as XAD-2 [15,89,108,176], XAD-4 [223,224] or XAD-8 [80,81,225]. Small amounts of organic solvents are usually sufficient to elute the adsorbed surfactants in a concentrated solution. These extracts are normally further processed by chromatographic steps.

5.   Desulfonation/Distillation

In the quantitative determination of LAS via desulfonation followed by GC analysis, the linear alkylbenzenes formed as products of the desulfonation reaction mixture (concentrated phosphoric acid) can be collected directly after distillation of the organic compounds. No further sample preparation is generally necessary before GC analysis using this special analytical technique [14,15,106, 187,189].

6. Chromatographic Methods

(a) Ion Exchange. For concentrating and purifying surfactants after primary extraction, ion exchange columns are most popular in isolation and preconcentration procedures documented in the literature [11,15,24,25,27,48,80,94,98, 103,104,106,108,110,176,192,203,214,216,220,225,226].

Ion exchange columns are used to isolate one single class of surfactants while discarding all others by isolating cationics on a cation exchanger, anionics on an anion exchanger, or nonionics from the eluate of a mixed-bed ion exchanger incorporating both cation and anion exchanger. Alternatively, a combination of anion and cation exchanger in series can be employed to isolate all three classes of surfactants by successive elution of the various ion exchange columns (for example, Ref. 194). Ion exchange columns, however, usually have a polymeric hydrophobic backbone (for example, a styrene-divinylbenzene copolymer similar to XAD resins) that can lead to hydrophobic interactions between the backbone structure and surfactants not usually trapped on this type of exchange column. This should be especially true when trace levels of surfactants are handled. Care must be taken to avoid these nonspecific interactions.

Nevertheless, ion exchange chromatography has been the workhorse of surfactant analysis for a long time, especially when only a single class of surfactant is analyzed during environmental monitoring. It is also incorporated into the standardized nonspecific procedures, such as MBAS or DSBAS [11,27].

(b) Electrophoresis. Although not a chromatographic procedure in the strict sense, electrophoresis can be utilized to effect separation of surfactants into different classes (anionic, cationic, or nonionic) depending on their charge in much the same way as ion exchange chromatography. To the best of the author's knowledge, only one such method has been described in the literature, in 1977, effecting separation by electrophoresis [227,228].

(c) Thin-Layer Chromatography. Concentration and purification of surfactants can be performed by thin-layer chromatography. Most often, SiO_2 was the stationary phase. The chromatographic properties of most surfactants were covered fairly comprehensively, both qualitatively and quantitatively, quite some time ago [220,227,228].

Purification of surfactants is possible using primary extracts [16]; more often, however, the separation was effected using the anionic surfactant/methylene blue complex [190,192] or the precipitate formed from nonionic or cationic surfactants and $Ba(BiI_4)_2$ [74,229]. After visualizing the fractions on the plate, they are removed by scraping, extracted, and used for quantitative analysis, most often in conjunction with molecular spectroscopy.

(d) Column Chromatography. Rather similar to thin-layer chromatography, column chromatography was employed to separate surfactants. Al_2O_3 [28,42,86,

108,156,204,217,229] was utilized more often than SiO_2 [78,181,204,224]. Both materials have also been used in solid-phase extraction cartridges to separate surfactant classes [230]. The separation was effected in these cases by interaction of the polar head group with the stationary phase. Apart from separating surfactant classes, nonpolar compounds that potentially interfere with the final analytical methods can be eliminated at this stage (for example, Ref. 217). Usually, no auxiliary reagents are necessary to improve separation at this point. In comparison to thin-layer chromatography, column chromatographic separations need less operator maintenance and are more amenable to automatization. Thus, for routine analytical methods, column chromatographic steps have a distinct advantage.

7. Complete Isolation and Preconcentration Procedures (Examples)

From the methods presented earlier as potential building blocks for isolation and preconcentration procedures, a minimum number is usually assembled in a modular way to obtain a sample fraction that can be quantitatively analyzed using a method as specific as possible. More steps are generally necessary to prepare sludge or sewage samples for final analysis than for receiving water samples. It is also true that for the more compound-specific analytical methods involving the application of internal standards as well as selective and sensitive detection systems (like HPLC or GC), less sample preparation effort is necessary for obtaining meaningful data than is the case with the nonspecific summary parameter methods.

For example, Fig. 11 shows the flow sheet for HPLC analysis of LAS, both from water samples and from sludge, sediment, or soil samples. Three preparatory steps are involved before final HPLC quantification [108]. Similarly, the flow sheet for the HPLC analysis of AEO nonionic surfactants from wastewater is shown in Fig. 12. Significantly more preparatory steps are necessary for this class of compounds before final HPLC quantification [176]. Similar schemes for the isolation and preconcentration of cationic surfactants from various matrices for HPLC analysis can be found in the literature [144].

The examples presented here aim at specifically isolating and concentrating one single class of surfactants for analysis while discarding all others. Often, auxiliary surfactants (of classes or structures not analyzed in the analytical method) are added to minimize adsorption losses. Thus, for a general surfactant monitoring exercise, a large number of samples must be handled under these circumstances. Only a few attempts have been documented in the literature tackling the challenging task of isolating all surfactant classes from a single water sample [226,231]. More work must be invested toward this end.

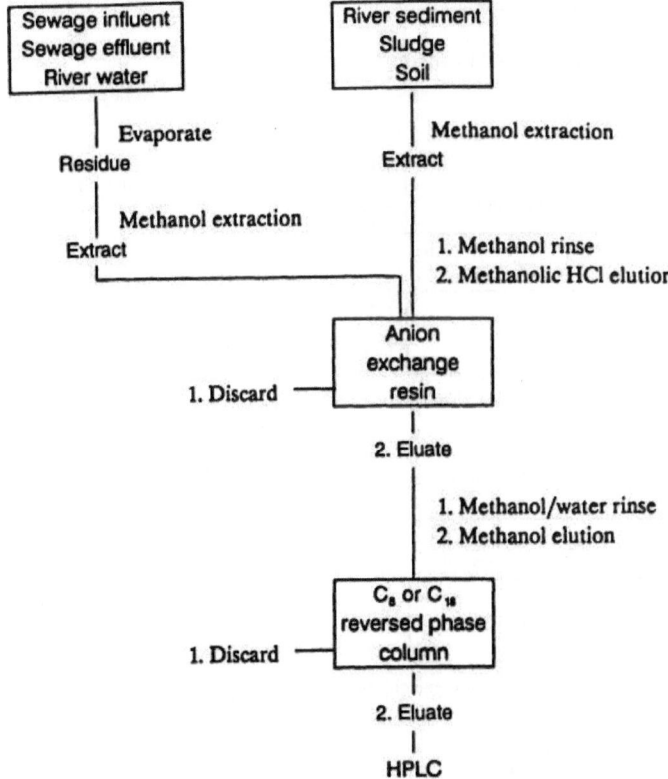

FIG. 11 Flow sheet for sample preparation for the HPLC analysis of LAS from environmental matrices. (Reprinted from Ref. 108 by permission of Carl Hanser Verlag.)

E. Summary

Although the number and variability of analytical procedures described in the literature for potential application to surfactant analysis at concentrations relevant to environmental occurrence is impressive, the actual number of procedures validated for environmental monitoring is rather limited. Often, environmental monitoring is performed using nonspecific methods, such as MBAS, DSBAS, BiAS, or CTAS. Given the limitations of these approaches, their use can be justified only for screening purposes, that is, to define samples that contain significant amounts of surfactants. These must be quantified using more specific, usually chromatographic methods to obtain valid data for risk assessment. For environmental matrices, validated chromatographic methods are available for LAS, APEO, and DSDMAC, respectively. Other compounds increasingly used

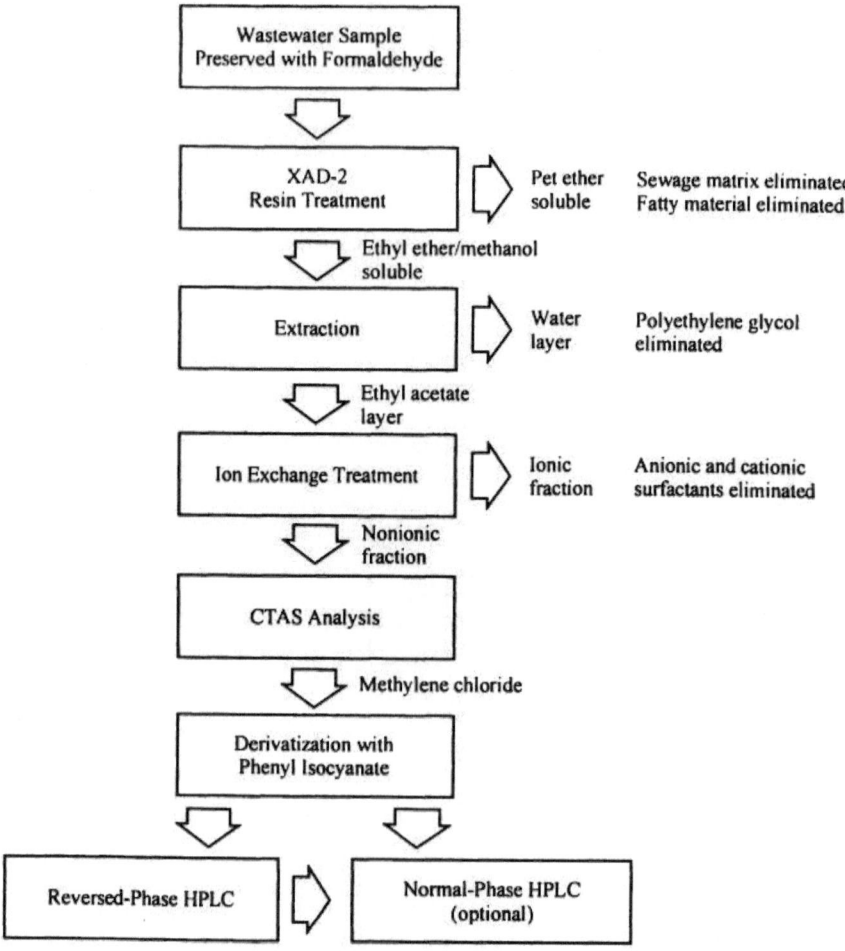

FIG. 12 Flow sheet for sample preparation for the HPLC analysis of AEO's from waste water. (Reprinted from Ref. 176 by permission of the American Oil Chemist's Society.)

in detergent formulations, such as AEO, alkanesulfonates, alkyl sulfates, and various cationic surfactants that are easily hydrolyzable, can be analyzed in concentrations relevant to environmental occurrence from standard solutions. More development effort must be invested, however, before a validated quantitative method for these compounds from environmental samples is available. Efforts toward this end are in progress in various groups active in the field.

F. Final Remarks

Obviously this is not the first review of the subject of trace analysis of surfactants. Recently, two books covering surfactant analysis have been published that also deal with trace analysis [232,233]. Recent reviews of the subject can be also found in Refs. 234 and 235. The coverage of surfactant analysis in the application reviews of *Analytical Chemistry* was discontinued in 1983 [236]. The topic, however, is now partly covered in the chapter on water analysis [237,238].

III. COMPLEXING AGENTS

A. General Remarks

Following discussions on the effect of detergent phosphates on eutrophication of the aquatic environment, bans or limitations of detergent phosphate contents have been enforced in various industrialized countries. As a consequence, alternative detergent builders have been introduced or considered phosphate substitutes, namely citrate, aminopolycarboxylic acids, organic phosphonates, zeoliths, or polycarboxylates [239,240]. All these compounds are ionic hydrophilic substances that serve the purpose of removing alkaline earth cations (water hardness) by complexation or ion exchange. Because of their hydrophilic nature, they are not expected to be adsorbed on sludges or sediments (with the exception of zeoliths, which are minerals themselves): they are expected to stay in solution and to enter the aquatic environment if not biodegraded. In receiving waters they may be detrimental by remobilizing heavy metals from suspended particles or sediments owing to their complexing power [241,242].

The concentrations of complexing agents in receiving waters must therefore be monitored to avoid the buildup of concentrations reaching effect levels. Numerous methods for determination of phosphate or citrate at trace levels are known (and are not covered in this review). Both compounds, however, are ubiquitous in nature from biogenic or geogenic sources. For zeoliths, phosphonates, and polycarboxylic acids, no methods are available as yet for the quantitative determination at concentrations relevant to expected environmental occurrence [239]. For phosphonates and polycarboxylic acids, however, attempts are made to approach this end; these are reviewed shortly. The main topic of this review, however, covers the various analytical approaches for aminopolycarboxylic acids, such as nitrilotriacetic acid (NTA) or ethylenediaminetetraacetic acid (EDTA), both of which are expected to be present in surface waters in the low µg/liter range.

B. Aminopolycarboxylic Acids

EDTA and NTA are amphoteric compounds containing a tertiary amine base and carboxylic acid residues; because of these properties they are soluble in water at all pH values. They cannot be extracted from water because their solubility in immiscible organic solvents is far less than in water. They have high complex formation constants with some heavy metals, especially Bi, Fe, and Cu; therefore, the presence of significant concentrations of these cations interferes with some methods.

Colorimetric, polarographic, and HPLC determination methods have been proposed for EDTA and NTA analysis in water samples needing only a small amount of sample preparation apart from metal complex formation. GC determination methods have been developed that need more detailed sample preparation (derivatization) procedures. Far more sensitive detectors can be used for GC compared with the other methods; furthermore, specificity is also higher because of the superior separation power obtained on GC.

The analysis of NTA was reviewed quite some time ago [241,243,244]. Colorimetric methods generally have detection limits of 500 µg/liter and are therefore not overly suited to environmental analysis [244]. They are consequently not reviewed here.

1. Polarography

In polarography, the different reduction potentials of metal complexes of NTA and EDTA are utilized for quantitative determination [241,244–248]. Hg [245] and Bi [244,246–248] were employed as the complexed cation to be reduced. Generally, EDTA can be determined in the presence of NTA, and vice versa. As sample preparation, water samples are acidified to liberate complexed cations from the complexing agent; a cation exchange column then efficiently removes potentially interfering cations. After addition of an excess of Bi and pH adjustment, samples are usually ready for measurement [248]. Because suitable internal standards are not available, standard addition is employed in quantitative analysis [241,248].

The advantage of the polarographic method is that it tolerates fairly large concentrations of salt in the aqueous matrix. The detection limits, however, are higher than needed for sensitive environmental analysis. Although lower detection limits have been claimed in the literature (to 15 µg/liter [245]), the German standard procedure limits the method to concentrations higher than 100 µg/liter [248].

2. HPLC and Ion Chromatography

Among chromatographic methods for hydrophilic polar nonvolatile compounds such as NTA and EDTA, HPLC and ion chromatography (IC) are obvious choices for separation and quantification of these compounds. Detectability, however, is a problem with compounds containing only aliphatic residues. Direct detection methods usually use conductivity [249,250] or electrochemical detection [251,252]; indirect detection methods use metal complexation with a colorimetric (UV) [253–257] or atomic absorption spectroscopic finish [258]. Some of the proposed methods have been applied only to commercial formulations [253] or standard solutions [249,258], but not to environmental samples. Thus, potential matrix interferences associated with these procedures cannot be assessed.

Water sample pretreatment before HPLC or IC separation is usually fairly simple: filtration of the sample to remove particulates, addition of the metal ions for complexation, and pH adjustment for indirect detection by UV [254–257]. For conductivity detection following IC of samples from biological matrices, a protein precipitation step was found to be sufficient sample pretreatment [250]. For electrochemical detection, either filtration [251] or filtration followed by removal of nonpolar organic compounds on a C18 cartridge [252] produced solutions ready for chromatographic determination.

For analysis by indirect (UV) detection, separation of the Fe-EDTA and Fe-NTA complexes (which are anionic in charge) was effected by either chromatography on an anion exchange column [254] or by ion pair separation on C18 RP columns using tetrabutylammonium as the counterion [255–257]. Separation for conductimetric detection was performed on a combination of column materials with an ion exclusion column as the step governing resolution. Thermostatting the detector system efficiently was mandatory for sensitive detection [251]. For electrochemical detection, chromatographic separation was effected on a C18 RP column or a styrene-divinylbenzene column with trichloroacetic acid or acetic acid as electrolytes/eluents [251] or on an ion chromatography (ion exchange) column using strongly acid (6 M HNO_3) conditions [252]. In all cases, NTA and EDTA were clearly separated from one another. Detection limits, were usually higher than 100 µg/liter for each of these compounds, however, limiting the application of HPLC or IC to samples with a higher content of complexing agents, such as process waters or sewage influents.

3. GC

Capillary GC has an inherently higher resolution for organic compounds than LC. Compounds to be analyzed by GC, however, must be volatile under the conditions envisaged for chromatographic separation. For such hydrophilic polar ionic organic compounds as NTA or EDTA, chemical derivatization before GC separation is mandatory to obtain sufficiently volatile derivatives. Consequently, sample preparation for GC analysis is a more complex and tedious procedure than for HPLC.

On the other hand, highly sensitive and specific detection systems like the nitrogen selective detector (NPD) or a mass spectrometer (MS) allow reliable quantification of EDTA and NTA at trace levels in the presence of organic matrix components.

The obvious choice for a derivatization method for carboxylic acids like

EDTA or NTA is an esterification reaction. All published methods have used this approach, preparing methyl esters [259,260], propyl esters [260,261], trimethylsilyl esters [262], or butyl esters [241,244,263–269]. Because derivatization reactions cannot be expected to have quantitative chemical yields, especially not for aminopolycarboxylic acids, the use of an internal standard is mandatory to correct for nonquantitative yields (as well as for incomplete recovery during sample preparation). In some method development investigations, radioactive (^{14}C)EDTA or NTA was used for this purpose [261,264]. During routine analysis, however, this is not a viable solution. Some procedures only incorporate standards that control the injection volume, such as dibutyl phthalate [264] or alkyl nitrile (17 and 18 carbon chain length) [260,266–269]. Heptadecanoic acid has been employed [261], but its properties are too different from EDTA and NTA for a reliable internal standard. Aminopolycarboxylic acids seem to be the internal standard most similar in behavior to EDTA and NTA; compounds used for this purpose are cyclohexandiamine-(1,2)-tetraacetic acid [259], nitrilotripropionic acid (mainly for NTA) [241,244], and 1,2-diaminopropane-N,N,N',N'-tetraacetic acid (DPTA) [270]. They are true internal standards because they cover the whole sample preparation procedure.

Sample preparation is still comparatively easy: usually the formaldehyde-preserved sample is acidified, internal standards (when applicable) are added, and the water is gently evaporated to dryness using N_2 or a heating bath [241,262, 270]. Some methods alternatively use an anion exchange column [241,260, 263,266] to separate EDTA and NTA from matrix components. A small amount of concentrated formic acid is sufficient for elution. After concentration, derivatization reagents (butanol/HCl or butanol/acetyl chloride) are added [241,263,266, 270] to effect derivatization. After addition of alkyl nitrile control standards, liquid-liquid extraction is used to clean EDTA and NTA esters further by partioning between water and apolar organic solvents [260,266,270]. After drying, the organic phase is concentrated and injected for GC analysis.

Historically, packed column GC was used for separation of EDTA and NTA esters in conjunction with flame ionization detection. Nonpolar phases, such as OV1, OV17, and SE54, were used for this chromatographic problem [259,263, 264]. More specific detection was possible after the introduction of the nitrogen selective detector (NPD), increasing both sensitivity and selectivity in the determination of aminopolycarboxylic acids [265]. After the advent of capillary column GC, their superior resolution was also utilized for the determination of NTA and EDTA. Glass capillary columns coated with nonpolar phases [260,266,268], as well as fused silica columns with a bonded nonpolar phase, such as DB-1 or DB-5 [267,269,270], were employed for this purpose in conjunction with NPD (see Fig. 13). Mass spectrometric detection can be expected to increase specificity further.

The method is in the process of being standardized in Germany [270]. Two

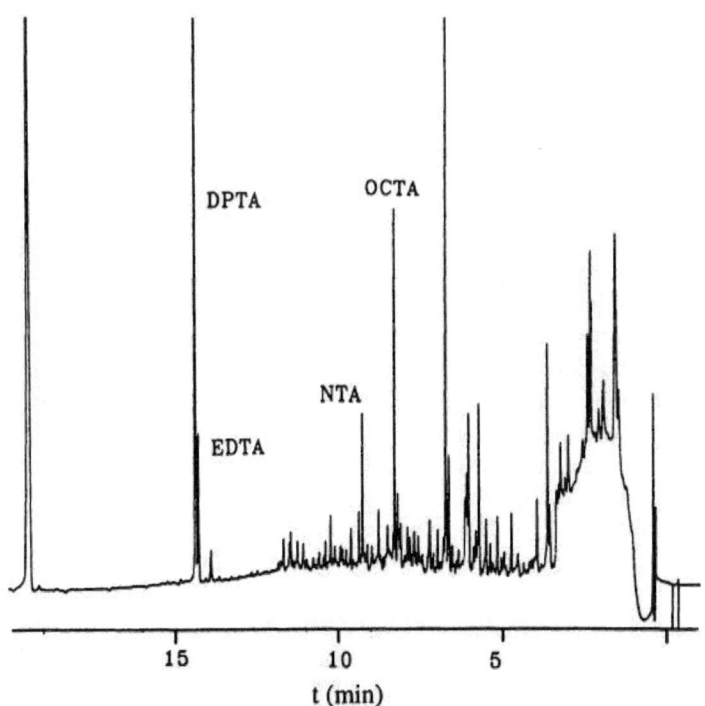

FIG. 13 GC determination of NTA and EDTA (both derivatized to form their isopropyl esters) from surface water at concentrations of 6 µg/L for NTA and 8 µg/L for EDTA, respectively, using the proposed German Standard procedure [237]; DB 5 column: 25 m × 0.32 mm ID (0.26 µm film thickness); carrier gas: He at 4 ml/minute; splitless injection; NPD; temperature program: 1 minute hold time at 80°C, ramping to 180°C at 20°C/minute, ramping to 260°C at 8°C/minute, further ramping to 300°C at 14°C/minute; final hold time at 300°C: 8 minutes. Octa: octadecyl nitrile (standard for injection volume control).

problems remain: in the presence of large concentrations of electrolytes [267], problems arise in the derivatization step. Second, the EDTA tetrabutyl ester, as well as the internal standard DPTA tetrabutyl ester are eluted from the GC column at very high oven temperatures; losses for unknown reasons among these two compounds compared with the control standards have irregularly been noticed.

Limits of detection of the proposed methods are significantly lower than for polarography and HPLC: for the packed column GC/FID analytical method, a method detection limit of 25 µg/liter was reported [244], whereas for the capillary column GC/NPD combination, method detection limits as low as 0.2

μg/liter were reported [266]. The proposed German standard method aims at a method detection limit of 1 μg/liter to increase reproducibility and reliability [270].

C. Organic Phosphonates

Organic phosphonates were recently introduced as complexing agents in detergents. Their molecular structure is much the same as that of corresponding aminopolycarboxylic acids; every carboxylate residue is substituted by a phosphonate residue [239]. Phosphonate residues differ from carboxylates by the fact that they are potentially bivalent anions compared with monovalent carboxylates. Consequently, metal phosphonate complexes potentially useful for polarography have different net charges compared with EDTA or NTA complexes. Quantitative esterification of phosphonates is inherently more difficult (if not impossible) than for EDTA or NTA for the same reason. GC is therefore not a viable option in the environmental analysis of phosphonates.

Consequently, only HPLC or IC separation has been investigated for the analysis of organic phosphonates at low concentrations. Anion exchange columns have been employed to separate individual phosphonates using isocratic elution conditions [271–276]. Detection of separated phosphonates is either by postcolumn digestion to form phosphate subsequently transformed to molybdatophosphate [271,275,276] or by metal complex formation with Fe(III) [272–274]. In both cases absorption measurements are the final detection method. Although one method has been applied to monitor degradation experiments [274], they are not amenable for environmental analysis, mainly because the detection limits obtained (>10 mg/liter) are totally insufficient compared with the expected environmental concentrations (low μg/liter) [239]. These methods, however, may be the basis for improvements and method development to achieve detection limits sufficient for environmental analysis.

D. Polycarboxylates

Polycarboxylates, mainly copolymers of acrylic and maleic acid of various molecular weight distribution, are used as cobuilders in phosphate-free detergents in fairly substantial quantities. Whereas size exclusion chromatography was used for determination of polycarboxylates in detergent formulations [277], only a titrimetric procedure has been published for polycarboxylates at trace concentrations. Polyelectrolyte titration uses a polymeric cationic compound that forms complexes with the polycarboxylates to be determined. Anionic dyes are used to detect an excess of the polycation titrant by a metachromatic shift upon adsorption of this dye on the polycation [278]. Using this method, all high-molecular-weight anionic compounds are indirectly detected. Interference by humic acids, for example, must be expected for environmental samples. Permanganate oxidation was proposed to remove this interference selectively [278]. For environmental analysis, however, more specific as well as more sensitive methods are urgently needed.

E. Summary

For quantitative determination of complexing agents at concentrations relevant to environmental occurrence, only a very limited number of methods is available. NTA and EDTA can be analyzed in highly polluted samples by polarography, HPLC, or GC. At the lower concentrations expected for surface waters, only the more sensitive and specific GC method achieves the detection limits required for this task. For organic phosphonates and polycarboxylic acids, no method has been reported that has the required sensitivity and specificity to monitor them at concentrations relevant to environmental occurrence. Method development for these compounds is urgently needed.

REFERENCES

1. P. Gerike, in *Surfactants in Consumer Products* (J. Falbe, ed.), Springer Verlag, Berlin, 1987, pp. 450–474.
2. W. K. Fischer and K. Winkler, Vom Wasser *47*:81 (1976).
3. W. K. Fischer, Tenside Deterg. *17*:250 (1980).
4. P. Gerike, K. Winkler, and W. Jakob, Tenside Surf. Det. *26*:270 (1989).
5. Bundesgesetzblatt (Teil I), 2613: (1990).
6. *Die Analytik der Tenside*, Chemische Werke Hüls AG, Marl, 1976.
7. H. Stache, and K. Kosswig, (eds.), *Tensid-Taschenbuch*, 3rd ed. Hanser-Verlag, München-Wien, 1990.
8. S. R. Epton, Trans. Faraday Soc. *44*:226 (1948).
9. J. Longwell and W. D. Maniece, Analyst *80*:167 (1955).
10. H. L. Webster and J. Halliday, Analyst *84*:552 (1959).
11. DIN 38409, Teil 23 (1980).
12. V. W. Reid, G. F. Longman, and E. Heinerth, Tenside *4*:292 (1967).
13. J. Waters, Vom Wasser *47*:131 (1976).
14. J. Waters and J. T. Garrigan, Water Res. *17*:1549 (1983).
15. Q. W. Osburn, J. Am. Oil Chem. Soc. *63*:257 (1986).
16. H. Hellmann, Tenside Surf. Det. *28*:111 (1991).
17. W. Huber, Tenside Surf. Det. *28*:106 (1991).
18. S. Motomizu, S. Fujiwara, A. Fujiwara, and K. Toei, Anal. Chem. *54*:392 (1982).
19. L. Shaopu, H. Zhigui, and J. F. Ying, J. Surface Sci. Technol. *6*:223 (1990).
20. E. Orthgiess and B. Dobias, Tenside Surf. Det. *27*:226 (1990).
21. H. Kubota, M. Katsuki, and S. Motomizu, Anal. Sci. *6*:705 (1990).
22. H. Pobiner and H. T. Hoffmann, Jr., Anal. Chim. Acta *141*:419 (1982).
23. Y. Shimoishi and H. Miyata, Fresenius' J. Anal. Chem. *338*:46 (1990).

24. J. Waters and W. Kupfer, Anal. Chim. Acta 85:241 (1976).
25. W. Kupfer, Tenside Deterg. 19:158 (1982).
26. Q. W. Osburn, J. Am. Oil Chem. Soc. 59:453 (1982).
27. DIN 38409, Teil 20 (1989).
28. H. Hellmann, Z. Wasser-Abwasser-Forsch. 22:131 (1989).
29. H. Klotz, Münch. Beitr. Abwasser-, Fisch.-Flussbiol. 44:205 (1990).
30. M. B. Simon, A. D. E. Cozar, and L. M. P. Diez, Analyst 115:337 (1990).
31. U. Denter, H.-J. Buschmann, and E. Schollmeyer, Tenside Surf. Det. 28:333 (1991).
32. S. Marhold, E. Koller, I. Meyer, and O. Wolfbeis, Fresenius' J. Anal. Chem. 336:111 (1990).
33. S. Motomizu, M. Oshima, and Y. Hosoi, Mikrochim. Acta 106:57 (1992).
34. S. Motomizu, M. Oshima, and Y. Hosoi, Mikrochim. Acta 106:67 (1992).
35. P. Berndt and J. Richter, Pharmazie 20:359 (1965).
36. R. Wickbold, Tenside Deterg. 9:143 (1972).
37. R. Wickbold, Tenside Deterg. 10:148 (1973).
38. C. Hennequin and A. Lerenard, Analusis 3:177 (1975).
39. J. Waters and G. F. Longman, Anal. Chim. Acta 93:341 (1977).
40. R. Wickbold, Vom Wasser 33:229 (1966).
41. W. Giger, P. H. Brunner, and C. Schaffner, Science 225:623 (1984).
42. M. Ahel and W. Giger, Anal. Chem. 57:1577 (1985).
43. R. A. Greff, E. A. Setzkorn, and W. D. Leslie, J. Am. Oil. Chem. Soc. 42:180 (1965).
44. A. Nozawa, T. Ohnuma, and T. Sekine, Analyst 101:543 (1976).
45. N. T. Crabb, and H. F. Persinger, J. Am. Oil. Chem. Soc. 45:611 (1968).
46. A. L. Huddleston and R. C. Allred, J. Am. Oil. Chem. Soc. 42:983 (1965).
47. N. H. Anderson and J. Girling, Analyst 107:836 (1982).
48. S. L. Boyer, K. F. Guin, R. M. Kelley, M. L. Mausner, H. F. Robinson, T. M. Schmitt, C. R. Stahl, and E. A. Setzkorn, Environ. Sci. Technol. 11:1167 (1977).
49. H. Jehring, Elektrosorptionsanalyse mit der Wechselstrompolarographie, Akademie-Verlag, Berlin, 1974.
50. D. R. Canterford, J. Electroanal. Chem. 111:269 (1980).
51. E. Bednarkiewicz, M. Donten, and Z. Kublik, J. Electroanal. Chem. 127:241 (1981).
52. M. Bos, J. H. H. G. van Willigen, and W. E. van der Linden, Anal. Chim. Acta 156:71 (1984).
53. M. Bos, P. van Marion, and W. E. van der Linden, Anal. Chim. Acta 223:387 (1989).
54. B. Cosovic and M. Branica, J. Electroanal. Chem. 46:63 (1973).
55. Z. Kozarac, B. Cosovic, and M. Branica, J. Electroanal. Chem. 68:75 (1976).
56. B. Cosovic and D. Hrsak, Tenside Deterg. 16:262 (1979).
57. Z. Kozarac, D. Hrsak, B. Cosovic, and J. Vrzina, Environ. Sci. Technol. 17:268 (1983).
58. B. Cosovic and V. Vojvodic, Marine Chem. 22:363 (1987).
59. A. Szymanski and Z. Lukaszewski. Anal. Chem. Acta 260:25 (1992).

60. W. Selig, Fresenius' Z. Anal. Chem. *312*:419 (1982).
61. W. A. Straw, Anal. Proc. *22*:142 (1985).
62. N. Buschmann and R. Schulz. Tenside Surf. Det. *29*:128 (1992).
63. J. Chlebicki and W. Garncarz, Tenside Deterg. *17*:1 (1980).
64. J. Courtot-Coupez and A. Le Bihan, Anal. Lett. *2*:567 (1969).
65. A. Le Bihan and J. Courtot-Coupez, Anal. Lett. *10*:759 (1977).
66. A. Le Bihan and J. Courtot-Coupez, Analusis *6*:339 (1978).
67. A. Le Bihan and J. Courtot-Coupez, Analusis *6*:346 (1978).
68. Australian Standard AS 3506-1987 (1987)
69. M. J. Spencer, G. G. Wallace, P. T. Crisp, and T. W. Lewis, Anal. Chim. Acta *244*:197 (1991).
70. K. J. Irgolic and J. E. Hobill, Spectrochim. Acta *42B*:269 (1987).
71. H. Hellmann, Z. Wasser-Abwasser-Forsch. *23*:62 (1990).
72. H. Hellmann, Z. Wasser-Abwasser-Forsch. *24*:178 (1991).
73. D. Hummel, *Analyse der Tenside*, Hanser Verlag, München, 1962.
74. H. Hellmann, Fresenius' Z. Anal. Chem. *300*:44 (1980).
75. H. Hellmann, Vom Wasser *55*:249 (1980).
76. H. Hellmann, Z. Wasser-Abwasser-Forsch. *16*:174 (1983).
77. H. Hellmann, Vom Wasser *64*:29 (1985).
78. H. Hellmann, Vom Wasser *66*:111 (1986).
79. R. A. Hearmon, Anal. Proc. *22*:147 (1985).
80. E. M. Thurman, T. Willoughby, L. B. Barber, Jr., and K. A. Thorn, Anal. Chem. *59*:1798 (1987).
81. J. A. Leenher, R. L. Wershaw, P. A. Brown, and T. I. Noyes, Environ. Sci. Technol. *25*:161 (1991).
82. E. Schneider, K. Levsen, P. Dähling, and F. W. Röllgen, Fresenius' Z. Anal. Chem. *316*:277 (1983).
83. E. Schneider, K. Levsen, P. Dähling, and F. W. Röllgen, Fresenius' Z. Anal. Chem. *316*:488 (1983).
84. F. Ventura, J. Caixach, A. Figueras, I. Espalder, D. Fraisse, and J. Rivera, Water Res. *23*:1191 (1989).
85. J. Rivera, F. Ventura, D. Fraisse, J. Caixach, A. Figueras, and I. Espadaler, in *Organic Contaminants in Waste Water, Sludge and Sediment: Occurrence, Fate and Disposal* (D. Quaghebeur, I. Temmerman, and G. Angeletti, eds.), Elsevier, London, 1989, pp. 94–102.
86. J. R. Simms, T. Keough, S. R. Ward, B. L. Moore, and M. M. Bandurraga, Anal. Chem. *60*:2613 (1988).
87. J. R. Simms, D. A. Woods, D. R. Walley, T. Keough, B. S. Schwab, and R. J. Larson, Anal. Chem. *64*:2951 (1992).
88. A. J. Borgerding and R. A. Hites, Anal. Chem. *64*:1449 (1992).
89. F. Ventura, D. Fraisse, J. Caixach, and J. Rivera, Anal. Chem. *63*:2095 (1991).
90. A. Nakae, K. Tsuiji, and M. Yamanaka, Anal. Chem. *52*:2275 (1980).
91. A. Nakae, K. Tsuji, and M. Yamanaka, Anal. Chem. *53*:1818 (1981).
92. D. E. Linder and M. C. Allen, J. Am. Oil Chem. Soc. *59*:152 (1982).
93. M. Kikuchi, A. Tokai, and T. Yoshida, Water Res. *20*:643 (1986).

94. H. De Henau, E. Matthijs, and W. D. Hopping, Int. J. Environ. Anal. Chem. 26:279 (1986).
95. A. Marcomini, S. Capri, and W. Giger, J. Chromatogr. 403:243 (1987).
96. A. Marcomini and W. Giger, Anal. Chem. 59:1709 (1987).
97. E. Kunkel, Tenside Surf. Det. 24:280 (1987).
98. E. Matthijs and H. De Henau, Tenside Surf. Det. 24:193 (1987).
99. P. D. Brunner, S. Capri, A. Marcomini, and W. Giger, Water Res. 22:1465 (1988).
100. J. Waters, M. S. Holt, and E. Matthijs, Tenside Surf. Det. 26:129 (1989).
101. P. Gerike, K. Winkler, W. Schneider, and W. Jakob, Tenside Surf. Det. 26:136 (1989).
102. A. Marcomini, S. Stelluto, and B. Pavoni, Int. J. Environ. Anal. Chem. 35:207 (1989).
103. M. A. Castles, B. L. Moore, and S. R. Ward, Anal. Chem. 61:2534 (1989).
104. M. S. Holt, E. Matthijs, and J. Waters, Water Res. 23:749 (1989).
105. M. T. Garcia Ramon, I. Ribosa, J. Sanchez Leal, and F. Commelles, Tenside Surf. Det. 27:118 (1990).
106. R. A. Rapaport and W. S. Eckhoff, Environ. Toxicol. Chem. 9:1245 (1990).
107. I. Fujita, Y. Ozasa, T. Tobino, and T. Sugimura, Chem. Pharm. Bull. 38:1425 (1990).
108. E. Matthijs and E. C. Hennes, Tenside Surf. Det. 28:22 (1991).
109. A. Marcomini, F. Cecchi, and A. Sfriso, Environ. Technol. 12:1047 (1991).
110. Y. Yokoyama and H. Sato, J. Chromatogr. 555:155 (1991).
111. A. Di Corcia, M. Marchetti, R. Samperi, and A. Marcomini, Anal. Chem. 63:1179 (1991).
112. J. L. Berna, A. Moreno, and J. Ferrer, J. Chem. Tech. Biotechnol. 50:387 (1991).
113. H. DeHenau, E. Matthijs, and E. Namkung, in Organic Contaminants in Waste Water, Sludge and Sediment: Occurrence, Fate and Disposal (D. Quaghebeur, I. Temmerman, and G. Angeletti, eds.), Elsevier, London, 1989, pp. 5–18.
114. M. S. Holt, E. Matthijs, and J. Waters, in: Organic Contaminants in Waste Water, Sludge and Sediment: Occurrence, Fate and Disposal (D. Quaghebeur, I. Temmerman, and G. Angeletti, eds.), Elsevier, London, 1989, pp. 161–182.
115. M. Stalmans, E. Matthijs, and N. T. de Oude, Water Sci. Technol. 24:115 (1991).
116. H. Takada, N. Ogura, and R. Ishiwatari, Environ. Sci. Technol. 26:2517 (1992).
117. G. R. Bear, J. Chromatogr. 459:91 (1988).
118. J. Weiss, J. Chromatogr. 353:303 (1986).
119. J. B. Li and P. Jandik, J. Chromatogr. 546:395 (1991).
120. M. Kudoh and K. Tsuji, J. Chromatogr. 294:456 (1984).
121. J. R. Larson, J. Chromatogr. 356:379 (1986).
122. D. J. Pietrzyk, P. G. Rigas, and D. Yuan, J. Chromatogr. Sci. 27:485 (1989).
123. G. Liebscher, G. Eppert, H. Oberender, H. Berthold, and H. G. Hauthal, Tenside Surf. Det. 26:195 (1989).
124. G. Eppert and G. Liebscher, J. Chromatogr. Sci. 29:21 (1991).
125. S. A. Maki, J. Wangsa, and N. D. Danielson, Anal. Chem. 64:583 (1992).
126. C. P. Terweij-Groen, J. C. Kraak, W. M. A. Niessen, J. F. Lawrence, C. E. Werkhoven-Goewie, U. A. T. Brinkman, and R. W. Frei, Int. J. Environ. Anal. Chem. 9:45 (1981).

127. F. Smedes, J. C. Kraak, C. E. Werkhoven-Goewie, U. A. T. Brinkman, and R. W. Frei, J. Chromatogr. 247:123 (1982).
128. A. Nakae and K. Tsuji, Comun. J. Com. Esp. Deterg. 14:133 (1983).
129. M. Kanesato, K. Nakamura, O. Nakata, and Y. Morikawa, J. Am. Oil Chem. Soc. 64:434 (1987).
130. C. De Ruiter, J. H. Wolf, U. A. T. Brinkman, and R. W. Frei, Anal. Chem. Acta 192:267 (1987).
131. M. Schoester and G. Kloster, Fresenius' J. Anal. Chem., in press.
132. P. Arpino, Mass Spectrom. Rev. 8:35 (1989).
133. R. Arpino, Mass Spectrom. Rev. 9:631 (1990).
134. P. Arpino, Mass Spectrom. Rev. 11:3 (1992).
135. H. F. Schröder, Vom Wasser 73:111 (1989).
136. H. F. Schröder, J. Chromatogr. 554:251 (1991).
137. J. J. Conboy, J. D. Henion, M. W. Martin, and J. A. Zweigenbaum, Anal. Chem. 62:800 (1990).
138. R. E. A. Escott and D. W. Chandler, J. Chromatogr. Sci. 27:134 (1989).
139. A. L. Rockwood and T. Higuchi, Tenside Surf. Det. 29:6 (1992).
140. L. B. Clark, R. T. Rosen, T. G. Hartman, J. B. Louis, I. H. Suffet, R. L. Lippincott, and J. D. Rosen, Int. J. Environ. Anal. Chem. 47:167 (1992).
141. K. Balasanmugam and D. M. Hercules, Anal. Chem. 55:145 (1983).
142. V. T. Wee and J. M. Kennedy, Anal. Chem. 54:1631 (1982).
143. V. T. Wee, Water Res. 18:223 (1984).
144. H. Klotz, Tenside Surf. Det. 24:370 (1987).
145. E. Matthijs and H. DeHenau, Vom Wasser 69:73 (1987).
146. M. Emmrich and K. Levsen, Vom Wasser 75:343 (1990).
147. L. Nitschke, R. Müller, G. Metzer, and L. Huber, Fresenius' J. Anal. Chem. 342:711 (1992).
148. J. R. Larson and C. D. Pfeiffer, Anal. Chem. 55:393 (1983).
149. C. J. Dowle, W. C. Campbell, and B. G. Cooksey, Analyst 114:883 (1989).
150. A. J. Wilkes, G. Walraven, and J.-M. Talbot, J. Am. Oil Chem. Soc. 69:609 (1992).
151. J. Kawase, Anal. Chem. 52:2124 (1980).
152. J. Kawase, Y. Takao, and K. Tsuji, J. Chromatogr. 262:293 (1983).
153. C. De Ruiter, J. C. H. F. Hefkens, U. A. T. Brinkman, R. W. Frei, M. Evers, E. Matthijs, and J. A. Meijer, Int. J. Environ. Anal. Chem. 31:325 (1987).
154. M. Schoester and G. Kloster, Vom Wasser 77:13 (1991).
155. M. Schoester and G. Kloster, in 11. Königsteiner Chromatographietage 1991 (Millipore GmbH, ed.), GIT Verlag, Darmstadt, 1991, pp. 357–362.
156. M. Ahel and W. Giger, Anal. Chem. 57:2584 (1984).
157. Y. Mengerink, H. C. J. De Man, and S. van der Wal, J. Chromatogr. 552:593 (1991).
158. S. van der Wal, LC-GC Int. 5:36 (1992).
159. T. Okada, J. Chromatogr. 609:213 (1992).
160. M. S. Holt, E. H. McKerrell, J. Perry, and R. J. Watkinson, J. Chromatogr. 362:419 (1986).

161.   A. Marcomini, P. D. Capel, T. Lichtensteiger, P. H. Brunner, and W. Giger, J. Environ. Qual. *18*:523 (1989).
162.   B. B. Sithole and L. H. Allen, J. Assoc. Off. Anal. Chem. *72*:273 (1989).
163.   E. Kubeck, and C. G. Naylor, J. Am. Oil. Chem. Soc. *67*:400 (1990).
164.   P. Varughese, M. E. Gangoda, and R. K. Gilpin, J. Chromatogr. *499*:469 (1990).
165.   P. Jandera, J. Urbanek, B. Prokes, and J. Churacek, J. Chromatogr. *504*:297 (1990).
166.   I. Zeman, J. Chromatogr. *509*:201 (1990).
167.   C. Zhou, A. Bahr, and G. Schwedt. Anal. Chim. Acta *236*:273 (1990).
168.   A. Marcomini, B. Pavoni, A. Sfriso, and A. A. Orio, Marine Chem. *29*:307 (1990).
169.   U. Zoller, E. Ashash, G. Ayali, S. Shafir, and B. Azmon, Environ. Int. *16*:301 (1990).
170.   C. G. Naylor, J. P. Mieure, W. J. Adams, J. A. Weeks, F. J. Castaldi, L. D. Ogle, and R. R. Romano, J. Am. Oil Chem. Soc. *69*:695 (1992).
171.   A. Aserin, M. Frenkel, and N. Garti, J. Am. Oil Chem. Soc. *61*:805 (1984).
172.   H. Brüschweiler, Mitt. Gebiete Lebensm. Hyg. *68*:46 (1977).
173.   A. Nozawa and T. Ohnuma, J. Chromatogr. *187*:261 (1980).
174.   P. L. Desbene, B. Desmazieres, J. J. Basselier, and A. Desbene-Monvernay, J. Chromatogr. *461*:305 (1989).
175.   M. C. Allen and D. E. Linder, J. Am. Oil Chem. Soc. *58*:950 (1981).
176.   T. M. Schmitt, M. C. Allen, D. K. Brain, K. F. Guin, D. E. Lemmel, and Q. W. Osburn, J. Am. Oil Chem. Soc. *67*:103 (1990).
177.   T. Okada, Anal. Chem. *62*:327 (1990).
178.   T. Okada, Anal. Chem. *63*:1043 (1991).
179.   M. Benning, H. Locke, and R. Ianniello, J. Liq. Chromatogr. *12*:757 (1989).
180.   H. G. Nadeau, D. M. Oaks, Jr., W. A. Nichols, and L. P. Carr, Anal. Chem. *36*:1914 (1964).
181.   E. Stephanou and W. Giger, Environ. Sci. Technol. *16*:800 (1982).
182.   G. Czichocki, W. Gerhardt, and D. Blumberg, Tenside Surf. Det. *25*:169 (1988).
183.   B. D. Bhatt, J. V. Prasad, G. Kalpana, and S. Ali, J. Chromatogr. Sci. *30*:203 (1992).
184.   C. Wahlberg, L. Renberg, and U. Wideqvist, Chemosphere *20*:179 (1990).
185.   M. Ahel, T. Conrad, and W. Giger, Environ. Sci. Technol. *21*:697 (1987).
186.   E. C. Jennings, Jr., and H. Mitchner, J. Pharm. Sci. *56*:1590 (1967).
187.   R. D. Swisher, J. Am. Oil Chem. Soc. *43*:137 (1966).
188.   W. T. Sullivan and R. D. Swisher, Environ. Sci. Technol. *3*:481 (1969).
189.   H. Leidner, R. Gloor, and K. Wuhrmann, Tenside Deterg. *13*:122 (1976).
190.   J. McEvoy and W. Giger, Naturwiss. *72*:429 (1985).
191.   J. McEvoy and W. Giger, Environ. Sci. Technol. *20*:376 (1986).
192.   H. Hon-Nami and T. Hanya, J. Chromatogr. *161*:205 (1978).
193.   M. L. Trehy, W. E. Gledhill, and R. G. Orth, Anal. Chem. *62*:2581 (1990).
194.   H. Kataoka, N. Muroi, and M. Makita, Anal. Sci. *7*:585 (1991).
195.   J. J. Kirkland, Anal. Chem. *32*:1388 (1960).
196.   H. Kataoka, T. Okazaki, and M. Makita, J. Chromatogr. *473*:276 (1989).
197.   L. K. Ng and M. Hupé, J. Chromatogr. *513*:61 (1990).

198. N. J. Fendinger, W. M. Begley, D. C. McAvoy, and W. S. Eckhoff, Environ. Sci. Technol. 26:2493 (1992).
199. E. L. Sones, J. L. Hoyt, and A. J. Sooter, J. Am. Oil Chem. Soc. 56:689 (1979).
200. R. P. Eganhouse, Int. J. Environ. Anal. Chem. 26:241 (1986).
201. C. C. Raymundo and M. R. Preston, Marine Pollut. Bull. 24:138 (1992).
202. H. Takada and R. Ishiwatari, Environ. Sci. Technol. 21:875 (1987).
203. M. S. Holt and S. L. Bernstein, Water Res. 26:613 (1992).
204. P. Fernandez, M. Valls, J. M. Bayona, and J. Albaigés, Environ. Sci. Technol. 25:547 (1991).
205. D. E. Knowles, L. Nixon, E. R. Campbell, D. W. Later, and B. E. Richter, Fresenius' Z. Anal. Chem. 330:225 (1988).
206. P. R. Geissler, J. Am. Oil Chem. Soc. 66:685 (1989).
207. N. Pericles and A. Giorgetti, J. Chromatogr. Sci. 28:218 (1990).
208. G. Krusche, Tenside Surf. Det. 27:122 (1990).
209. S. Brossard, M. Lafosse, and M. Dreux, J. Chromatogr. 591:149 (1992).
210. A. H. Silver and H. T. Kalinoski, J. Am. Oil Chem. Soc. 69:599 (1992).
211. P. L. Desbène, C. Rony, B. Desmaizières, and J. C. Jacquier, J. Chromatogr. 608:375 (1992).
212. C. Tribet, R. Gaboriaud, and P. Gareil, J. Chromatogr. 609:381 (1992).
213. C. S. Weiss, J. S. Hazlett, M. H. Datta, and M. H. Danzer, J. Chromatogr. 608:325 (1992).
214. J. A. Field, L. B. Barber, II, E. M. Thurman, B. L. Moore, D. L. Lawrence, and D. A. Peake, Eviron. Sci. Technol. 26:1140 (1992).
215. J. A. Field, D. J. Miller, T. M. Field, S. B. Hawthorne, and W. Giger, Anal. Chem., in press.
216. T. A. Neubecker, Environ. Sci. Technol. 19:1232 (1985).
217. H. Hellmann, Z. Wasser-Abwasser-Forsch. 22:4 (1989).
218. H. Klotz, Fresenius' Z. Anal. Chem. 326:155 (1987).
219. S. B. Hawthorne, D. J. Miller, D. D. Walker, D. E. Whittington, and B. L. Moore, J. Chromatogr. 541:185 (1991).
220. E. R. Michelsen, Tenside Deterg. 15:169 (1978).
221. K. Inaba and K. Amano, Int. J. Environ. Anal. Chem. 34:203 (1988).
222. R. A. Menges, T. S. Menges, G. L. Bertrand, D. W. Armstrong, and L. A. Spino, J. Liq. Chromatogr. 15:2909 (1992).
223. P. Jones and G. Nickless, J. Chromatogr. 158:87 (1978).
224. P. Jones and G. Nickless, J. Chromatogr. 158:99 (1978).
225. J. A. Field, J. A. Leenheer, K. A. Thorn, L. B. Barber, II, C. Rostad, D. L. Macalady, and S. R. Daniel, J. Contam. Hydrol. 9:55 (1992).
226. A. Neufahrt, K. Hofmann, and G. Täuber, Comun. J. Com. Esp. Deterg. 15:123 (1984).
227. D. Frahne, S. Schmidt, and H. G. Kuhn, Fette Seifen Anstrichm. 79:32 (1977).
228. D. Frahne, S. Schmidt, and H. G. Kuhn, Fette Seifen Anstrichm. 79:122 (1977).
229. H. Hellmann, Fresenius' Z. Anal. Chem. 335:265 (1989).
230. N. Buschmann, A. Kruse, and R. Schulz, Comun. J. Com. Esp. Deterg. 23:317 (1992).

231. G. Kloster, M. Schoester, and H. Prast, in preparation
232. M. R. Porter (ed.), *Recent Developments in the Analysis of Surfactants*, Critical Reports on Applied Chemistry, Vol. 32, Elsevier, London, 1991.
233. T. M. Schmitt, *Analysis of Surfactants*, Surfactant Science Series, Vol. 40, Marcel Dekker, New York, 1992.
234. P. T. Crisp, in *Nonionic Surfactants-Chemical Analysis* (J. T. Cross, ed.), Surfactant Science Series, Vol. 19, Marcel Dekker, New York, 1987, pp. 77–116.
235. R. D. Swisher. *Surfactant Biodegradation*, 2nd ed., Surfactant Science Series, Vol. 18, Marcel Dekker, New York, 1987, pp. 47–146.
236. R. A. Llenado and T. Neubecker, Anal. Chem. *55*:93R (1983).
237. P. MacCarthy, R. W. Klusman, and J. A. Rice, Anal. Chem. *61*:269 R (1989).
238. P. MacCarthy, R. W. Klusman, S. W. Cowling, and J. A. Rice, Anal. Chem. *63*:301R (1991).
239. N. T. de Oude, (ed.), *Detergents, The Handbook of Environmental Chemistry* (O. Hutzinger, series ed.), Vol. 3, Part F, Springer-Verlag, Berlin, 1992.
240. M. Dwyer, S. Yeoman, J. N. Lester, and R. Perry, Environ. Technol. *11*:263 (1990).
241. H. Bernhardt, (ed.), *NTA—Studie über die aquatische Umweltverträglichkeit von Nitrilotriacetat*, Verlag Hans Richarz, St. Augustin, 1984.
242. Projektträger Wassertechnologie und Schlammbehandlung (ed.), *Aquatische Umweltverträglichkeit von Nitrilotriessigsäure (NTA)*, Kernforschungszentrum Karlsruhe, 1992.
243. P. W. W. Kirk and J. N. Lester, Rev. Anal. Chem. (Tel Aviv) *5*:207 (1981).
244. P. W. W. Kirk, R. Perry, and J. N. Lester, Int. J. Environ. Anal. Chem. *12*:293 (1982).
245. Z. Stojek and J. Osteryoung, Anal. Chem. *53*:847 (1981).
246. A. Voulgaropoulos, P. Valenta, and H. W. Nürnberg, Fresenius' Z. Anal. Chem. *317*:367 (1984).
247. F. Guerrieri and G. Bucci, Anal. Chim. Acta *167*:393 (1985).
248. DIN 38413, Teil 5 (1990).
249. S. G. Chen, K. L. Cheng, and C. R. Vogt, Mikrochim. Acta *1983*:473 (1983).
250. R. P. Schneider, F. Zürcher, T. Egli, and G. Hamer, Anal. Biochem. *173*:278 (1988).
251. J. Dai and G. R. Helz, Anal. Chem. *60*:301 (1988).
252. W. Buchberger, P. R. Haddad, and P. W. Alexander, J. Chromatogr. *546*:311 (1991).
253. C. C. T. Chinnick, Analyst *106*:1203 (1981).
254. J. Harmsen and A. van den Toorn, J. Chromatogr. *249*:379 (1982).
255. A. Yamaguchi, A. R. Rajput, K. Ohzeki, and T. Kambara, Bull. Chem. Soc. Jpn. *56*:2621 (1983).
256. D. L. Venezky and W. E. Rudzinski, Anal. Chem. *56*:315 (1984).
257. W. Huber, Acta Hydrochim. Hydrobiol. *20*:6 (1992).
258. E. B. Milosavljevic, L. Solujic, J. L. Hendrix, and J. H. Nelson, Analyst *114*:805 (1989).
259. L. Rudling, Water Res. *6*:871 (1972).

260. S. Schürch and G. Dübendorfer, Mitt. Gebiete Lebensm. Hyg. *80*:324 (1989).

261. Y. K. Chau and M. E. Fox, J. Chromatogr. Sci. *9*:271 (1971).

262. R. J. Stolzberg and D. N. Hume, Anal. Chem. *49*:374 (1977).

263. W. A. Aue, C. R. Hastings, K. O. Gerhardt, J. O. Pierce, H. H. Hill, and R. F. Moseman, J. Chromatogr. *72*:259 (1972).

264. R. A. Larson, J. C. Weston, and S. M. Howell, J. Chromatogr. *111*:43 (1975).

265. D. T. Williams, F. Benoit, K. Muzika, and R. O'Grady, J. Chromatogr. *136*:423 (1977).

266. C. Schaffner and W. Giger, J. Chromatogr. *312*:413 (1984).

267. F. H. Frimmel, R. Grenz, E. Kordik, and F. Dietz, Vom Wasser *72*:175 (1989).

268. A. C. Alder, H. Siegrist, W. Gujer, and W. Giger, Water Res. *24*:733 (1990).

269. E. Kordik-Kolb, Münch. Beitr. Abwasser-, Fisch-, Flussbiol. *44*:493 (1990).

270. Fachgruppe Wasserchemie in der Gesellschaft deutscher Chemiker, DIN Working Party NAW 14/UA5/AK8, *Bestimmung von EDTA und NTA mittels Gaschromatographie*; in preparation.

271. H. Waldhoff and P. Sladek, Fresenius' Z. Anal. Chem. *320*:163 (1985).

272. G. Tschäbunin, P. Fischer, and G. Schwedt, Fresenius' Z. Anal. Chem. *333*:111 (1989).

273. G. Tschäbunin, P. Fischer, and G. Schwedt, Fresenius' Z. Anal. Chem. *333*:117 (1989).

274. G. Tschäbunin, G. Schwedt, and P. Fischer. Fresenius' Z. Anal. Chem. *333*:123 (1989).

275. H. Wagner, G. Metzner, and R.-R. Wagner, Münch. Beitr. Abwasser-, Fisch-, Flussbiol. *44*:218 (1990).

276. E. Frigge and E. Jackwerth, Anal. Chim. Acta *254*:65 (1991).

277. G. Krusche, J. Illert, and H. Mandery, Chromatographia *31*:17 (1991).

278. U. Schröder, D. Horn, and K.-H. Wassmer, Seifen-Öle-Fette-Wachse *117*:311 (1991).

II
Zeolites

4

Zeolites in the Environment

CLAUS PETER KURZENDÖRFER R & D Physical Chemistry, Henkel KGaA, Düsseldorf, Germany

PETER KUHM R & D Automotive Industry, Henkel KGaA, Düsseldorf, Germany

JOSEF STEBER Department of Ecology, Henkel KGaA, Düsseldorf, Germany

I. FEATURES OF SYNTHETIC ZEOLITES

A. History

Mineral zeolites were discovered by the Swedish amateur mineralogist Baron von Cronstedt in 1756. They were first exploited industrially exactly 140 years later, as ion exchangers in the sugar industry. The zeolites owe their name to the fact that they release large amounts of water on heating. Von Cronstedt derived the name from the Greek *zeo* (I boil) and *lithos* (stone). The term "molecular sieves" now used in many sectors as a synonym for industrially manufactured zeolites was introduced by McMain. The suitability of zeolites for separation processes at the molecular level had been discovered by Weigel and Steinhoff in 1925, and 1 year later McMain succeeded in finding a practical application (sorption) for this property [1,2].

From 1938 zeolite synthesis was studied systematically at Oxford University, initially by Barrer. Industrial research commenced in 1945. An important development in this context was the introduction of x-ray diffraction analysis, which enabled zeolite structures to be clearly differentiated and identified for the first time. From 1954, UCC approved samples of the three zeolite types, A, X and Y, which are still preeminent today, for use in the petrochemical industry. Since that time zeolites have been the subject of more than 20,000 scientific publications and almost as many printed patent specifications. In the industrial sphere the zeolites initially started their triumphal advance as sorbents. However, as early as the 1960s there were reports of their use as catalysts, and the industrial breakthrough was made after only a few years. By the early 1990s, annual sales of zeolitic sorbents amounted to more than 10 billion US$, the annual growth rate was about 10%, and world consumption was about 100,000 t. Consumption of zeolitic catalysts was almost as high; such catalysts are used preferentially for catalytic cracking in fluidized beds. Applications for the catalytic production of fine chemicals have also gained in importance in recent years. Since 1975, the most important application in terms of quantity has been the use of zeolite NaA as ion exchanger for alkaline-earth metals in detergents and cleaning agents. As early as 1981, annual consumption of zeolite as a substitute for condensed phosphates reached 150,000 t, and by 1992 it was more than 1,000,000 t worldwide, primarily in Europe (530,000 t), followed by North America (350,000 t) and the East Asia (190,000 t). This is therefore by far the biggest field of application for zeolite and will remain so indefinitely.

B. Structure

Zeolites are natural or synthetic framework aluminosilicates. They contain water of crystallization and have a characteristic pore and cavity structure. They can be reversibly dehydrated without undergoing significant structural change.

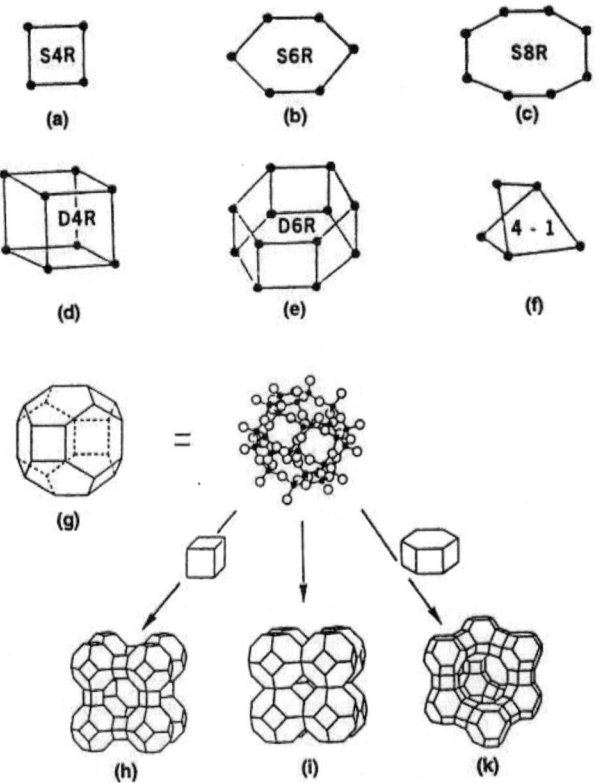

FIG. 1 Structure and building units of zeolites. (a–f), secondary building units of zeolites. Corner points = positions of tetrahedral silicon and aluminum atoms. (g), "Sodalite cage," tertiary building unit. (h), Zeolite A. (i), Sodalite. (k) Zeolite X, Y.

The general formula for a zeolite is

$$M_{2/n}O/Al_2O_3/xSiO_2/yH_2O$$

where M is an exchangeable cation belonging to the alkali metals, alkaline-earth metals, transition metal, or a quaternary ammonium ion and n is the valence. This method of representing the oxide relates the chemical composition to $Al_2O_3 = 1$, so that x is the molecular ratio $SiO_2:Al_2O_3$, which has a considerable influence on the chemical properties of the zeolites. The basic structural units are SiO_4 and AlO_4 tetrahedra, which are linked to so-called secondary building units (SBU) (Fig. 1). The modern classification of the approximately 50 mineral zeolites and more than 200 synthetic zeolites [3,4] into 11 structural groups is based on the following 16 SBUs: single rings S4R, S5R, S6R, and S8R of four, five, six, or eight tetrahedra; double rings D4R (cube), D6R (hexagonal prism), and D8R (octagonal prism); and nine structural units comprising complexly linked rings. Combinations of these 16 SBUs form polyhedral tertiary building units. Silicon and/or aluminum ions may be isomorphically replaced by network modifiers, such as P, Ge, Ga, B, Fe, Co, Ti, As, or Zn, although these "special cases" have yet to attain any industrial importance.

Zeolites A, X, and Y all contain so-called sodalite or β cages as tertiary building units. These are flattened cubooctahedra with the Si or Al atoms at the corners (Fig. 1). The feldspathoid sodalite is made up of densely stacked sodalite cages; every 8 cubically arranged sodalite cages enclose a cavity, which itself has the shape of a sodalite cage. Hydroxysodalite is the basic form of sodalite, which may be formed as an undesired secondary product during zeolite synthesis. In zeolite A, groups of 8 sodalite cages are linked to each other by D4R building units. Each such group encloses a large cavity with eight-ring pores. In faujasite and the structurally analogous synthetic forms X and Y, an SBU consisting of a double six-membered ring links to sodalite cages, which enclose the so-called faujasite cage.

The channels and cavities (e.g., the faujasite cage just mentioned) in the zeolites form a three-dimensional sieve with mesh sizes in the molecular range (between 200 and 1400 pm, hence the term "molecular sieve"). The mesh width of the anionic sieve is considerably influenced by the associated cations. Their lattice position, size, and charge (and therefore the dimensions of the hydrate shell) determine the pore diameter. In the sodium form of zeolite A, for example, the pore diameter is 420 pm (molecular sieve 4A), but this narrows to 200 and 300 pm if the cations are replaced by cesium and potassium, respectively. The smaller number of cations in the calcium form of zeolite A results in a pore diameter of about 500 pm (molecular sieve 5A). In general, the type and position of the cations influence the charge distribution in the zeolite and therefore its sorption and cation exchange properties and, to a large extent, its catalytic activity.

In total contrast to the ideal composition of the elementary cell, the Si/Al ratio of a lattice type can exhibit a wide range of variation. Zeolite NaA, the sodium form of lattice type A, exhibits a Si/Al ratio of 1, and its idealized formula is $Na_2O/Al_2O_3/2SiO_2/4.5H_2O$. The high cation exchange capacity of zeolite NaA is a result of its high aluminum content or, rather, the charge of the resulting anion lattice. This is because aluminum is only trivalent, so that, per incorporated aluminum atom, one univalent cation must be accommodated in lattice cavities. The proportion of aluminum in the lattice is limited by the so-called Loewenstein rule. This states that the Si/Al ratio cannot fall below 1 if tetrahedra are to remain as primary structural units. Only then does each aluminum ion have the opportunity of becoming linked to a silicate tetrahedron via an oxygen bridge. Direct condensation to the neighboring aluminum tetrahedra is unstable.

The industrially most important zeolites A, X, and Y, and many zeolites that are used less extensively, are first manufactured in their sodium form; cation modifications are formed from this by means of subsequent ion exchange. Types of lattice that do not exist in the synthesis field $Na_2O/Al_2O_3/SiO_2/H_2O$ can be synthesized by the additional or exclusive use of other cations, especially quaternary ammonium ions. In recent years it has proved possible to manufacture especially silicate-rich zeolites with the help of quaternary ammonium compounds. Tetramethylammonium hydroxide, for example, has been used to manufacture analogs of the A, X, and Y types (known as N-A, N-X, and N-Y), as well as alpha, omega, phi, TMA-offretite, and a few ZSM types. For the most important member of the ZSM family, that is, ZSM 5, which is used industrially for catalytic purposes, there are a whole series of specifications for synthesis in the presence and absence of quaternary ammonium compounds, in both aqueous and nonaqueous systems. These zeolites are characterized by their high Si/Al ratios, which can be increased to values in excess of 10,000 by subsequent dealuminization. The final link in this chain is silicalite, which is completely aluminum free and structurally identical to ZSM 5 and has pore widths of about 600 pm [5]. For further literature on the structure and systematics of zeolites see Refs. 6–15.

C. Physical and Chemical Properties

Zeolites are generally characterized in terms of their structure type and chemical composition (type and number of cations, Si/Al ratio, water content, and network modifier). The zeolite structure can be clearly determined by means of x-ray diffraction analysis and comparison with the literature. Chemical analysis is very costly but can be replaced by energy-dispersive x-ray microanalysis. The dimensions, tracht, and habit of the crystals, as well as overgrowths, can be determined by scanning electron microscopy (Fig. 2). The individual crystals usually have diameters in the range 0.1–3 μm. Larger crystals with a length of about 100 μm can be grown only if the synthesis parameters are specially selected for this purpose. The size of individual crystals (primary particles) cannot be determined from the grain size distribution, because this only provides information about the diameter of single crystals, crystal aggregates, and agglomerates (so-called secondary particles). An arbitrary limit of 50 μm was chosen, and grains whose size exceeded this limit were classified as grit. Zeolites undergo continuous loss of weight during drying: that is, water of crystallization is continuously released as the temperature steadily rises. The thermal stability of the zeolites is comparatively high but depends on the structural type and the ions (alkali metal/alkaline-earth metal) in the lattice. For example, the detergent zeolite NaA transforms itself into a feldspathoid lattice at temperatures above approximately 700°C, and

FIG. 2 SEM of zeolite A.

CaA also goes through such a phase transition at above approximately 800°C. Thermal stability increases as a function of the Si/Al ratio of the lattice.

The Si/Al ratio has a considerable influence not only on the thermal but also the chemical properties of the zeolites. At pH < 9, the rate of degradation of so-called low-silicate zeolites (Si/Al ratio of 1–2; e.g., zeolite A, P, and X) is directly related to the proportion of Al in the lattice. In theory it is impossible for the ratio to be less than 1 (Löwenstein rule) because Al-O-Al- bridges are unstable in tetrahedral links. In practice these ratios can go down as far as 0.9. This is attributable to the aluminate solution that is retained in the zeolite pores and cages during rapid crystal growth.

More information on the chemical and physical behavior of the zeolites can be found in Refs. 16 and 17.

The basic principles of ion exchange at zeolites have been extensively described, especially for aqueous systems [1,6,18,19]. The exchange behavior of the cations that are not incorporated in the lattice depends on the cation type, the steric arrangement of the anion lattice (i.e., the zeolite type), the temperature, and the concentration of the cation in the solution (Sec. IV.A).

A treatment of the catalytic properties of the zeolites is beyond the scope of this work. Information on this subject can be found in Refs. 20–24.

D. Zeolite A in Detergents

Zeolite A is used in detergent formulations as a water-insoluble ion exchanger substituting the soluble complexing agent sodium triphosphate (STP). Not the least because of their differences in water solubility, the detergency properties of the two builders are in part quite different [25–28]. Similarly to STP, zeolite A exhibits a high binding capability for calcium ions (Sec. II.A.1.b) and heavy metal ions; in addition, their alkaline reaction (Sec. III.B.1) contributes to the detergency. In contrast to STP, zeolite A does not support soil removal and dispersion by specific adsorption onto soil pigments and fabrics. On the other hand, zeolite A promotes the washing process, preventing redeposition of soil by adsorption of molecularly disperse soil particles by heterocoagulation with colloidal soil pigments and by providing a crystallization substrate for precipitation reactions.

The cation exchange on zeolites is time dependent, and cation complexation by a complexing agent is spontaneous. There are also differences with respect to the temperature dependence of the water-softening process, that is, the binding of calcium and magnesium ions. The exchange of alkaline-earth ions by hydration reduction is increased at elevated washing temperatures, whereas complexation of these ions is decreased. The calcium ion exchange of zeolites is improved by employment of water-soluble cobuilders used in substoichiometric concentrations. These cobuilders are sequestering agents acting as carriers and accelerating the removal of polyvalent cations from solid surfaces (soil and fabrics) and transporting them to the zeolite. Because of the removal of alkaline-earth ions from the accumulated soil, the soil is loosened and thus can be washed off more easily. Substoichiometric threshold-active substances, such as polycarboxylates (polyacrylates and acrylate/maleinate copolymers) and phosphonates, are able to stabilize calcium carbonate forming in an amorphous-colloidal status and to inhibit the crystal growth of calcite, thus accelerating the dissolution of this amorphous material by the calcium ion exchange of zeolite A.

E. Manufacturing Process

Because there are a large number of different zeolite types and, inevitably, a variety of production methods, it is difficult to gain a clear overview of the literature on this subject. For this reason, only the detergent zeolite 4A and the structurally related types X and Y are dealt with here, because these three account for more than 95% of total zeolite production. The manufacturing methods are then limited to two different variants of the precipitation process.

The starting products for synthesizing zeolites by the precipitation process are sodium hydroxide solution, sodium aluminate solution, and "activa silica" as a source of SiO_2. The way in which the silica is activated constitutes the difference between the "acid-clay process" usually used in Japan [29] and the related process, involving precipitation with water glass, which is used in countries where water glass is easily obtainable (Central Europe and the United States). A factor common to both variants is the need to remain within defined concentration ranges in the four-component system $Na_2O/Al_2O_3/SiO_2/H_2O$ [30,31]. It must be taken into consideration that the processes involved in zeolite synthesis make up a system that is always in disequilibrium and is therefore disturbance prone. The A type in particular and the X and Y types to a lesser degree are thermodynamically metastable intermediate products that may change into feldspathoid hydroxysodalite or foreign zeolitic phases. In past years extensive studies have been carried out on the effects of varying the numerous synthesis parameters (e.g., stoichiometry of the reaction mixture, temperature curve during the synthesis, influence of foreign substances, energy input, and quality of the raw materials), and these are reflected especially in the patent literature [32–36].

The mechanism of hydrothermal zeolite formation is still not fully understood. There is, however, a useful theory: when silicate and aluminate components are mixed in a strongly alkaline environment, a short sol phase, the so-called precipitation delay period, is followed by the formation of an alkali-rich alkali aluminosilicate gel. This gel contains the secondary building units already mentioned. As a result of a permanent process of rearrangement of the primary and secondary building units via the sol phase (solubility of SiO_2 and Al_2O_3 is about 0.5%) and under the influence of dissolved alkali, the liquid phase becomes saturated with building units of low molecular weight. "Disordered" aluminosilicates tend to dissolve more readily than "ordered" varieties, that is, secondary particles consisting of alternating tetrahedra of silicate and aluminate. Solvated sodium ions function as starting points for crystallization (as templates), and the tertiary structural groups form around them. After an induction period, the so-called nucleation period, the rate of crystallization increases exponentially under the influence of autocatalysis. When crystallization is virtually complete, the rate of crystallization decreases just as dramatically. In extremely alkaline gels the nucleation phase can last for about 10 minutes, whereas actual crystallization is complete after 2 minutes. Information on the basic principles of zeolite synthesis can be found in Refs. 37–42.

In the precipitation process, more accurately described as the "hydrogel process", the gel is produced by mixing the liquid components. A so-called heterogeneous gel can first be produced by using solid aluminate and/or silicate. In general, however, zeolites formed from heterogeneous gels have a more inhomogeneous spectrum of properties than zeolites made from homogeneous gels. Homogeneous gels are of greater importance worldwide, whereas in Japan only heterogeneous gels are used, owing to the unusual raw materials situation there.

The three precipitation components water glass, aluminate solution, and sodium hydroxide solution must be used in very pure condition, because readily soluble impurities in particular can cause feldspathoid secondary products to be formed or can delay crystallization. This must be avoided at all costs in modern processes, which are designed for the maximum possible space-time yield and the minimum possible cycle times. When the educt components are thoroughly mixed, the gel forms suddenly after the precipitation delay period, causing a dramatic increase in viscosity. This can be moderated by adding one of the educts in portions. Although the precipitation of sodium aluminosilicate gel occurs at temperatures of 60–70°C, a temperature of just under 100°C is required for rapid and controlled crystallization. The crystallization time also depends on the alkali concentration and the solids content in the reaction mixture. The rounded edges typical of detergent zeolite, on material that otherwise crystallizes in cubic form, are attributable to rapid crystal growth after a controlled nucleation period and increased redissolving because of the high alkalinity of the reaction mixture after crystallization has taken place (Fig. 2).

After crystallization the solid material is separated from the mother liquor and washed; if necessary it is also dried, milled, or granulated. Despite the finely particulate nature of the solid material, filtration presents no insurmountable problems for products with a high degree of crystallization and a uniform particle size of about 3 μm. Filter cakes with about 50% solids content (calculated with regard to absolutely dry zeolite, that is, zeolite containing no water of crystallization) are obtained. The filter cake itself is strongly thixotropic and is pumpable after a short period of shearing stress. The zeolite is made fit for storage by drying it until it has a residual water content (water of crystallization) of 20%, when it takes the form of a dry, free-flowing powder, or by adding a stabilizer. Stabilizers are usually nonionic surfactants. The hydrophilic group of the niotenside molecule absorbs on the surface of the zeolite, and the hydrophobic hydrocarbon chain prevents the individual crystals from agglomerating. In this way it is possible to keep zeolite suspensions with a solid content of 50% as pumpable liquids for a period of weeks [43].

Figure 3 is a general scheme for the industrial manufacture of zeolite by precipitation. Gel precipitation and zeolite filtration are performed continuously, but crystallization still always occurs in batch operation. The educts are mixed in separate streams to avoid sudden viscosity changes during precipitation. A stirred cascade effects the gradual homogenization of the gel [44–47].

The zeolitization of natural aluminosilicates is repeatedly advanced as an alternative to the industrially preferred precipitation synthesis. Kaolinitic clays are most suitable for this. This clay process is especially advantageous if pure raw materials are available in sufficient amounts and the product is to be used an ion exchanger or sorbent (lower as quality demands) [48]. In kaolinite the molecular ratio SiO_2/Al_2O_3 is approximately 2, so that no more silicate or aluminate must be added to facilitate conversion to zeolite A. This is currently the only process available for converting kaolinite into zeolite; it involves the calcination

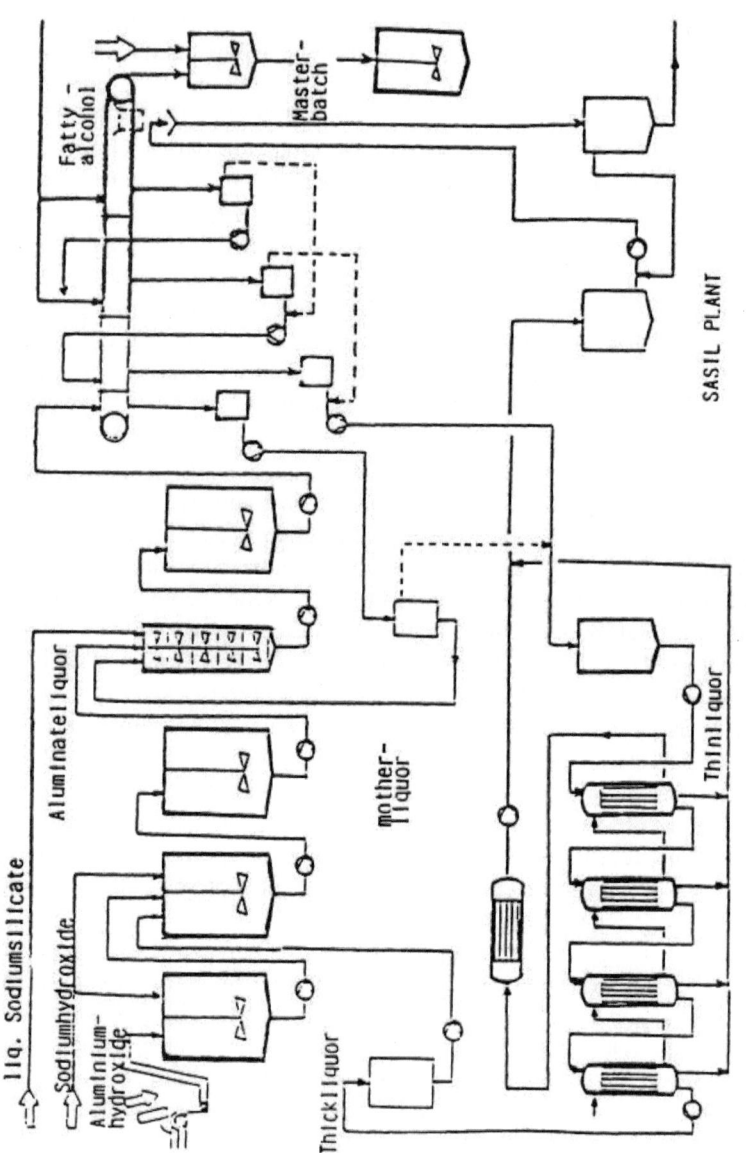

FIG. 3 Continuous production of zeolite NaA according to the precipitation/hydrogel process.

of elutriated fine kaolinites in a rotary kiln at approximately 600°C. X-ray amorphous metakaolinite is formed, which is then heated in dilute sodium hydroxide solution at more than 80°C to convert it to zeolite NaA. Because kaolinite has an iron content of >0.3% Fe_2O_3, the zeolites manufactured in this way are not suitable for use in detergents (destabilization of persalts and loss of active oxygen).

II. ENVIRONMENTAL DISTRIBUTION OF ZEOLITE A

A. Zeolite A in Sewage

Conventional detergents contain mainly soluble components, that is, molecularly disperse substances and association colloids (surfactants). With the advent of the insoluble ion exchanger Na-zeolite A, large amounts of a water-insoluble substance were included in detergents for the first time. The particulate structure of Na-zeolite A is such that the particle size of the zeolite suspended in the wash liquor must be limited for technical and ecological reasons [49,50]. As zeolite particle size decreases, calcium ion exchange rates increase, so that the formation of calcium carbonate precipitates in particular is more strongly inhibited and zeolite depositions on textiles decrease [51–54]. The initial exchange rates of about 130 mg CaO/g zeolite/minute [54] or 2 grains Ca/gallon/g zeolite/minute [53] needed to remove calcium ions sufficiently quickly from wash liquors and to allow the zeolite particles to pass the mesh sizes of the order of 10 μm encountered in the most closely meshed textiles can be achieved only by zeolite particles smaller than 10 μm. The minimum solid particle size for sedimentative removal of particles suspended in sewage is about 1 μm. The range of zeolite particle sizes required for technical and ecological compatibility is therefore 1 μm < particle size < 10 μm [53,54]. The poorly soluble pigment soil formed from molecularly disperse substances from tap water, textiles, and soil lies preferentially in the colloidal particle size range (<0.6 μm) and becomes more concentrated as the washing process progresses, reinforced by the dispersion process of laundry-active substances (surfactants) in the wash liquor [51].

From the washing machine the zeolite particles, which are primarily loaded with calcium, are discharged together with the wash liquor, which contains pigment soil, greasy soil, and surfactants, into domestic sewage pipes and are then carried along with domestic sewage through the communal sewerage into a clarification plant or surface waters. According to estimates (based on Stokes' law) of the sedimentation rate of discrete solid particles in a static water phase, industrially manufactured Na-zeolite A in the requisite particle size range of 1–10 μm and with the relatively high density of 2 g/cm^3 in sewage, can in principle be partially sedimented in a time that can be regarded as finite for practical purposes. By contrast, metastable colloidal solid particles are largely dispersion stable. If sodium triphosphate (STP) in detergents were to be completely replaced by zeolite, it is estimated that the zeolite concentration in domestic sewage would be of the order of magnitude of 15–30 mg/liter (Sec. II.C.3).

Model studies and practice tests must be carried out to determine the extent to which suspended zeolite particles are removed from static and flowing domestic sewage in the context of the sedimentation and deposition of zeolite in domestic sewage pipes, in the communal sewerage and in the primary settling stage of a clarification plant.

1. Particle Size Distribution and Sedimentation in Pure Water and Sewage

(a) Sedimentation Theory. Liquids have a viscosity that is demonstrated during flow and is attributable to the friction between shifting layers of liquid that is, to internal friction. Viscosity influences the sedimentation rate of solid particles in that the particles do not move with uniform acceleration but rather adopt a constant rate after a short time. In this state the internal friction is equivalent to the forces acting on the particle, including gravitation and lift. This is the basis of Stokes' law for small, spherical, solid particles in a static liquid, according to which the equilibrium sedimentation rate (fall height/time) of discrete particles is proportional to the square of the particle radius and to the density difference and is inversely proportional to the viscosity of the liquid.

The lower limit at which the law can be applied corresponds to the transition of a suspension of solids to a colloid solution (particle size < 0.6 μm), when so-called Brownian molecular movement inhibits the sedimentation of particles. The upper limit is dependent on not only the size of the particles (up to approximately 50 μm) but also their density and the physical properties of the liquid. The upper limit is defined in terms of the dimensionless Reynolds number Re < 1, when a sedimenting particle is in a laminar flow region (laminar flow is largely determined by internal friction). Furthermore, as particle concentration increases and mutual obstruction between the particles becomes significant, and under the influence of the counterflow speed of the displaced liquid, the sedimentation rate decreases [55,56].

Naturally occurring and industrially manufactured suspensions of solids are usually polydisperse: that is, the sizes of the particle are variable. The particle size distribution of such suspensions usually conforms to Gaussian frequency distribution, for the characterization of which various average particle diameters are used. In symmetrical distribution curves these average particle diameters are in agreement [57]. A distinction is made between the mode (passes through the peak of the relative percentage frequency curve, that is, the diameter at which the frequency is at a maximum), the median (diameter at 50% size on the cumulative frequency curve), and the mean diameter (diameter given by the number of particles, with the different diameters and the respective percentage proportions relative to the total number of particles). The half-value latitude of a distribution curve is a measure of the polydispersity of a suspension of solids (it is half the width of the relative percentage frequency curve at half the height of the frequency curve at the maximum). The mode and median diameters and half-value latitude can be determined graphically.

The particle size measurements carried out to determine the particle size distribution were made with the sedimentation balance (measured particle size is proportional to the particles with the same sedimentation rate: Stokes' law) and the Coulter counter (measured particle size is proportional to the particles with the same volume: particle volume interferes with an electrical field).

(b) Characterization of Particulate Na-Zeolite A. Industrially manufactured Na-zeolite A is characterized in terms of the technically and ecologically relevant parameters of the particle size distribution (in pure water) and the calcium binding capability (at 30°d Ca, pH 10, with 1 g/liter of zeolite).

The calcium binding capability of suspended Na-zeolite A at pH 10 in water with a relatively high level of hardness (30°d), that is, in water with an excess of calcium ions compared with its theoretical capacity, is preferably in the range 100–200 mg CaO/g zeolite after an ion exchange period of 15 minutes, depending on the zeolite synthesis conditions and the particle size distribution [49].

An aqueous suspension contains primary particles and agglomerates of primary particles, that is, secondary particles. Primary particles are distinctive individual crystallites with edge lengths of approximately 0.1–3 μm or intergrowths of individual crystallites. Suspensions of industrial zeolites are correspondingly polydisperse because of the various particle sizes and can be represented by Gaussian frequency distributions of the particle sizes (Fig. 4) [51]. Accordingly, Na-zeolites A exhibit differences in the mean diameters of 4.0, 7.9, and 10.4 μm, depending on the synthesis conditions. Most of the particles lie clearly above the colloidal particle size region.

The application of Stokes' law for the sedimentation of suspended zeolite particles is permissible, because for sewage-relevant—diluted—zeolite suspensions of 30 mg/liter with particle sizes smaller than 20 μm and a density of 2 g/cm^3, sedimentation rates of less than 0.2 mm/s and Reynolds numbers of less than 0.005 are expected, and these lie within the application range of Stokes' law for undisturbed sedimentation of discrete particles [58].

(c) Model Tests. Findings obtained from sedimentation analyses with solutions of detergent ingredients in pure water and with filtered synthetic sewage were used to illustrate work on the influence of water-soluble sewage ingredients on the particle size distribution and sedimentation of industrial zeolite A (mean diameter = 10.3 μm) [59]. Anionic surfactants and a conventional model detergent (without builders and NaCl) containing a mixture of anionic and nonionic

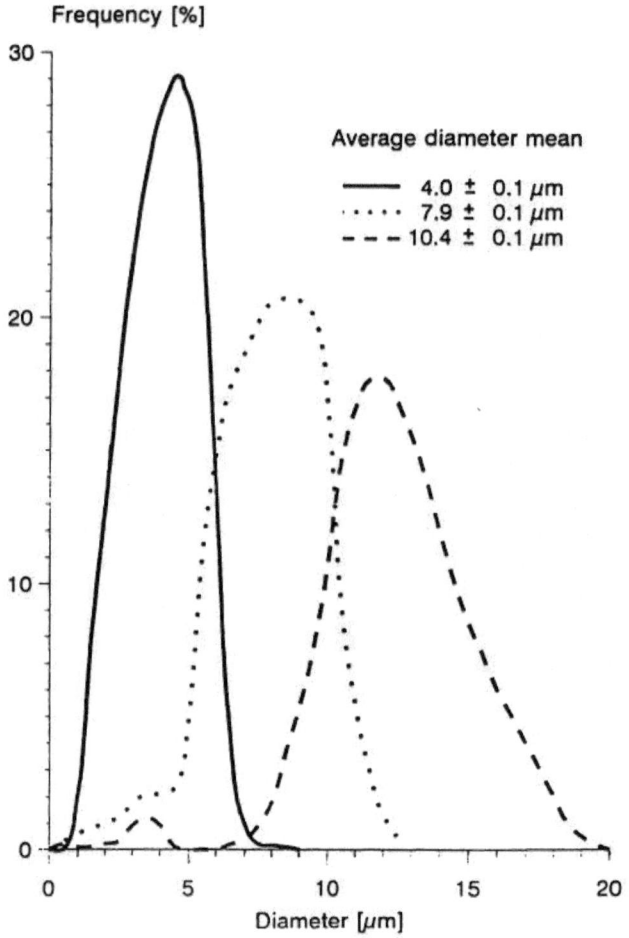

FIG. 4 Particle size distribution of Na-zeolites A [51]. Test method: sedimentation analysis. Apparatus: sedimentation balance, 25°C (Cahn).

surfactants have no significant influence on the sedimentation rate (half-life time) and the parameters of the distribution curve of zeolite (Table 1). Anionic and nonionic surfactants are not adsorbed by negative zeolite (with Na^+ counterions) with a high negative ζ potential and are therefore not able to change the high repulsion potential between the zeolite particles or, consequently, the particle size distribution [60] (see also Ref. 146). In contrast, high salt concentrations promote a higher sedimentation rate (lower half-life) owing to the increased particle diameter and the increased polydispersity (Table 1). It follows that the

TABLE 1 Sedimentation (Half-Value Time) and Particle Size Distribution of
Na-Zeolite A (Sedimentation Balance, 25°C, Height of Fall 20 cm)

System	Concentration (g/liter)	Half-value time (minutes)	Average particle diameter mean (μm)	Average particle diameter mode (μm)	Half-value latitude (μm)
Water	—	21.8 ± 1	10.3	8.1	3.2
Alkylbenzene-sulfonate	0.48	21.0 ± 1	10.4	9.1	3.8
Sodium chloride	30.0	11.7 ± 0.6	14.5	13.8	5.5
Model detergent	6.40	21.4 ± 1	11.4	9.4	3.6

Source: From Ref. 59.

addition of salt increases the counterion concentration at the zeolite and reduces
the repulsion potential between the zeolite particles, so that zeolite agglomerates
with large secondary particle diameters can form. The soluble substances in
synthetic sewage from which the turbidity substances have been removed have
the same effect as high salt concentrations in that they stimulate zeolite agglomeration to larger secondary particles and increased polydispersity (Fig. 5).

Practice-relevant studies of the sedimentation of industrial zeolite A (median
diameter = 4.2 μm) in concentrations similar to those found in sewage (30
mg/liter) in static pure water were carried out with the help of the static column
test. These demonstrated that the sedimentation rate of zeolite under the given
conditions approximates Stokes' law and that it is possible to remove almost 30%
of the zeolite from the water over a period of time (Fig. 6) [58]. Comparable
zeolite removal is obtained from the zeolite particle size distribution determined
with the Coulter counter method and evaluation with Stokes' law (Fig. 6) [58].
It follows that Stokes' law enables reasonably accurate predictions to be made
concerning the removal of zeolite from water under pilot conditions.

Analogous studies with natural sewage containing a high level of solids (direct
zeolite determination through aluminum analysis) indicate that 30 mg/liter of
zeolite only minimally reduces the strong percentage removal of solids suspended in sewage compared with solids removal from sewage in the absence of
zeolite (Fig. 7) [58]. It follows that zeolite, in the concentrations in which it
occurs in sewage, has practically no effect on the removal of solids by sedimentation. Furthermore, zeolite is removed to a lesser extent than the other solids
suspended in the sewage, especially in the initial stages of sedimentation (Fig. 7).
In comparison with the approximate 30% removal of zeolite from pure water
(Fig. 6), however, much higher levels (up to about 70%) of zeolite are removed
from sewage and at a faster sedimentation rate, especially in the initial stages of

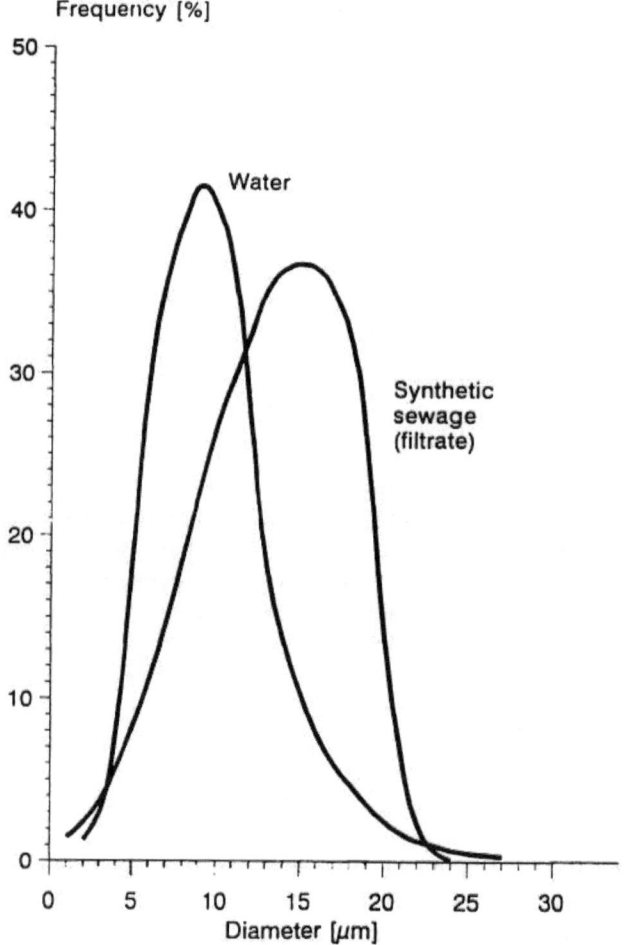

FIG. 5 Particle size distribution of Na-zeolite A in synthetic sewage filtrate (sedimentation balance, 25°C) [54].

sedimentation (Fig. 7). One possible explanation is that soluble substances in the sewage stimulate zeolite agglomeration (Fig. 5) [54,59] and heteroagglomeration of the relatively sparse zeolite with silicate solids [60].

Analogous studies by other authors [61] with the static column pair test and natural solids-containing sewage and 100 mg/liter of Na-zeolite A (median diameter = 3–4 μm) confirm that zeolite is removed to a lesser extent, albeit only slightly, than the other solids suspended in the sewage but show also that higher zeolite concentration can cause a temporary weak increase in solids removal.

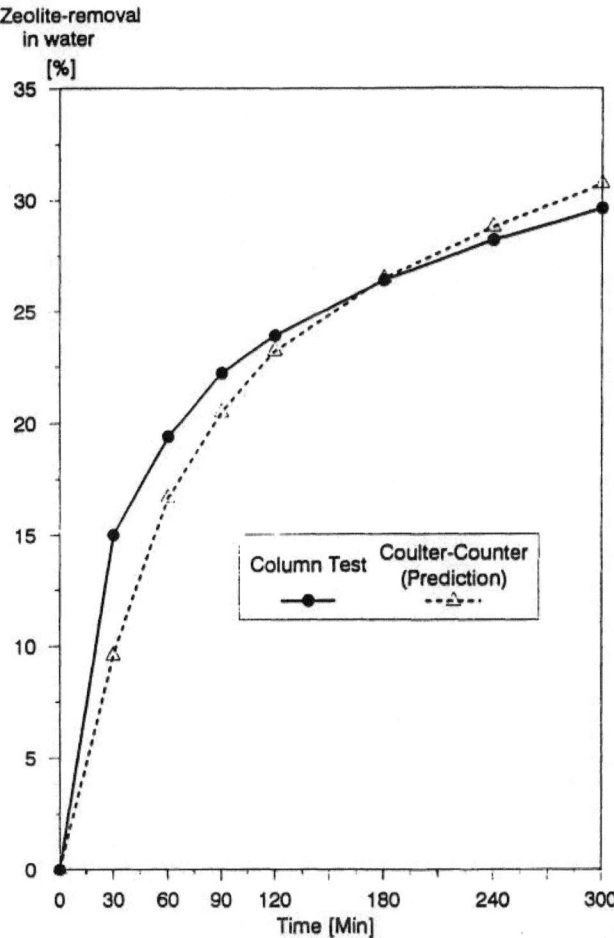

FIG. 6 Removal (%) of Na-zeolite A in water (test method: static column test and zeolite product HAB A40 with average diameter median = 4.2 μm) [58].

Studies of 20 mg/liter of zeolite in natural sewage, performed with the dynamic column test, which simulates the primary settling stage in a clarification plant, indicate that 77 and 62% of the zeolite in total and in the smallest 10% size fraction, respectively, are removed [61].

Laboratory studies of sedimentation in coarsely filtered natural sewage provide information on the influence of calcium-loaded Na-zeolite A on finely dispersed turbidity substances with only a slight tendency to sediment out in sewage [54]. With the help of continuous electrophotometric turbidity measure-

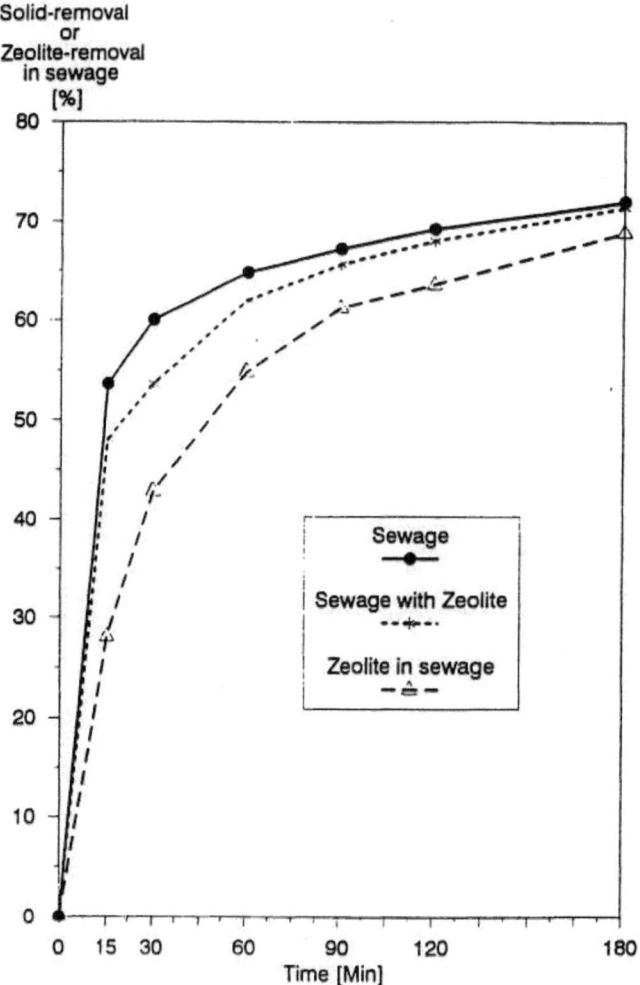

FIG. 7 Removal (%) of Na-zeolite A and suspended solids in raw sewage (test method: paired static column test and zeolite product HAB A40 with average diameter median = 4.2 µm) [58].

ments, the turbidity and clarification of sewage by zeolite can be determined by measuring the relative percentage change in the turbidity of zeolite suspended in sewage compared with the turbidity of the sewage in the absence of zeolite as a function of time in the static sewage (Fig. 8). With 100 mg/liter of Ca-Na-zeolite A of product I (mean diameter = 5.2 µm; proportion <3 µm = 6.6%) or product II (mean diameter = 4.4 µm; proportion <3 µm = 55.1%), turbidity initially

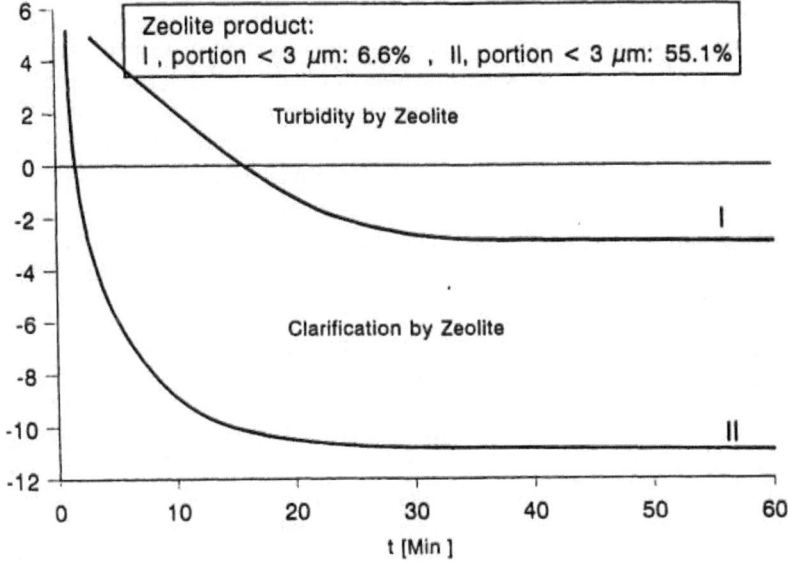

FIG. 8 Turbidity change (relative percentage) of coarse-filtered domestic sewage by addition of Ca-Na-zeolite A as function of time (test method: static settling test) [54].

increases after the zeolite is added and then decreases with the passage of time until the clear range is reached. Zeolite causes finely dispersed turbidity substances in the sewage to sediment out and increases their sedimentation rate.

The zeolite product II with the greatest proportion of smaller zeolite particles causes faster and stronger sedimentation of the turbidity substances, which therefore combine preferentially with smaller zeolite particles to form heteroagglomerates. Turbidity substances can therefore be removed with zeolite as heteroagglomerate sediment in a clarification process. However, zeolites cause clarification to occur only if other solids suspended in the sewage have not already done so. The clarification properties of zeolites are no better than those of other suspended solids.

Model tank studies of discrete particle sedimentation are difficult to scale up because it is almost impossible to achieve dynamic and hydraulic similarities. Dynamic similarity cannot be achieved between model and prototype sedimentation tanks; that is, it is not possible to have equal Reynolds and Froude numbers and velocity ratios simultaneously [58,62]. With raw sewage, predictions are even less reliable because its flocculant or agglomerate nature means that the depth of the tank cannot be neglected. As a result, retention time becomes a parameter of major importance because it affects the flocculation or agglomeration mechanisms in sedimentation of raw sewage. As a consequence of these difficulties, predictions for full-scale sedimentation design are often developed from column sedimentation tests. Typically, the proportion of suspended solids removed from sewage increases as a function of column depth, in sharp contrast to discrete sedimentation, for which depth is irrelevant. This is a result of the nonuniform sedimentation velocities of flocculant or aggregate suspensions. The amounts removed at a given depth in the column may be used for comparative purposes. In sedimentation and activated sludge tests involving the operation of two units in parallel, one as a control and the other containing the zeolite to be tested, comparative predictions can be made even if scale-up predictions are not completely accurate [58].

2. Transportation, Deposition, and Resuspension in Sewage Pipes

(a) Basics. As a result of extensive studies by various authors concerning natural and simulated water courses, a number of basic facts are known about the transportation and deposition of suspended materials and the resuspension (erosion) of deposited materials [63]. A distinction is made between monodisperse materials and the polydisperse materials encountered in practice, such as loam, clay, gravel, sand and silty loam, and clay, and studies are carried out involving particle sizes of 0.002–50 mm (2–5×10^3 µm) and flow velocities rates of 0.1–1000 cm/s. A strictly defined and practical starting point is obtained by assuming that the suspended material flows over a bed of loose material of the same particle size (monodisperse material) or the same particle size distribution (polydisperse material).

The curves for the lowest transportation velocity of suspended particles and the lowest resuspension velocity of deposited particles are shown on the graph of the flow velocity as a function of the particle size of monodisperse material (Fig. 9) [63]. In the region below the transportation curve material deposition occurs: that is, sediment forms on the bottom. In the region above the resuspension curve resuspension occurs from the deposited material. The resuspension curve passes through a minimum at the relatively small particle size of about 0.1–0.5 mm (100–500 µm), and the transportation curve ascends continuously as the particle size increases. The flow velocities for the resuspension of deposited material are higher than those for the transportation of material with corresponding particle size.

For materials containing all sizes of particle, such as those encountered in practice, the situation with regard to transported, deposited, and resuspended material is much more complicated [63]. The interval between the resuspension velocity (the velocity at which all particle sizes are set in motion) and the lowest transportation velocity (the velocity at which material starts to deposit) is known as the transportation interval and is narrower than that of a monodisperse material. If the velocity is kept constant at any value within the transportation

Flow velocity [cm/s]

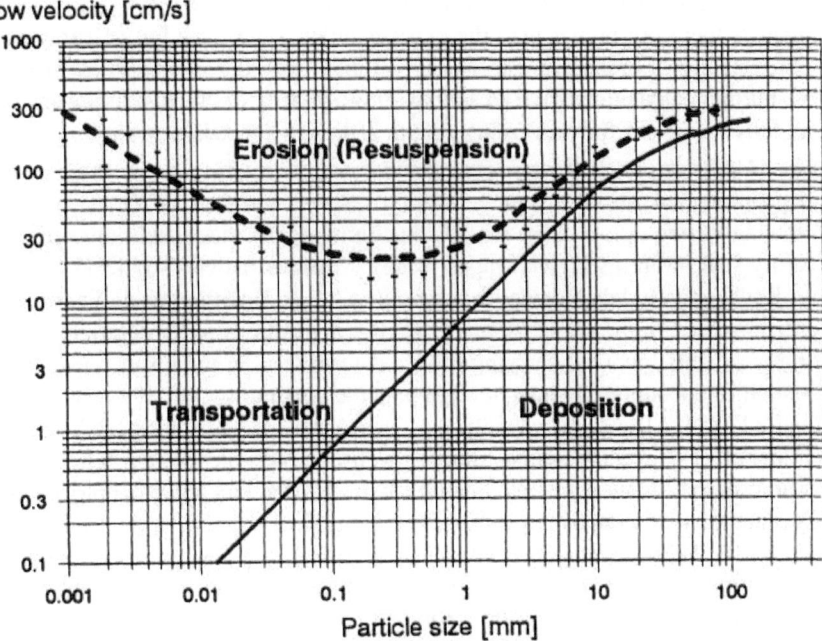

FIG. 9 Curves for erosion (resuspension) and transportation r.s.p. Deposition of monodisperse material on a bed of loose material of the same particle size [63].

interval, neither resuspension nor sedimentation occurs. The resuspension velocity varies with mixed grades of particles. The mixture is more readily resuspended if fine material is added to coarser and is resuspended most readily if the mixture contains an average of 75% finer material.

The experimental results obtained with monodisperse fluvial materials with a density of 2.65 g/cm^3 (Fig. 9) [63] also apply to industrially manufactured zeolite A, which is a similar silicate compound and has a comparatively narrower particle size distribution and a density of 2 g/cm^3. Suspended zeolite particles with a diameter close to the upper limit of 10 μm (0.01 mm) can therefore be fully transported without sediment formation at the very low flow velocity of, for example, 1.0 cm/s, but acquired depositions of zeolite cannot be resuspended (Fig. 9). It is expected, however, that at flow velocities observed in sewage practice [61,64], approximately 40–60 cm/s, zeolite suspended in sewage will be transported without hindrance and initial resuspension of zeolite depositions in sewage pipes will occur (Fig. 9).

Calculations for zeolite (particle diameter 4 μm) give a lowest transportation velocity of 3.7 cm/s for zeolite in sewage pipes [65]. The calculated lowest transportation velocity of zeolite particles with a diameter of 4 μm in sewage pipes is therefore higher than the experimentally determined lowest transportation velocity of monodisperse fluvial materials with the same particle size (Fig. 9). Further calculations give an average flow velocity of 30 cm/s in sewage pipes as fully sufficient for preventing zeolite sediment formation [66].

(b) Model Studies.   Tests were carried out on domestic sewage to observe the deposition and resuspension behavior of suspended Na-zeolite A in sewage flowing through pipes; pipe models that simulate practice conditions on a laboratory scale were used, with pipes made of various materials [54,61].

Radioactively labeled zeolite (chromium 51) [67] with a mean diameter of 8 μm was used in the study to enable zeolite depositions to be determined quantitatively; this involved carrying out short-term tests in the lower range of flow velocities encountered under sewage practice conditions (18–39 cm/s) (Figs. 10 and 11) [54]. The flow velocity was increased and then decreased in stages. Cumulative irreversible depositions as well as reversible (resuspendable) depositions of zeolite were detected. With clean cast iron and VA steel pipes, zeolite was cumulatively deposited irrespective of the flow velocity of domestic sewage (Fig. 10). With clean polyvinyl chloride (PVC) and glass pipes some zeolite deposition occurred at the beginning at low flow velocities, but these depositions were fully resuspended and transported in the sewage as the flow velocity increased. When the flow velocity was then reduced again, small amounts of zeolite were cumulatively deposited. This deposition and resuspension behavior, which is presumably attributable to the different adhesion mechanisms of zeolite on different materials (surface roughness of the pipe walls and electrostatic and van der Waals forces) indicates that in principle not only sedimented zeolite but also adhesion mechanisms contribute to the degree of zeolite deposition.

As an initial coating of the iron pipes with material from the sewage increases (sewage with suspended unlabeled zeolite), subsequent tests result in ever weaker zeolite depositions (Fig. 11). When the initial coating of the iron pipes was at its strongest (curve III), no additional cumulative zeolite deposition occurred at flow velocities of some 27 cm/s and higher. In iron pipes with considerable coatings, such as those found in sewage pipes under real conditions, tests showed that zeolite depositions formed at very low flow velocities are fully resuspended at flow velocities of about 30 cm/s and higher and that the deposition and resuspension of zeolite are independent of the pipe material. It follows that no significant deposition of zeolite is expected at the flow velocities of approximately 40–60 cm/s [61,64] encountered in sewage practice.

Studies carried out by other authors involve the use of a pipe model for long-term tests with visual assessment of zeolite depositions and qualitative determination of zeolite; zeolite-containing wash liquors from a washing machine (median zeolite particle diameter 3–4 μm) are passed continuously

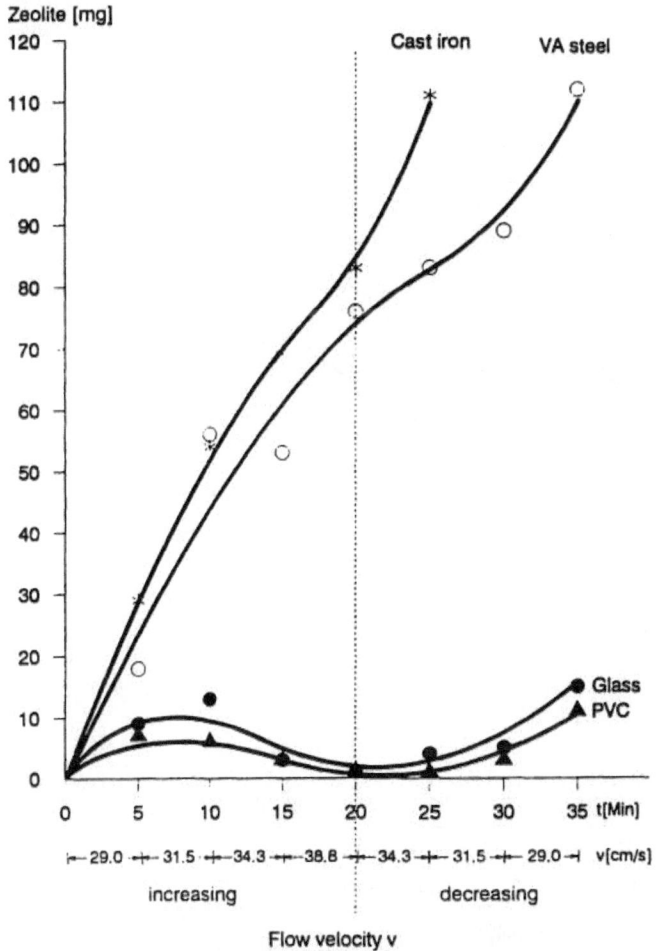

FIG. 10 Total Na-zeolite A deposition (pipe model) in raw domestic sewage with the clean pipe materials cast iron, VA steel, glass, and PVC [54].

through pipes made of different materials. At the end of the test the pipes are examined for depositions. Increasing zeolite depositions in the pipe material sequence [61] PVC < copper < iron confirm that deposition on iron pipes is greater than that on PVC in the low deposition range [54]. Surface roughness of the pipes is thought to cause differences in deposition, because visible zeolite depositions occur only at certain locations on the pipe walls.

(c) Practical Tests. The model calculations and model studies carried out on a laboratory scale indicate that the use of zeolite-containing detergents is asso-

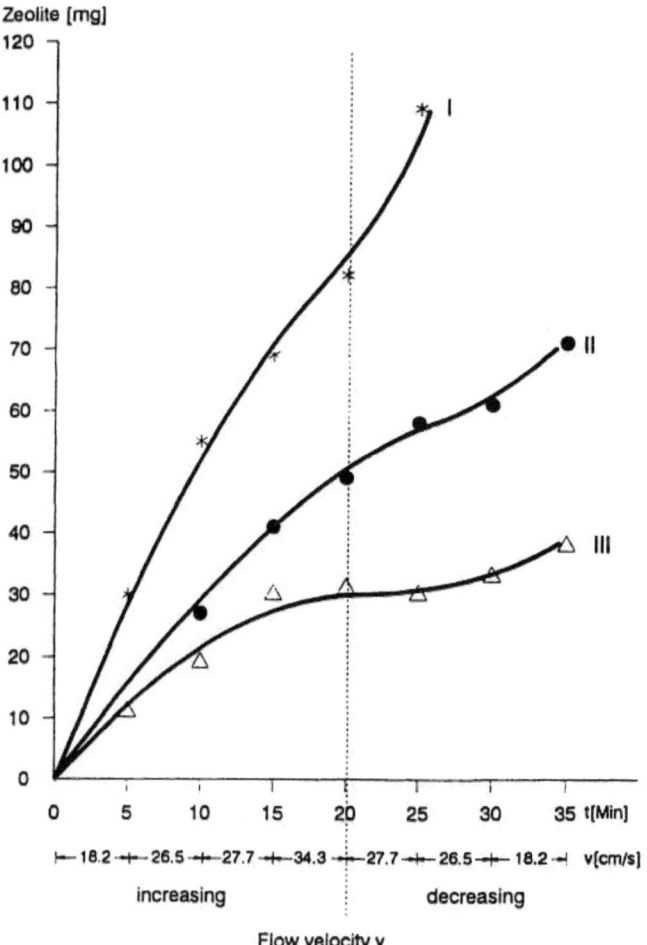

FIG. 11 Total Na-zeolite A deposition (pipe model) in raw domestic sewage with increasing precoating of cast iron pipes (I, II, III) [54].

ciated with only minimal risks with regard to the deposition of Na-zeolite A in the sewage pipes of domestic and communal sewerage. Some findings concerning field tests that were carried out to substantiate these findings are discussed here [61,64,67–69].

The use of zeolite A-containing detergents for practical tests was started in 1974 when the Na-zeolite A (charge HAB A40) provided by Henkel and Degussa (Germany) was used for large-scale tests in the Düsseldorf and Stuttgart areas (Germany) [68]. The use of zeolite-containing detergent over a period of 19–33 months neither caused stoppages nor resulted in increased accumulation of depositions in the domestic sewage pipes of residential blocks of up to 180 households in the Henkel settlement in Düsseldorf-Holthausen. It was clear that the disposal of waste water through the domestic sewage pipes would not be disturbed after the introduction of zeolite-containing detergents [64]. To have a specific analysis method and quantitative deposition determination method available for Na-zeolite-A in certain discharge pipe sections the zeolite was labeled with indium (zeolite particle size range from 2 to 20 μm) and determined by means of neutron activation analysis [67], and segment-like metal strips with the same area were inserted in the sewage pipes and removed for quantitative analysis. Sampling locations were horizontal pipes, down pipes, pump sumps, and drainage ditches.

It was found that the amount of deposition in horizontal pipes was the same in the absence (control) and the presence of zeolite; inorganic depositions remained constant after a threshold value was established, with a maximum value of approximately 5% zeolite. The composition of the depositions fluctuated within narrow limits but did not change significantly. Over a relatively long period of time only minor zeolite deposition occurred at the variable flow velocities prevailing in household sewage pipes, and the zeolite had no decisive influence on the extent or composition of the deposition of the solids suspended in the sewage. In the horizontal pipes in particular, it was clear that temporary zeolite depositions may form and then become resuspended (see the proven resuspension effects on deposited zeolite in model studies [54]). Zeolite sediments at locations where flow velocity is low, such as drainage ditches, in which sediments of all suspended solids accumulate.

A 1 year large-scale study was carried out in Stuttgart-Büsnau. This involved 960 households in which only detergents containing indium-labeled zeolite A were used. Tests carried out in the communal sewerage, extending as far as the clarification plant inflow, confirmed that there was no long-term deposition of zeolite in the sewage pipes of the municipal sewerage [68]. Four shafts were selected in the Büsnau sewerage. Checks were made on the basis of visual assessment in the visible part of the sewer. The zeolite content in the laundry wastewater had no permanent effect on the formation of depositions in the sewers of the municipal sewerage. Depositions tend to form sporadically, mainly from other more coarsely grained materials present in the sewage; the stagnations thus created then encourage the formation of more finely particulate zeolite depositions. However, these finely particulate depositions disappear just as sporadically, together with the other depositions, as a result of the rinsing effect of stronger sewage discharges and rainwater [68] (see the findings mentioned in the basics section, that a mixed deposition of fine and coarsely particulate material is, overall, easier to resuspend [63]).

Tests carried out in the communal sewerage after a 4 month period during which a zeolite-containing detergent was used in 75 households in a small Belgian town also showed no significant long-term accumulation of depositions and no impairment of the flow behavior of the sewage in the municipal sewerage [61]. Analytical SiO_2 determinations in the upper layers of depositions in the sewerage pipes before and after heavy rain showed that the zeolite concentration in the upper deposition layers was reduced by the rain from about 6.7 to 0.9%. Zeolite depositions in sewerage pipes are only temporary and can be resuspended by increased flow velocities.

To summarize, all of the zeolite suspended in the wash liquor is discharged into domestic and communal sewage pipes when the wash liquor is pumped off. In the long term only a very small amount of the zeolite is removed from the sewage by deposition, so that almost all of the zeolite is transported to the clarification plant.

B. Elimination of Zeolites in Sewage Treatment Plants

Zeolites used in detergents ultimately drain and in most cases enter either a municipal sewage treatment plant or an individual home treatment system. To evaluate the extent of zeolite A elimination under these conditions, a number of laboratory tests, including simulation model studies and field trials, have been carried out. The fate and the resulting concentrations of zeolite A were determined in these studies on the basis of Al and/or Si measurements and in some cases by use of a zeolite A doped with indium [67], which is accessible to characterization and quantification by neutron activation analysis. Because of the very favorable outcome of the investigations into the ecological behavior of zeolite A [70], there was thought to be no need for the development of an analytical method for the technical material that could be applied in environmental samples.

1. Mechanical Treatment

Because of the particulate structure, water insolubility, and inorganic nature of zeolite A, it can be expected that the most relevant process for the removal of this material is the sedimentation process (Sec. II.A). A particle diameter of 1 μm is frequently regarded as the minimum size removed by sedimentation during primary treatment [71]. Thus, detergent-use zeolite A with a particle size of 1–10 μm should be removed efficiently in the first stage of the waste water treatment process in most municipal sewage plants, that is, in the mechanical stage.

Investigations by King et al. [61] into the removal of zeolite A and other suspended solids in static settling tests suggested that the removal of the zeolite from waste water by primary treatment is similar or slightly higher than that of other suspended solids. Carrondo et al. [58] noticed a marginally lower removal of zeolite A than of the other suspended solids in raw sewage when testing the zeolite in settling columns at a concentration of 30 mg/liter (Sec. II.A.1.c). The removal rates of the zeolite were 55 and 69% at retention times of 1 and 2 h, respectively.

Data for zeolite A removal from a dynamic sedimentation test were provided by King et al. [61] using a 6 liter tank supplied with raw sewage containing 20 mg/liter of zeolite. The removal of bulk zeolite A was 77% at a 2 h retention time and 62% for the smallest 10% size fraction of the zeolite; the corresponding removal of suspended solids was in the range 33–41%. Thus, it could be concluded that smaller particles of zeolite A are not selectively discharged in waste water treatment plant effluents and that zeolite is better removed by sedimentation than other suspended solids. The same conclusion was drawn by Rossin et al. [72], who measured zeolite A and suspended solids removal in a continuous primary sedimentation pilot plant (volume 0.9 m^3). At concentrations of 30 and 60 mg/liter of zeolite A in the influent raw sewage, a normally accepted range of suspended solids removal in primary sedimentation of 50–60% was observed, which was not enhanced in the presence of the zeolite. The zeolite A removal was 66% at a retention time of 0.76 h and reached 76% at variable flow conditions. The higher removal of zeolite A compared with ordinary suspended solids, which was found in dynamic but not static settling systems, is explained by the higher turbulence and, hence, more intensive contact between zeolite A and sewage particles aiding the agglomeration of the zeolite particles.

The zeolite elimination prognoses deduced from model tests were confirmed and partly exceeded by the results obtained under practical conditions. Data from a municipal activated sludge plant [73] with an influent doped with indium-labeled zeolite A yielded 50% removal of zeolite in the sand trap and a 85% decrease after passing the mechanical stage of the sewage plant. In a large 13 month field trial in the village of Büsnau (Germany) comprising more than 3000 people, a 63% removal of zeolite A was observed after mechanical treatment of the raw waste water in the sewage plant; this elimination rate was the result of a removal of 21% in the sand trap and 42% in the primary settler [68]. The available data from realistic trickling filter model plants showed a 30% removal of zeolite A by primary treatment at an influent concentration of about 10 mg/liter; the suspended solids removal was 48% at the same time [74]. A considerably higher removal rate of zeolite A (81%) was measured in the primary settler of a technical scale trickling filter [75] when 60 mg/liter of Ca-zeolite was added to the waste water influent; the difference in the results from these two trickling filter plants may be explained by the differing zeolite A concentrations in the plant influents.

2.   Biological Treatment

Having passed the mechanical purification stage of sewage treatment, the dissolved or suspended material contained in the effluent enters the biological stage.

Biological waste water treatment is effected in most cases by the activated sludge process or by trickling filters.

In a 10 day semicontinuous activated sludge test (SCAS test) involving a fill and draw operation with a daily cycle of 23 h aeration followed by 0.5 h settling and discarding the clarified supernatant, the zeolite elimination was 99 and 97% for the bulk zeolite and the smallest 10% size fraction, respectively [61]. A 72 day study conducted in the same test at the same zeolite concentration (20 mg/liter) yielded an elimination rate of about 80% [61]. The same authors found a zeolite A removal of 77–87% when feeding a continuous activated sludge model plant with raw municipal waste water containing 20 mg/liter of zeolite. Carrondo et al. [76] found zeolite A concentrations of 2–4 mg/liter in the effluent of a model activated sludge plant fed with waste water and containing 15 or 30 mg/liter of zeolite in the influent; this corresponds to an elimination rate of 73–93%. In a further study by Carrondo et al. [58] conducted in the same test system under good clarifying conditions, the zeolite A removal averaged 88%. The authors concluded that removal rates >90% are possible in waste water treatment plants comprising primary settling and secondary treatment by activated sludge.

The elimination rates of zeolite A obtained in model tests could be verified in a number of realistic field studies. Hopping [77] determined varying removals between 80–90% in a package activated sludge plant serving a 60-household area; the removal rates were not influenced by the influent zeolite concentrations of about 8–17 mg/liter. Similar rate values were obtained in a field study with aerobic home treatment systems [78]: the zeolite A removal was 88% at an influent concentration of about 23 mg/liter.

The high removal of zeolite A under practical conditions was also indicated by the findings of Malz and Jendreyko [73] who only detected traces of Indium-labelled zeolite A in the effluent of an activated sludge plant; up to 85% of the added zeolite was already removed in the mechanical stage of the plant. Ultimately, a field trial in the village of Büsnau (Germany) conducted in 1976–1977 [68] revealed a total elimination rate of indium-labeled zeolite A of 96% in a mechanical-biological sewage treatment plant; this rate is caused by 63% removal by mechanical and 33% by the biological treatment corresponding to a relative elimination rate of 89% in the activated sludge process.

Zeolite A removal rates are also reported for waste water treatment in trickling filters. In a laboratory-scale trickling filter experiment, the influent zeolite A concentration of 20 mg/liter was reduced by 89% during the 56 day test period [74]. Baumann et al. [74] found a 74% removal of the zeolite in a trickling filter under field conditions; this rate was considerably higher than the suspended solids elimination (57%) measured in the secondary treatment stage. The total elimination rate of zeolite A including mechanical and biological treatment in this trickling filter plant was 81% compared to 78% of the suspended solids.

Field investigations by Scherb [79] showed, in addition, that the small percentage of zeolites not yet removed in the mechanical treatment of a trickling filter plant (about 20%) is reduced by 75% in the biological stage, resulting in a total elimination of approximately 95%.

In sum, the data from laboratory simulation and field tests form a sound basis for the prognosis that the zeolite A concentrations in waste waters are reduced in mechanical-biological sewage treatment plants by at least 80% in trickling filters and by ≥90% in activated sludge plants.

3. Anaerobic Sludge Treatment

Sewage sludges originating from primary and secondary treatment of waste waters are usually stabilized in anaerobic digesters, resulting in the reduction in their water content and the removal of those organic materials that are accessible to anaerobic biodegradation. Because the overwhelming part of detergent-based zeolites end up in the primary and secondary sludge of sewage treatment plants, a relatively high concentration of this material can be expected in sludge digesters. It was shown in a number of model tests that even zeolite concentrations far above realistic values have no undesirable influence on the drainability of sewage sludges, on the ability of digester sludges to remove heavy metals, or on the performance of anaerobic degradation processes [73,78,80,81]. Based on aluminum analyses, 85–92% of the added zeolite was recovered in the sludge of laboratory digester units fed semi-continuously with synthetic waste water plus a zeolite suspension (78 and 780 mg/liter) over a period of 20 weeks; the zeolite content in the suspended solids of the supernatant was considerably less than in the digester sludge, indicating a much better sedimentation of zeolite A than normal suspended solids in the digester [78].

In the laboratory model of a septic tank representing individual anaerobic waste water home treatment systems, the elimination of suspended solids and zeolite A was compared after continuous dosing of zeolite-containing sewage (up to 55 mg/liter) over a 110 day study period. At a hydraulic retention time of 3 days in the septic tank, the zeolite removal was ≥97% compared with 57–77% removal of total suspended solids, confirming once again the excellent removal of zeolites under waste water treatment conditions [78]. Ultimately, the results from a 7 month septic tank field study by Holman and Hopping [78] revealed that the use of detergents containing about 20% of Na-zeolite A (equivalent to a mean of 26 mg/liter in the influent) had no effect on the performance or metal removal of the home waste water treatment systems; the zeolite removal was 81% on average. Based on these data it was estimated that the zeolite could concentrate up to approximately 30% of suspended solids in a septic tank, and the zeolite concentration in the tank effluent corresponded to about 2% of the suspended solids. Additional experiments by the same authors showed that the zeolites had little or no effect on the settled volume of septic tank sludge. Thus, the use of zeolite A in detergents was not expected to increase the rate of sludge accumulation in or the frequency of necessary clean-out of the septic tanks.

4. Tertiary and Quaternary Treatment

Although mechanical-biological waste water purification in two-stage sewage treatment plants represents the principle most often used to reduce insoluble contents and the organic load of waste waters, tertiary treatment is increasingly applied to remove inorganic phosphorus by chemical treatment. The elimination of phosphates is mainly effected either by direct addition of inorganic flocculants, such as ferric chloride, aluminum salts, or lime, into the aeration basin of the activated sludge plant (simultaneous precipitation) or by subsequent addition of these chemicals to the biostage effluent.

King et al. [61] investigated the elimination extent and the possible influence of zeolite A on the efficacy of the phosphorus removal. In a jar experiment, zeolite A was added to raw municipal waste water (0–20 mg/liter), mixed with doses of 50–250 mg/liter of ferric chloride, alum, or lime, stirred, and allowed to settle for 30 minutes. Analysis of the supernatants showed that the zeolites did not appreciably affect the concentrations of total phosphate or orthophosphate, turbidity, Fe, or Al; the removal of zeolite A was nearly 100% in comparison with 33–72% observed in the control (without chemical treatment). Thus, it can be concluded that zeolites are almost quantitatively eliminated by tertiary treatment of waste waters.

In certain cases, if exacting concepts of waste water purification are to be realized, before ingress into lakes or dam waters, for example, even quaternary treatment by filtration techniques may be carried out. As expected from the particulate structure of zeolite A, the application of a zeolite-containing secondary sewage effluent to a dual-media filter column (containing about 30 cm sand and 40 cm anthracite coal) resulted in 92% removal of the zeolite A and the removal of BOD_5 and total suspended solids was 46 and 78%, respectively [74].

C. Zeolite Concentrations in Raw and Treated Waste Waters and Sludges

As pointed out earlier, no sufficiently specific and sensitive analytical method is available to measure the concentrations of detergent-based zeolites in the various areas of the aquatic and soil environment. Nevertheless, the usage figures for detergents and their ingredients allow realistic calculation of the zeolite content of raw sewage. Taking the previously discussed elimination mechanisms and rates into account, the concentration of zeolites in sewage treatment plant effluents and in the sewage sludges produced can also be estimated. Ultimately, these calculated figures can be compared with experimentally determined data obtained from realistic field trials by application of the special efforts discussed here, which require/analytical methods.

1. Raw Waste Water

In most cases a conservative estimate of zeolite concentrations in raw sewage is based on the per capita consumption of detergents, the zeolite content of detergents, and the per capita drinking water consumption. Corresponding calculations for detergent and zeolite usage conditions in the United States led to the prognosis of approximately 10 mg zeolite/liter in raw sewage [77,82]. Because of the lower per capita flow rate of waste water from strictly domestic sources, the zeolite concentrations determined in the influents of individual home treatment systems were estimated as approximately two times higher; this was confirmed in a 7 month field study yielding mean influent concentrations of 23–26 mg/liter of zeolite [78].

With the low water consumption and greater hardness of European waters, a higher zeolite concentration in sewage was expected. Fischer et al. [83] anticipated a maximum of 50 mg/liter in typical German waste water. Similarly, Roland [84] calculated a mean zeolite concentration of 40 mg/liter in raw sewage based on a 20% zeolite content in detergents, a weekly per capita consumption of 200 g detergents and 1000 liters water.

Recent estimations based on the current zeolite A usage in European countries (in 1000 ton/year: Germany 160, France 65, Italy 105, Spain 50, Great Britain 70, Netherlands 20, and Austria 15) [85] and an average per capita water consumption of 200 liters/day predict a zeolite concentration in raw sewage in the range of 15–30 mg/liter. Thus, the previously calculated average zeolite concentrations in raw waste waters are somewhat high compared with the present situation.

2. Treated Waste Waters

As discussed in Sec. II.B.2, the elimination of zeolite A is at least 90% in mechanical-biological sewage treatment plants. Because the overwhelming part of waste waters in Central, Western, and Northern Europe is treated in such two-stage sewage plants, it can be anticipated that the effluents contain zeolites at concentrations up to 1.5–3 mg/liter. These figures correspond fairly well with previous predictions by Fischer and Gode [86], who expected a zeolite A concentration of 1–2 mg/liter in sewage plant effluents. Virtually the same average concentration of zeolites in treated waste water, that is, 2 mg/liter, was predicted [87] by calculations applying a generic environmental fate model (HAZCHEM 2) based on the fugacity concept of Mackay et al. [88].

3. Sewage Sludges

Based on the data for zeolite consumption by detergent use and zeolite removal in sewage treatment plants (Sec. II.B.2), the zeolite content of sewage sludges can be calculated. For the usage conditions of zeolite-containing detergents in the United States, the zeolite A concentration in sludges from two-stage sewage treatment plants was estimated as 2–3% of total suspended solids [77,78]. The digested sludge from a trickling filter field experiment contained zeolites corresponding to 6% of total dry solids [74]. Sludges from home treatment systems, such as septic tanks, are expected to contain higher zeolite contents, up to 30% of suspended solids [78].

Calculations based on the German consumption figures of zeolite-containing detergents forecast somewhat higher zeolite sludge concentrations: Roland and Schmid [80] estimated a per capita consumption of about 30 g/day of detergents containing 20% zeolite A. It was calculated from this that zeolites contribute approximately 6 g per inhabitant and per day to the sewage sludge dry matter; this is equivalent to about 7.5% of the per capita sludge production of a sewage treatment plant (80 g/day). Malz and Jendreyko [73] estimated an increase in sludge dry matter of about 13% as a result of zeolite usage. Unfortunately, the published data from the 13 month field trial in the village of Büsnau [68] do not provide figures for the zeolite content of the sewage sludges. However, based on the measured elimination rates of about 8 kg/day of zeolite A and 108 kg/day of BOD_5 and on the ground of a per capita waste water production of 60 g BOD_5/day and of 55 g digested sludge dry matter/day [89], it can be calculated that the plant produced about 110 kg sludge dry matter/day. Thus, the zeolite content of the digested sludge amounted to around 7% of the sludge dry matter. Similar conclusions can be drawn from current German usage data: the specific zeolite A load in raw sewage is 5.5 g per capita × day; the specific sewage load is 15 g per capita × day for sedimenting inorganic materials and 30 g per capita × day for sedimenting organic matter [90]. Thus, the specific sewage load of sedimenting materials is 45 g per capita × day. Because the major part of these materials is eliminated in the mechanical stage of a sewage plant, it can be concluded that the zeolite content of primary sludge is approximately 12%. Recent calculations on the basis of a generic environmental fate model (HAZCHEM 2) predicted a zeolite concentration of about 18% in primary sludge [87].

In summary, zeolite A is almost quantitatively eliminated from waste waters in sewage treatment plants and contributes around 10% to the mass of sludge produced during sewage treatment. However, field tests show that the sewage sludge volumes do not increase when zeolite-containing detergents are used [68,75]. On the contrary, the presence of zeolite A ameliorated the sedimentation and density properties of the sludges, resulting in a lower sludge volume index compared with zeolite-free sludges [75,83].

D. Zeolite A in Soils

The prevailing part of zeolite A contained in detergents ends up in sewage sludge after the primary and secondary treatment of waste waters. In most cases, these sludges are stabilized by anaerobic treatment in a digester and subsequently disposed of in landfills or incinerated or spread on agricultural land for fertilizing purposes. The agricultural use of sludges is quite different in European contries, e.g. 20% in Italy, 30% in West Germany and France, 45% in Great Britain and Denmark, 60% in The Netherlands and Spain, and 80% in Switzerland [91].

1. Concentrations

The agricultural use of zeolite-loaded sludges results in an introduction of this compound into the soil compartment. Based on the zeolite content of digested sludges (range 5–15% of sludge dry matter) and the standard conditions for sludge application to soil (rate 0.2 kg/m^2 per year; incorporation of the sludge to a soil depth of 0.15 m; soil bulk density: 1200 kg/m^3), the predicted zeolite concentration in sludge-amended soils is in the range of 70–200 mg/kg of soil. Because of the weathering of zeolite A in natural waters by hydrolysis, forming natural alumosilicates [92], it can be anticipated that comparable processes may bring about the transformation of the synthetic zeolite A in natural soils. Natural alumosilicates (clays) constitute a substantial component of soils, so that the contribution of zeolite A to the mineral content of the soils is marginal.

2. Mobility

Because of the utilization of zeolite-loaded sewage sludges for agricultural purposes, the migration behavior of radioactively labeled (Cr^{51}) zeolite A was investigated in a laboratory test simulating a 2 year precipitation period [81]. After application of labeled zeolite (52–67 mg) on the top of three 30 cm length glass columns containing different standard soils, the columns were percolated with drinking water for 24 days (10 ml/h). No radioactivity could be detected in the column effluents, showing that these particulate alumosilicates (having a diameter <3 μm to about 60% and <10 μm to about 99%) do not penetrate the soil and will not contaminate groundwaters.

Similar conclusions were drawn from experimental results concerning the retention of zeolites in slow sand filters used for raw drinking water preparation [93]. Zeolite A was retained like other suspended solids. Zeolite particles with a diameter <1.5 μm may pass the sand filter but are very well eliminated by flocculation because of their higher kinetics. Thus, the removal of zeolite A by the drinking water preparation process is anticipated to be higher than 90%.

Even a complete decomposition of zeolites in the soils would not result in an impairment of the groundwaters that form the source of drinking water in many cases. As discussed by Roland and Schmid [81], the formation of water-insoluble aluminohydroxides and soluble silicates would not influence significantly the SiO$_2$ concentrations usually found in German drinking waters.

E. Zeolite A in Surface Waters

Although zeolites used in detergents are largely eliminated during waste water treatment, a certain portion enters receiving waters together with the sewage treatment plant effluent. Of course, the amounts of zeolites reaching surface waters are considerably higher in those areas where waste water treatment is poor and raw sewage is introduced directly into rivers. However, it can be expected that this situation will improve in the countries of the European Community within a relatively short time because the corresponding EEC regulations have already been implemented.

1. Concentrations in River Waters

As discussed previously (Sec. II.C.2), the zeolite A concentration in sewage treatment plant effluent is in the range of 1.5–3 mg/liter. The zeolite concentration in rivers is mainly dependent on the dilution ratio between the volumes of the plant effluents and the receiving waters. Although very low dilution factors, such as 1:3 and lower, can be found in particular cases, it is generally accepted that calculations based on a dilution factor of 10 are suited for a realistic conservative estimation of substance concentrations in rivers if the substance is introduced via treated waste waters [94]. Taking this into account, the predicted environmental concentrations of zeolite A in river waters is in the range of 150–300 µg/liter. Comparable figures are expected when the German zeolite consumption figures are used for the calculation of environmental concentrations according to the fugacity fate model HAZCHEM 2 [87]. The predicted environmental concentration (PEC) was 287 µg/liter, assuming that no decomposition of zeolite A takes place in river water. However, if a slow abiotic decomposition of zeolite A in natural waters corresponding to a half-life of 1–2 months [95] is taken into account, the PEC of particulate zeolite A in rivers is 57 µg/liter.

2. Relevance to the Silicon and Aluminum Balance in Rivers

Occasionally, detergent-derived zeolites have been debated as materials that may considerably influence the loads of aluminum and silicon present in rivers; this might disturb the natural mass balance of these chemical elements in the aqueous environment and may lead, ultimately, to undesirable consequences, such as increased aluminum concentrations in rivers and such eutrophication phenomena as the mass growth of algae (e.g., diatoms) along the sea-coast.

A few simple calculations can be made to evaluate the possible contribution of detergent-based zeolites to the concentrations of dissolved silicon and aluminum in surface waters. The median concentration of dissolved silicon in fresh waters is in the range of 300 µm corresponding to more than 8 mg/liter [96]. As discussed, the zeolite concentration in rivers is predicted to be in the range of 150–300 µg/liter. Assuming that all zeolites hydrolyze completely and form water-soluble decomposition products, the silicon content of zeolite A (approximately 15%) would clearly result in a contribution of less than 1% of the total Si content of rivers. This is a completely insignificant value compared with the variation in the Si concentrations in river flows. Calculation of the standard deviation of annual mean silicon concentrations (within a 7 year period) at various sampling points along the Rhein [97] showed a variation of 5–13%.

In principle, similar conclusions could be made for aluminum concentrations in surface waters. If detergent-based zeolites present in river waters hydrolyzed completely and formed water-soluble compounds, the contribution would be 22–44 µg aluminum/liter (Al content of zeolite A is 14.8%). However, the total aluminum concentrations in surface waters (<0.1–740 µM, corresponding to <3–20,000 µg/liter) show an extremely wide range depending on the pH value, the presence of particulate matter and organic or inorganic ligands, temperature, and ionic strength; not the least, analytical methods applied to the determination of aluminum concentrations and its species are also responsible for obvious inconsistencies in the reported data [98]. Because of the relatively low solubility of natural aluminum minerals at around neutral pH values [solubility of $Al(OH)_3$ is 6–12 µg/liter] and low concentrations of complexing ligands, the concentrations of dissolved aluminum are generally low in most natural waters. Median values of 0.4 µM (corresponding to around 10 µg/liter) for terrestrial waters [96] and an average concentration of 9 µM (corresponding to around 240 µg/liter) for fresh water, including bogs [99], have been reported. These conflicting figures highlight the problem of obtaining a sound basis for the intended calculations and show, on the other hand, that the formal assumption of a complete transformation of zeolite-derived aluminum into water-soluble products is unrealistic. This conclusion is supported by the fact that even in acidic lake water the water column aluminum is suggested to be largely in particulate form. Stream sediments may release aluminum directly to the water column in waters that are undersaturated while serving as zones of Al deposition in other circumstances [98]. Therefore, it is evident that the concentrations of dissolved and, thus, more bioavailable aluminum in surface waters are independent of the formation of water-soluble decomposition products of zeolite A. Because aluminum is the most abundant metallic element within the lithosphere and aluminum inputs to the aqueous environment are largely associated with particulate minerals [98], it is obvious that quantification of the Al contribution of synthetic zeolites to surface waters lacks a scientifically sensible background.

III. CHEMICAL STABILITY OF ZEOLITE A IN THE AQUATIC ENVIRONMENT

A. Analytical Methods

Zeolite A either is not detectable or is not sufficiently detectable in samples taken from the environment. This is because it contains no rare elements and exhibits no characteristic properties. Thanks to its typical cubic shape it can be immediately recognized using an electron microscope. It is true that its typical line spectrum is easily recognizable in x-ray diffraction analysis, but this technique requires zeolite A to be present in concentrations of more than 2%, and such concentrations have yet to be observed in nature (sediments or digested sludge). Zeolite A can be doped with indium [100] (quantification by neutron activation analysis) or radioactive ^{51}Cr or reacted with terbium chloride [101] (observation by fluorescence microscopy), but these methods are only suitable for limited studies. As a result of the generally positive results obtained about the behavior of zeolite A in piping systems, sewers, and clarification plants and its bioaccumulation in living organisms, for many years no other methods were developed for tracing the behavior of zeolite A in the environment. The latest research findings are concerned with the measurement of zeolite-derived aluminum concentrations of about 10 µg/liter and the detection of decomposition products by scanning electron microscopy, transmission electron microscopy on ultrathin sections, energy-dispersive x-ray microanalysis, and electron diffraction [102,103]. Infrared spectroscopy can be used as a "fingerprint" method.

B. Stability in the Aquatic Phase

1. Laboratory Investigations

Owing to its low Si/Al ratio, zeolite A undergoes hydrolysis in the pH range below 9. This is confirmed by the fact that, if fully demineralized water is added, the pH of the suspension rises, depending on the zeolite concentration, to 10–10.5 [104]. This rise in pH is caused primary by a fast ion exchange process (some minutes to hours, sodium cations versus protons). Technical zeolite A also contains a certain amount of free sodium aluminate (up to 3%), which causes the pH jump in the first seconds. Several authors have commented on the impossibility of converting zeolite A to the H form, that is, replacing sodium cations completely by protons, without causing the lattice to break down [104–106]. Zeolite A dissolves in strong acids, leaving no residue [107].

In the early 1980s, when zeolite-containing detergents were introduced, several publications appeared concerning the degradation or hydrolysis behavior of zeolite A. One group of authors [92] described the hydrolysis of zeolite NaA and CaA in the pH range 3–9 in distilled water, model environmental systems, and natural sewage media in closed and flow-through systems. The hydrolysis products were studied by scanning electron microscopy, infrared spectroscopy, density measurements, oxygen adsorption measurements, electron diffraction experiments, x-ray diffraction methods, and x-ray fluorescence.

The authors concluded that the hydrolysis behavior was incongruent. Initially the cations in the lattice were replaced by protons from the surrounding solution, and then the samples increasingly lost Si and Al. The resulting end product could not be identified with certainty, although there were numerous indications that gibbsite ($Al_2O_3/3H_2O$) and halloysite [$Al_2Si_2O_5(OH)_4$] were present. Identification of the hydrolysis end products was complicated by their completely amorphous nature, although in many experiments (especially at pH < 6) the external form of the zeolite crystal remained unchanged. The study showed weaknesses in the choice of the tested zeolite concentrations (because 0.5 and 10 g/liter of zeolite are not relevant to environmental conditions, but 5–20 mg/liter of zeolite are expected). Although zeolite concentrations of 7–17 mg/liter, for example, were regarded as realistic at clarification plant inflows and were measured [105], 1500 mg/liter of zeolite was used in one study [104]. After a few hours the concentrations of hydrolysis products were so high that the solubility limits for aluminum hydroxide were exceeded and the subsequent reactions were hardly comparable with hydrolysis processes occurring in natural waters. Acid uptake, which represents the first step in hydrolysis, has been measured using 0.5 g/liter deionized water containing a suspension of zeolite A in at pH value 5–7 in a closed system. It was shown that the proton exchange of zeolite CaA is slower than that of zeolite NaA: obviously, Ca ions have a higher affinity for the zeolite lattice. Therefore, the proton exchange of zeolite NaA in hard water containing Ca ions is relatively slow as well. In general, the rate of proton exchange decreases with rising pH value and with rising concentrations of Ca ions within the lattice or in the surrounding solution.

Similar investigations in flow-through systems containing 10 g/liter of zeolite NaA at pH 7 verify this statement. The proton exchange is delayed or slowed by a high concentration of salts, especially sodium and lead salts, for example in natural waste water. In the cationic exchange of zeolites, there is competition between the cations mentioned earlier and the protons.

The effect of metal cations on the proton exchange of zeolite NaA is reduced by natural or synthetic complexing agents. In natural water and in waste water, these compounds form stable, soluble complexes, in particular with multivalent cations.

Drummond et al. [108] encountered the phenomenon of zeolite NaA hydrolysis when they used this zeolite to measure ion exchange kinetics. They also made the important observation that, in the pH range 6.5–9.0, the rate of release of sodium ions is independent of the particle size of the zeolite samples used. This confirms the observation of these authors that Na is rapidly replaced by H at the start of hydrolysis. As a result of this exchange, which is complete after 1 h at pH 9 and after 5 minutes at pH 5, the zeolite lattice breaks down and the coordination of the aluminum changes from tetrahedral to octahedral [109]. The then soluble aluminum (as aluminate) is therefore available to participate in subsequent reactions. These subsequent reactions depend on the pH and the foreign ions available in the surrounding solution.

For many years the most comprehensive study of the hydrolysis of zeolite CaNaA in natural and synthetic surface waters containing water hardness, soluble silicate, and other salts, including heavy metals, was that by Allen et al. [95]. The samples were continuously shaken in polyethylene bottles. "Synthetic" water, that is, model waters, as well as water samples of natural origin, served as a medium for the hydrolysis experiments. Half-life values for zeolite degradation were given for the first time (1–2 months for neutral waters at pH 7.7–8.4). Allen et al. give a relatively detailed description of the composition of the water samples used. The degradation rate was measured by determining the amounts of aluminum and silicon released.

A further investigation in synthetic water at pH 6.3, 7.1, and 8.3 in glass containers by mixing with carbon dioxide-nitrogen gas mixtures showed that the decomposition rate of the zeolite increases as the pH decreases (decomposition after 3 days: 44% at pH 8.2 and 72% at pH 6.3). There were no significant differences in degradation rate at concentrations between 0.25 and 5 mg/liter, and the measured rates of release of aluminum and silicon increased only slightly at higher concentrations. The hydrolysis of the zeolite was more rapid under the conditions of this experiment than for samples contained in polyethylene bottles and mixed by shaking. The authors explained this by ascertaining that the rate of hydrolysis varies with the experimental conditions. The decomposition rate was measured indirectly via the amounts of silicate and aluminum released. Therefore, if the substances released reacted further to form insoluble derivatives, then in this concentration range alone this would simulate a low or almost nonexistent decomposition rate [110]. The decomposition rate of zeolite A increases as the pH decreases, and it is improbable whether decomposition is inhibited as the zeolite concentration increases.

In this context, new research results by Chappell et al. [111,112] showed that aluminosilicate complexes are more stable than corresponding aluminum-citrate complexes at pH > 7.4. One obvious conclusion is that, after protonation of the lattice, zeolite decomposition initially proceeds via soluble aluminosilicate complexes. New investigations of the hydrolytic decomposition of zeolite A are considering, in contrast to the literature just cited, the phosphate concentration of natural waters [102]. Although experimentally verified hydrolysis rates vary with the experimental conditions, there is no comparison between synthetic phosphate-free waters and natural waters (phosphate concentration about 1 mg/liter, as expected for zeolite concentrations in river water). Therefore, under environmental conditions, comparable results can be created only by using phosphate-containing synthetic waters.

Current studies in both closed and flow-through (laboratory) systems with synthetic water according to the composition of the Rhein have demonstrated that amorphous basic calcium aluminum phosphates (silicates), with compositions corresponding to crandallite, are formed as thermodynamically stable hydrolysis end products in the presence of sufficient amounts of calcium and phosphate ions [102]. Although the amorphous material apparently does not undergo change until the temperature is above 60°C under normal conditions [113], microcrystalline crandallite zones are present in at least the decomposition products. One fact in favor of this theory is that in hypereutrophic lakes more than 50% of the total phosphorus content is bonded to calcium [114] in forms that include whitlockite, a mineral closely related to crandallite. Now we can propose a two-step hydrolysis mechanism for zeolite A. In the first step, a more or less complete cation exchange versus protons, deriving from the surrounding water, takes place. Step 2 is the mobilization of aluminosilicate molecules. Depending on the chemical conditions of the aquatic system, the elimination of the aluminosilicates via calcium aluminum phosphates or via the formation of insoluble gibbsite/halloysite (in the absence of calcium/magnesium and phosphate) occurs.

If heavy metal cations are present in the first stages of the degradation of zeolites, multivalent cations at least should have a significant influence on the degradation process (for the ion exchange of zeolite A with heavy metal ions, Sec. IV.B.3). Analogy to the chemical variety of minerals from the apatite or the wavellite lazulite group suggests that large amounts of heavy metals can be incorporated into the lattices of the corresponding minerals.

In the course of the decomposition of zeolite A in phosphate-containing water, crystalline or amorphous compounds are formed that belong to the two groups of minerals just mentioned. In combination with multivalent cations, these kinds of aluminosilicates are insoluble. The author is thus convinced [110] that large amounts of heavy metals are eliminated by cation exchange at zeolites and decomposition of these zeolites thereafter, forming insoluble basic phosphates. This way, heavy metals are deposited.

2. Results from Field Trials

Only one true field test of the hydrolytic decomposition of zeolite A is reported in the literature because of the invincible experimental and analytical problems (sampling and rate of recovering). As long ago as 1979, Müller [101] reported field tests involving a permeable bag containing suspensions of 8 g/liter of zeolite CaA, which was left to float freely for 2 months in the Elsenz (a tributary of the Neckar). No x-ray photographic changes were found to have occurred, nor were there any changes in the Si/Al ratio, during the entire duration of the test. Because these results are not in agreement with the degradation rates obtained in various laboratory experiments and the author reported immense difficulties caused by blocking of the membranes and contamination of the sample material by suspended clay detritus, it can at least be suspected that hydrolysis of the zeolite was suppressed during these tests by encrustations on the sample material and inadequate water exchange. Müller assumed that the synthetic CaA zeolite behaves in much the same way as naturally occurring zeolites, which are much richer in silicate, however, and therefore more stable. Silverman [115] gave only indirect data for the degradation behavior of zeolite A in an article on the biological and organochemical degradation of silicates. This article dealt in particular with the biologically related weathering of natural silicates. With the help of numerous examples it was demonstrated which organisms can cause natural silicates to decay in what period of time (mostly by producing carboxylic acids). Numerous sources for biologically related weathering of natural zeolites were cited. Especially notable are the aluminum release rates from such stable minerals as biotite and muscovite, both of which are micas and eminently comparable to silicate-rich zeolites. According to this, up to 1% of the aluminum content of the minerals can be removed from silicates by biological action.

The hydrolytic decomposition of type A zeolites used in detergents can be summarized as follows. The hydrolysis of type A zeolites is incongruent, so it is a multistep process:

The hydrolytic decomposition of NaA zeolites is initiated by proton exchange. The rate decreases with a rising pH value and with a high concentration of salts. Proton exchange is accompanied by a loss of crystallinity of the zeolites. In general, the rate of deprotonation depends on the experimental conditions, including chemical composition of the water, hydrodynamics of the zeolite suspension, pH value, and temperature.

The second step is the detachment of soluble aluminosilicates from the amorphous zeolite particles. These aluminosilicates can undergo different secondary reactions, depending on the composition of the surrounding water. All these reactions give insoluble products.

Laboratory experiments, conducted with environmentally relevant zeolite concentrations of 1–10 mg/liter in neutral, phosphate-free, synthetic, or natural water show that the half-life of zeolites is about 1–2 months, depending on the factors mentioned earlier. Quoting the authors [95], however: "It seems unlikely that the level of zeolite or degradation products in the environment would be sufficiently great to prevent its hydrolysis."

Laboratory results based on investigations with high zeolite concentrations in phosphate-free synthetic water indicate that amorphous gibbsite and halloysite are the final products of zeolite degradation.

The half-life of zeolites in phosphate- and calcium-containing water (synthetic or natural water, zeolite concentration 1–10 mg/liter) has been determined as 12–30 days, depending on pH (7–8.5).

Under ordinary environmental conditions (phosphate concentration in the range of 1 mg/liter, calcium concentration ≥ 10 mg/liter, and no elevated concentrations of complexing agents), an amorphous basic calcium-aluminum-silicate-phosphate is rapidly formed as the end product of decomposition. It is extremely insoluble in water (<50 µg/liter).

IV. IMPACT OF ZEOLITE A ON THE HEAVY METAL DISTRIBUTION IN WATERS

During the washing process, Na-zeolite A becomes loaded with Ca ions from calcium hydrogen carbonate in the tap water and from calcium-containing soil pigments and textiles [116]. The calcium-loaded zeolite remains suspended in the wash liquor, which is discharged into the domestic and municipal sewers; it then passes together with industrial sewage into clarification plants or directly into streams. One important aspect in this context is the exchange of heavy metal ions (transition metal ions) at the calcium-loaded zeolite during its transportation in the sewer network, during primary settling and subsequent biological treatment in the clarification plant, and in streams. From an ecological point of view, the extent to which the various heavy metal ions can be removed from sewage by ion exchange at the zeolite (decontamination) and the extent to which the heavy metals become enriched in the zeolite (heavy metal loading) are of considerable importance. The extent to which natural and synthetic complexing agents in sewage can remobilize the heavy metals captured by the zeolite is also of importance, as is the question of whether zeolite can influence the heavy metal ion equilibrium between biomass (sludge) and sewage in the biological treatment stage of clarification plants and between river sediment and river water.

A. Ion Exchange Theory and the Environmental Relevance of Heavy Metal Ion Exchange

Cation exchange at Na-zeolite A, that is, the ion exchange of alkaline earths and heavy metals in exchange equilibrium, kinetically controlled by film and grain diffusion, depends not only on the chemical composition and hollow space structure of Na-zeolite A but also on the characteristics of the exchanged cations, such as ion size, ion charge, atomic weight, hydration tendency, and polarizability [1,117–119]. Cations are preferentially exchanged if they have high valency, high atomic weight, and a low tendency to undergo hydration, are readily polarizable, and exhibit a marked tendency to form ion pairs, and if their cationic compounds are strongly dissociated in aqueous solution. The ion exchange equilibrium of Me2 cations at Me1-zeolite A, with the release of equivalent amounts of Me1 cations to maintain the electroneutrality of the ion lattice, can be described by the equation [6,117]

$$Me1 \ Z + Me2^{n+} = Me2 \ Z + Me1^{n+}$$

The equilibrium constant corresponds to the selectivity coefficient S of the zeolite for a specific cation Me2 and can be calculated with the help of the following formula: $S = QMe2 \times qMe1/QMe1 \times qMe2$ (where Q and q are the equivalent fractions of the cations in the zeolite and in the solution, respectively).

S depends on the strength of the bond between cation and negative lattice ion and the ease with which the cations can access the lattice ions.

A comprehensive selectivity series of individual cations at zeolite A in pure water in exchange equilibrium, in order of decreasing selectivity, is as follows (6):

$$Ag^+ > Cu^{2+} > H^+ > Zn^{2+} > Sr^{2+} > Ba^{2+} > Ca^{2+} > Co^{2+} \; Au^{3+} > K^+ > Na^+ >$$
$$Ni^{2+} > Cd^{2+} > Hg^{2+} > Li^+ > Mg^{2+} \qquad \text{selectivity coefficient}$$

The selectivity series is valid for high initial cation concentrations in the solution, under which conditions zeolite-A becomes highly loaded with the cations. The selectivity coefficient of divalent cations changes with the loading of zeolite A, that is, with the equivalent fraction Q of the cation in the zeolite; the selectivity coefficients are especially large when loading levels are low [117]. This explains the differences in the literature with regard to the cation sequences in selectivity series [120].

The comprehensive selectivity series in pure water given in the literature do not include the heavy metals Pb^{2+} and Cr^{3+}. However, these metals can be expected to exhibit a tendency toward high selectivity coefficients on account of the high atomic weight of Pb and the high valency of Cr. At this point mention should be made to the frequent references in the literature to cation series based on percentage ion exchange values, such as percentage reduction in cation concentration, that is, the percentage of cations removed from solutions, lakes, and rivers and the percentage cation loading of zeolite A. Selectivity series of individual cations based on selectivity coefficients [117] or separation factors [1,118] can be compared with cation series based on percentage ion exchange values only if the initial molar concentrations of individual equivalent cations are equal, if the zeolite concentration is constant, if kinetic exchange delays are excluded so that cation exchange equilibria are reached, and if interfering competitive exchanges with other cations are excluded.

Environmentally relevant models of heavy metal ion exchange at Na-zeolite A have been carried out on individual heavy metals in solution as well as on heavy metal mixtures corresponding to the heavy metal content encountered in practice in sewage, lakes and rivers. Because, in the context of the environment, interest in heavy metal ion exchange at zeolite A is primarily directed towards the removal of heavy metals from sewage, lakes, and rivers, attention has been focused on determining reductions in percentage heavy metal ion concentrations. This has led to determination of the level of heavy metal ion exchange with zeolite as a percentage of the hypothetical case in which all heavy metal ions in the solution are removed. The heavy metal analysis is performed by atomic absorption spectroscopy.

A large number of influencing factors must be taken into account in the design of environmentally relevant model tests for determining heavy metal ion exchange. These include the following:

There is a wide range of very low heavy metal ion concentrations.

The duration of the test must be specified in view of kinetically controlled ion exchange.

Heavy metals in heavy metal mixtures compete among each other to participate in ion exchange.

In water that is hard or contains salts, heavy metals must compete with alkaline earths and alkalis to participate in ion exchange.

Heavy metal cations must compete with H^+ ions to participate in ion exchange at various pH values.

The heavy metals Fe^{3+}, Al^{3+}, and Cr^{3+} may preferentially precipitate out as hydroxide in the environmental pH range 6–8, whereas heavy metal bonds are dependent on the pH and heavy metal concentration.

Ca loading occurs during the washing process.

Increasing the temperature can reduce the hydration tendency of cations and thus improve cation exchange.

Heavy metals may preferentially form bonds with the synthetic and natural complexing agents present in sewage.

Initial heavy metal ion concentration decreases as heavy metal ions are exchanged at zeolite in a closed system (simulation of stagnant waters) or is kept constant in stream simulations.

B. Model Tests of Heavy Metal Ion Exchange, Desorption, and Remobilization

1. Exchange of Individual Heavy Metals and Mixtures in Pure and Hard Water

Calcium and magnesium ions compete with heavy metals ions to participate in ion exchange at Na-zeolite A, as has been demonstrated by comparative equilibrium tests in pure and synthetic 16°d hard water (Ca/Mg = 5:1) with 100 mg/liter of zeolite and equimolar heavy metal solutions of 2×10^{-4} M [121]. Because Ca has a higher selectivity than Mg, mainly owing to the stronger hydration shell of Mg in the temperature range 0–30°C, Ca is the main rival to the heavy metals with regard to participation in ion exchange at zeolite A [6,117]. The selectivity series of individual heavy metals in pure water, in decreasing order of selectivity [121] Cu > Zn > Ni > Cd (separation factor), is also valid for hard water, although the percentage exchange and the exchange rate of the individual heavy metal ions are reduced by competitive exchange by alkaline-earth ions (Table 2) [121]. Exchange of the comparatively less selective heavy

TABLE 2 Separation Factor (Selectivity), Heavy Metal Loading, and Percentage Heavy Metal Exchange of Na-Zeolite A (100 ppm) in 0 and 16°d Hard Water (Ca/Mg = 5:1)[a]

Metal ion Me^{2+}	Loading[b] (mol/g zeolite/liter) pure water	Exchange[c] (%)		Separation factor[d] pure water
		Pure water	Hard water	
Cu	1.966×10^{-3}	98.2	71.5	110
Zn	1.908×10^{-3}	95.4	18.0	38.5
Ni	1.175×10^{-3}	58.7	15.0	0.95
Cd	4.644×10^{-4}	23.3	3.5	0.06

[a]Test conditions: pH 6.7–7.2; 23°C; duration until approaching equilibrium; equimolar initial heavy metal concentration of 2×10^{-4} M.
[b]Concentrations converted to loading (Me analysis of the solution).
[c]Related to total initial heavy metal concentration (Me analysis of the solution).
[d]Separation factor [1,118].
Source: From Ref. 121.

metals Zn, Ni, and Cd is reduced more strongly than that of the much more selective Cu.

In binary mixtures of equimolar heavy metal ions, the selectivity of the heavy metals is the decisive factor in determining which heavy metals are preferentially exchanged at Na-zeolite A (100 mg/liter) in pure water [121]. The weaker the selectivity of a heavy metal and the stronger the selectivity of its rival in the binary mixture, the more it is disadvantaged in terms of its participation in ion exchange. In a Cd-Ni mixture, with its weakly selective heavy metals, the exchange of the comparatively weaker Cd is clearly reduced by the presence of the slightly stronger Ni, whereas exchange of Ni ions is practically unaffected. In Zn-Cu mixtures, in which both metals are strongly selective, the exchange values of both heavy metals are only slightly reduced. For both binary mixtures, the exchange rates of all heavy metal ions are lower than those measured in solutions containing only individual heavy metals.

In binary mixtures of equimolar heavy metals in hard water, the reduction in exchange is even greater because the effect of competition between the two heavy metals is reinforced by the effect of competition from the Ca ions present (Fig. 12) [121]. All in all, there is only a slight reduction in exchange of the comparatively most selective Cu in hard water mixtures with more weakly selective heavy metals. The comparatively less selective Zn and Ni exhibit strongly reduced exchange levels in mixtures with strongly selective heavy metal. If Cd is mixed with strongly selective equimolar heavy metals, its exchange activity is completely suppressed.

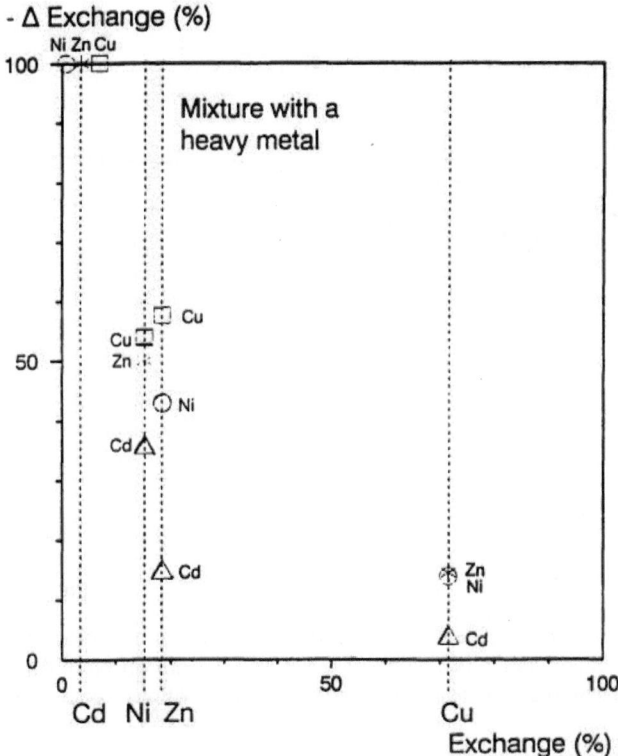

FIG. 12 Percentage reduction in heavy metal exchange at equilibrium ($-\Delta$ exchange) at Na-zeolite A in hard water when mixed with an equimolar heavy metal as a function of the percentage heavy metal exchange at equilibrium, that is, of the selectivity (the exchange at equilibrium reflects the selectivity if equimolar concentrations of divalent heavy metals are present) [121]. Test conditions: initial Me concentration, 2×10^{-5} M; zeolite concentration, 100 ppm; water hardness (Ca/Mg = 5:1), 16°d; temperature, 23°C; pH 6.7–7.2.

Tests of the exchange activity of a ternary mixture of the equimolar heavy metals Cu, Zn, and Ni at Na-zeolite A (100 mg/liter) in pure water demonstrated that selectivity remains the crucial factor in competitive exchange, as demonstrated by the following heavy metal series based on the percentage exchange of the equimolar heavy metals Cu, Zn, and Ni (the selectivity of the metals decreases in the sequence Cu > Zn > Ni) in the ternary mixture [121]: Cu (with Zn and Ni) > Zn (with Cu and Ni) > Ni (with Cu and Zn).

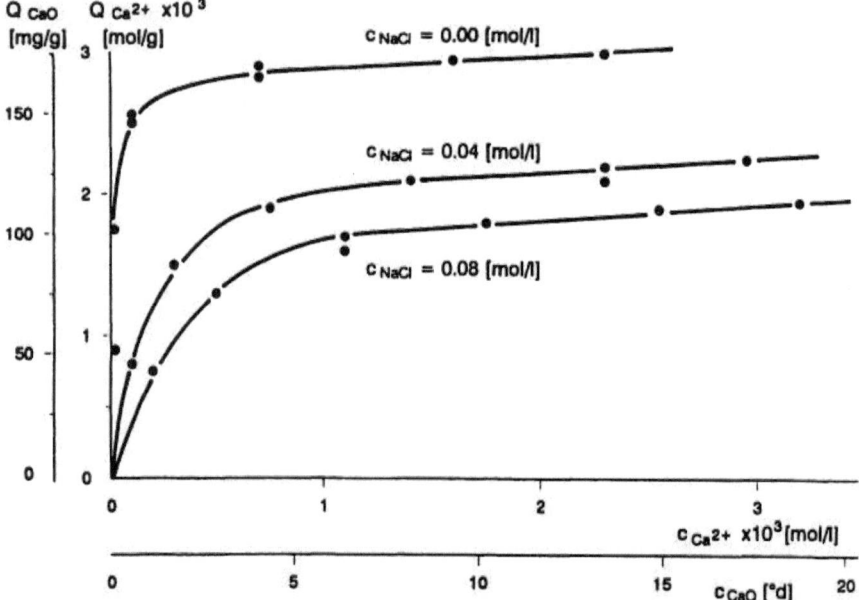

FIG. 13 Calcium loading Q of Na-zeolite A at different sodium concentrations as a function of calcium ion concentration (Ca hardness of the water). Comparison of calculated (curve) and the measured (points) isotherms [25]. Test conditions: temperature 22°C, duration 1 h.

2. Exchange of Individual Heavy Metals, Desorption, and Remobilization by Complexing Agents Under Environmental Conditions

Because Ca and the less selective Na compete to participate in ion exchange, the Ca loading of Na-zeolite A in tap water during the washing process depends on the concentrations of calcium and sodium in the wash liquor. Ca loading at exchange equilibrium decreases as the Na concentration increases, and it increases to an upper limiting value, which is dependent on the Na concentration, as the Ca concentration increases (Fig. 13) [25].

Despite the strong competition from Ca in an environmentally relevant system with 70% Ca-loaded Na-zeolite A in 16°D hard synthetic water (Ca/Mg = 5:1), the use of a closed agitator system ensures partial exchange of individual heavy metal ions that are present in low concentrations similar to those in the sewage, although the exchange equilibrium establishes itself more slowly (Table 3 and Fig. 14) [52,122].

At a relatively high zeolite concentration of 100 mg/liter and in the presence

TABLE 3 Percentage Heavy Metal Exchange in the Solution, Percentage Heavy Metal Loading of Zeolite A, and Selectivity Coefficient of Ca(70%)-Na-Zeolite A (100 ppm) Under Conditions That Apply in Sewage (16°d Ca/Mg = 5:1), pH 6.8–7.4, 23°C, Constant Initial Heavy Metal Concentration, Test Duration 40 days)

Metal ion Me^{2+}	Metal concentration in sewage[a] (ppm)	Initial metal concentration		Loading[b] (%)	Exchange[c] (%)	Selectivity coefficient[d] S
		ppm	M			
Pb	11.0	10.0	4.85×10^{-5}	38.5	87	150
Cu	2.4	2.5	3.60×10^{-5}	3.5	70	13
Zn	11.0	10.0	1.50×10^{-4}	3.5	32	3
Cd	<0.05	0.05	4.45×10^{-7}	0.13	63	17

[a]Maximum value in the influents to the clarification plants Hilden, Ohligs, Hochdahl, and Langenfeld.
[b]Related to total zeolite (Me analysis of the solution).
[c]Related to total heavy metal concentration (Me analysis of the solution).
[d]$S = Q_{Me}q_{Ca}/Q_{Ca}q_{Me}$.
Source: From Ref. 122.

of low concentrations of heavy metals (Table 3) [122], high percentages of Cu, Zn, and Cd participate in ion exchange in the solution (30–70%), producing low percentage loading at the zeolite (0.1–3.5%); in contrast, a higher percentage of the comparatively more selective Pb (higher selectivity coefficient) participates in ion exchange in the solution (87%), producing a higher percentage loading (38%). The selectivity coefficients obtained for the individual heavy metals Pb, Cd, and Cu with Ca-loaded zeolite A in hard water by applying the formula $S = QMeqCa/QCaqMe$ were 150, 17, and 13, respectively (Table 3); these values exhibit sufficient agreement with the selectivity coefficients of 810, 13, and 7 obtained under similar conditions [95,123].

Corresponding series were obtained for the percentage exchange of heavy metals in solution in tests carried out with the low initial heavy metal concentration of 1 mg/liter over a period of 1 week (sequence [52], Fig. 14: Pb > Cu > Cd > Zn > Ni %) and with various heavy metal concentrations approximating the concentrations present in sewage and maintained at a constant level over 40 days (sequence [122], Table 3: Pb > Cu > Cd > Zn %; r.s.p. sequence [122], Table 3: Pb > Cd ≥ Cu > Zn, selectivity coefficient).

This heavy metal series, obtained under conditions resembling those in sewage, can be compared accurately with the selectivity series of the heavy metals in pure water [6,121]: Cu > Zn > Ni > Cd (selectivity coefficient or separation factor) simply by referring to the heavy metal series based on selectivity coefficients [122]. This shows that, under conditions resembling those prevailing in

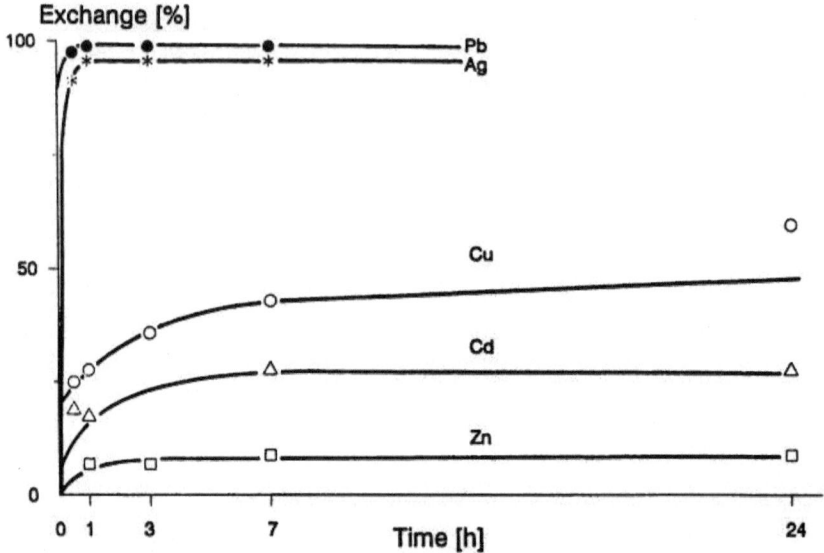

FIG. 14 Percentage heavy metal exchange (Pb, Ag, Cu, Cd, and Zn) at Ca-Na-zeolite A as a function of time (short-term test) under the following test conditions [52]: Initial Me concentration, 1 ppm; zeolite concentration, 100 ppm; Ca loading of zeolite, 72%; water hardness (Ca/Mg = 5:1), 16°d; temperature, 23°C; pH 6.6–7.1.

sewage, cadmium exhibits increased selectivity in comparison with the heavy metals Cu and Zn. This is probably attributable to the fact that the molar concentration of Cd in sewage is very low (10^{-7} M) and hence the Cd loading of zeolite A is also very low (Table 3); low loading levels result in increased selectivity coefficients [117]. The selectivity series of individual heavy metals at Na-zeolite A in pure water in a very low loading range [120] Cd > Cu > Zn > Ni (selectivity coefficient) confirms the comparatively higher selectivity of Cd in relation to Cu, Zn, and especially Ni and proves that, under conditions of low loading, Cd can be preferentially adsorbed in narrow hollow spaces with a high charge density in the zeolite A lattice [120].

Under environmentally relevant conditions, that is, initial heavy metal concentrations similar to those in sewage (1 mg/liter) and 70% Ca-loaded zeolite A, the percentage ion exchange of individual heavy metals in solution decreases significantly as the water hardness increases from 6 to 30°d (Fig. 15) [52]; as is the case with Na-zeolite A in hard water [121] (Sec. IV.B.1), the heavy metals that exhibit the greatest selectivity under the prevailing conditions are less inhibited with regard to ion exchange.

To summarize, model tests with Ca-loaded zeolite A in a closed agitator

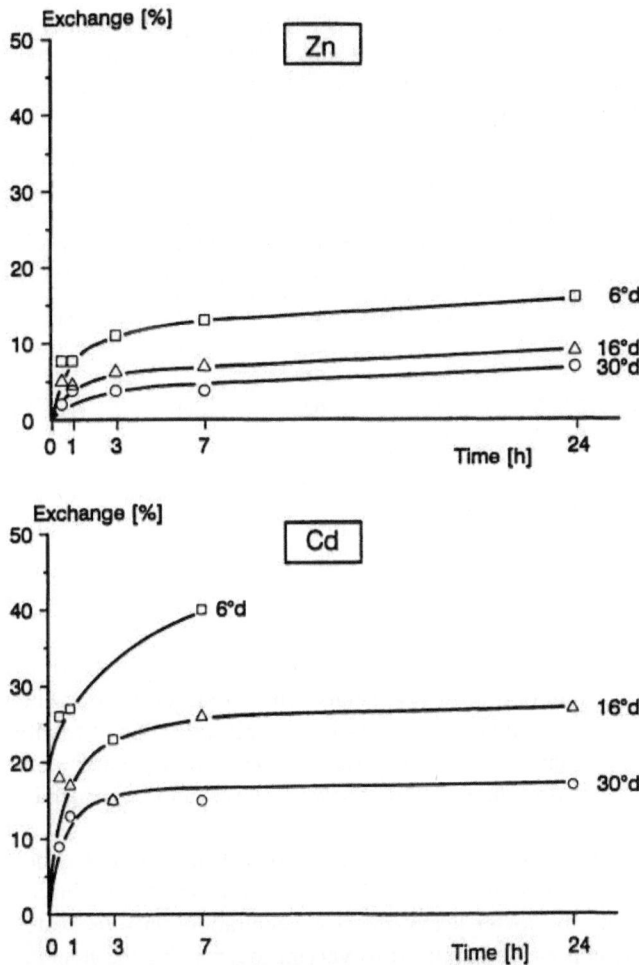

FIG. 15 Percentage heavy metal exchange (Zn and Cd) at Ca-Na-zeolite A at different water hardness as a function of time (short-term test) under the following test conditions [52]: initial Me concentration, 1 ppm; zeolite concentration, 100 ppm; Ca loading of zeolite, 72%; water hardness (Ca/Mg = 5:1), 6, 16, and 30°d; temperature, 23°C; pH 6.6–7.1.

system under conditions similar to those prevailing in sewage demonstrated that, depending on selectivity, a more or less strong percentage exchange of the individual heavy metals in solution takes place, and therefore there is partial removal of the heavy metal ions from the water. The degree of removal of heavy metal ions is ion specific and is dependent on concentration and water hardness.

TABLE 4 Percentage Desorption and Remobilization of Heavy Metals Me with 10 ppm EDTA and Calculated EDTA Remobilization Ability from Me-Ca-Zeolite A (100 ppm) Under Conditions That Apply in River Water (16°d Ca/Mg = 5:1, pH 6.8–7.4, 23°C, Constant Initial Heavy Metal Concentration, Test Duration 30 Days)

Metal ion Me^{2+}	Metal concentration in river water[a] (ppm)	Initial metal concentration (ppm)	Initial metal loading[b] (%)	Desorption[c] (%)	Remobilization with EDTA[c] (%)	EDTA remobilization ability[d] (%)
Pb	0.003 (0.3)	0.1	38.5	20	70	50
Cu	0.03 (0.11)	0.07	3.5	40	60	20
Zn	0.25 (0.66)	0.5	3.5	51	57	6
Cd	0.005	0.005	0.13	62	69	7
Hg	0.0005	0.005	0.17	88	88	0

[a]Average values in the Rhine (1972) and values in parentheses from other rivers.
[b]Related to total zeolite (Me analysis of zeolite).
[c]Related to total initial loading (Me analysis of zeolite).
[d]Difference between measured desorption and measured remobilization with EDTA.
Source: From Ref. 122.

This means that the concentration of heavy metals can be reduced by immobilization at suspended, poorly soluble zeolite A during the transportation of zeolite A in sewers, clarification plants, and surface waters. Therefore a positive contribution of zeolite A to the decontamination of sewage, lakes, and rivers can be expected.

However, the various heavy metals adsorbed at Ca-loaded Na-zeolite A under the conditions found in sewage are all partially desorbed under the conditions found in rivers, owing to the lower heavy metal concentrations in river water (Tables 3 and 4) [122] and are remobilized to varying degrees by the nonbiodegradable complexing agent EDTA (Table 4) [122]. The possible release of heavy metal ions from heavy metal-loaded zeolite by the partial hydrolysis of zeolite during these tests (closed agitator system) at pH 6.8–7.4 can be ignored owing to the relatively high zeolite concentration of 100 mg/liter and the relatively short test duration of 30 days [95].

The heavy metal series in order of increasing percentage desorption under river water conditions in a closed agitator system is (Tables 3 and 4) [122]: Pb > Cu > Zn > Cd > Hg (%); this corresponds to the heavy metal series in order of decreasing selectivity in pure water [6] Cu > Zn > Cd > Hg (selectivity coefficient). The degree of desorption of the heavy metal loading of zeolite A is inversely proportional to the selectivity of the heavy metals.

In a system containing zeolite and complexing agents, heavy metal ions and alkaline earth ions not only can form soluble complexes but can also participate in exchange at zeolite A. The dissociation equilibria of the complexes (i.e., the stability constants) and the exchange equilibria (i.e., the selectivity coefficients) compete for the heavy metal ions and alkaline-earth ions. The stability constants (complexes) and the selectivity coefficients (exchanges) of the heavy metals are mostly higher than those of the alkaline earths.

A laboratory test on Ca-zeolite A loaded with 3% individual heavy metals in the presence of the complexing agents EDTA and STP in synthetic 16°d hard water allows conclusions to be drawn about the basic influence of selectivity (exchange) and stability (complex) on the remobilization of heavy metals from zeolite A by complexing agents [122]. As selectivity increases, stability becomes the dominant factor influencing the remobilization ability of the complexing agents, and as selectivity decreases, selectivity becomes the dominant factor influencing the release of the heavy metals.

In a laboratory test under simulated river water conditions with the synthetic complexing agents EDTA and STP and the natural ·complexing agent peptone, the percentage remobilization ability of these complexing agents (difference between measured desorption and measured remobilization with complexing agent) is determined directly by carrying out an analysis of the heavy metal loading of zeolite [122]. The heavy metal series in order of increasing percentage EDTA remobilization ability under river water conditions at Ca-zeolite A loaded with individual heavy metals under sewage conditions is (Table 4) [122] Hg < Cd, Zn < Cu < Pb. The level of remobilization of the heavy metal loadings as a result of the action of the strongly complexing EDTA is directly proportional to the selectivity of the heavy metals. The heavy metal series in order of increasing EDTA remobilization ability is therefore the reverse of the heavy metal series in order of increasing desorption. This confirms that, under river water conditions, the effectiveness, that is, the stability of the complex, becomes the dominant factor influencing remobilization ability as selectivity increases. The more weakly complexing STP and peptone only exhibited significant remobilizing ability with regard to Pb and Cu, respectively. For example, the percentage remobilization of Pb with 10 ppm EDTA is greater than with 10 ppm STP because the stability constants of heavy metal complexes formed with EDTA are usually greater than those of heavy metal complexes formed with STP (Fig. 16) [122].

3. Exchange of Heavy Metal Mixtures and Desorption in Waters and Sewage

The content of zeolite in modern phosphate-free detergents results in an average concentration of 15–30 mg/liter in raw sewage and 0.15–0.3 mg/liter in surface waters (Sec. II.C.1). Laboratory tests with surface water were carried out on Ca-loaded zeolite A in the concentration range of a few milligrams per liter under

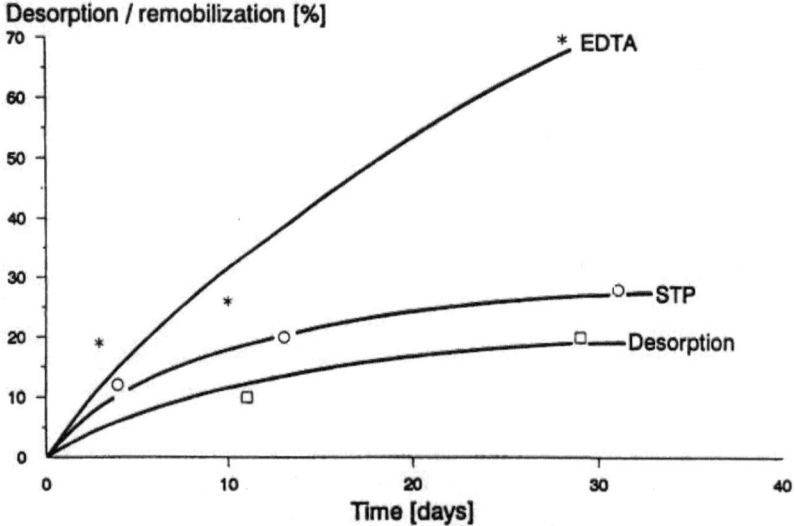

FIG. 16 Percentage desorption and remobilization with complexing agent of Pb from Pb-Ca-zeolite A as a function of time under the conditions that apply in river water (Table 4) [122]: initial Pb concentration 0.1 ppm; (kept constant), 0.1 ppm; zeolite concentration, 100 ppm; initial Pb loading of zeolite, 38.5%; complexing agent concentration, 10 ppm; water hardness (Ca/Mg = 5:1), 16°d; temperature, 23°C; pH 6.8–7.4.

simulated practice conditions. In a closed shake system with synthetic river water containing a mixture of heavy metals and with six samples of natural water containing heavy metal mixtures, the heavy metal ion exchange was measured, that is, loading and desorption as well as hydrolysis [95].

Long-term tests in synthetic river water were carried out over periods of up to 32 weeks with 1 mg/liter of zeolite at pH 8.2. Heavy metal ion exchange started relatively rapidly but decreased after about 12 weeks. This was attributable to the effect of continuously increasing zeolite hydrolysis with a half-life of about 2 months under partial release of heavy metal ions. The average values of the strongly fluctuating heavy metal loading values during the first 12 weeks gave the following heavy metal series in order of decreasing percentage loading (Table 5) [95]: Pb > Cd > Zn > Cu > Ni. The heavy metal loadings of zeolite were low for Cu, Zn, Ni, and Cd (in all cases ≤ 16%) and high for Pb (50%); these orders of magnitude are similar to those found when Ca-zeolite A at 100 mg/liter under sewage relevant conditions is loaded with individual heavy metals (Sect. IV.B.2) [122].

As the pH decreases in the range 8.2–6.3, the rate of hydrolysis of zeolite in synthetic river water increases; in short-term tests this has only a weak effect on

TABLE 5 Percentage Heavy Metal Loading and Rate of Hydrolysis of Ca-Na-Zeolite A (1 ppm) in Synthetic and Natural Waters

Waters	pH	Hydrolysis (half-life)	Loading[a] (%)				
			Pb	Cu	Zn	Ni	Cd
Synthetic river water	8.2	2 months	50	8	11	0.9	16
Lake Michigan	8.4	27 days	35	6	9	0.4	2
Lake Michigan + 15% secondary sewage effluent	8.3	31 days	30	4	6	0.6	3
Ohio River	7.7	31 days	53	3	3	0.1	5
Miami River	8.3	5 months	27	4	13	0.1	3
Raccoon Creek	4.3	Immediate hydrolysis	0	0	0	0	0
Raccoon River	8.2	Slow hydrolysis	21	4	16	0	2
Town River	6.5	Few days	29	8	2	1.0	3

[a]Average value (related to total zeolite/Me analysis of zeolite) of the equilibrated samples after 1 and 3 days and 2, 4, 8, and 12 weeks.
Source: From Ref. 95.

the heavy metal loading of zeolite and causes only very small amounts of heavy metal ions to be released. Hydrolysis products are apparently able to bind heavy metals (Sec. III.B.1) [92,95].

In short-term tests of synthetic river water in a closed shake system, zeolite loaded with a heavy metal mixture was added to synthetic river water that contains no heavy metals. The heavy metals Pb, Cu, Zn, Ni, and Cd were partially released from the zeolite by desorption within 2 weeks, whereas the proportion of heavy metals released by zeolite hydrolysis could not be determined [95]. The degree of desorption of the various heavy metals at equilibrium (characterized by constant residual loading) lies between 30 and 90%. By comparison, the degree of desorption from zeolite loaded with individual heavy metals under simulated sewage conditions when it is added under simulated river water conditions is 20–60%, that is, roughly the same order of magnitude (Table 4) [122].

Long-term tests were carried out with natural waters containing 1 mg/liter of Ca-loaded zeolite but differing in terms of the pH and the amounts of heavy metals and other substances present. During the first 12 weeks, when there was continuous zeolite hydrolysis with a half-life of 1–2 months (pH > 7), low loading was measured with Cu, Zn, Cd, and Ni (in all cases ≤16%) and high loading with Pb (20–50%), that is, results similar to those obtained with synthetic river water (Table 5) [95]. In this case the hydrolysis of zeolite was partially inhibited by the presence of high concentrations of calcium and silicate in the natural waters, but strong hydrolysis took place at pH values appreciably below 7 (Raccoon Creek, pH 4.3) and therefore no heavy metal bonds were formed by ion exchange. These laboratory tests with natural waters demonstrate that the enrichment of heavy metals in Ca-loaded zeolite A (present in concentrations similar to those in lakes and rivers, in partly hydrolyzed form) is only slight, with the exception of the most selective heavy metal Pb.

The exchange of heavy metal ions at Ca-loaded zeolite A in natural sewage flowing through a sewer network is simulated in an open flow system during a correspondingly very short test period with the help of a zeolite-filled pressure filter unit [124]. The filter unit is connected to the influent of a clarification plant. The sewage is drawn in with the help of a suction pump. For 2 h the sewage with additional heavy metals flows at a rate of 1000 liter/h/200 g zeolite (equivalent to 100 mg/liter of zeolite) through the zeolite bed, which is enclosed by a filter net (filter unit). The heavy metals in the zeolite are determined by direct analysis. The following heavy metal series in flowing natural sewage is obtained, arranged in order of descending percentage loading of zeolite at pH 7.5–8 and 10–14°C [124]: Cd > Pb, Zn > Cu > Hg. The respective heavy metal loading with 0.5–3.3% corresponds with one exception (Pb) to the orders of magnitude of the equilibrium loading of individual heavy metals obtained under sewage conditions with the analogous 100 mg/liter of zeolite in a closed agitator system, that is, 0.13–3.5% (Cu, Zn, and Cd) and 38.5% (Pb; Table 3) [122]. If, after each loading test, natural sewage containing no additional heavy metals is passed through the heavy metal-loaded zeolite bed during the same very short test period, only very weak heavy metal desorption occurs, in contrast to the equilibrium desorption of individual metals (20–88%) obtained under river water conditions in a closed agitator system (Table 4) [122]. The very weak desorption is probably attributable to the kinetic effects of the desorption from the zeolite bed during the short contact time.

The test results obtained from the open flow system, with only a short contact time with natural sewage, indicate that because of the short time available in practice with natural sewage flowing through a sewer network, the heavy metal enrichment in the zeolite bed is less than 5% (even Pb) and only very small amounts of heavy metals are desorbed if the heavy metal concentrations are reduced in the sewage.

4. Remobilization of Heavy Metal Mixtures from River Sediments by Zeolite A in Waters

If zeolite A finds its way into inland waters, the question must be asked whether this causes changes in the chemistry of these waters in terms of the distribution of heavy metals and the release of immobilized heavy metals in aquatic sediments. It is known that heavy metals can be released under changed environmental conditions if pH value and oxygen content decrease, salt content increases, or the synthetic complexing agent input increases [125]. In the same way that zeolite A has ion-exchanging ability, some natural sediment-forming substances are also able to adsorb cations from aqueous solutions and to release equivalent amounts of already adsorbed cations (especially clay minerals, such as kaolinite, illite, and montmorillonite [125]). The cation exchange of such natural substances is also pH dependent, because H^+ ions compete with the heavy metal ions and release heavy metal ions if the pH is low [125]. To standardize the practice-relevant determination of the remobilization potential of lake and river water in the pH range 7–8 usually encountered in such water, heavy metal-loaded zeolite A is used as a poorly soluble solid phase with a low heavy metal-desorbing blank value [126]. Therefore, a sediment-forming heavy metal-loaded zeolite A in inland water does not behave any differently in principle than sediment-forming natural heavy metal-loaded cation exchangers and, like these substances, will contribute to the stable deposition of heavy metals [125,127,128].

Basically it has been found that the various heavy metals are present not only in dissolved form in rivers and lakes but also in combined form in river sediments. The metal ion equilibria, which are dependent on the solubility products of the different metal compounds and the proportions of the dissolved and combined fractions of the different metals [125,129], can be disturbed by metal ion exchange at suspended zeolite A in a closed system (stagnant water) as a result of the temporary reduction in the metal ion concentrations in river water. The disturbed metal ion equilibria are restored in accordance with the solubility products by partial dissolution of the metal compounds of the sediment, that is, by partial remobilization of the metals until the initial metal concentrations are restored. Only in this sense can zeolite A be effective in remobilizing heavy metals.

Pilot studies of the influence of zeolite A on the heavy metal ion equilibria between membrane-filtered water from the Rhein and Neckar, containing mixtures of heavy metal ions, and sediment from the lower Rhein made up of mixtures of various heavy metal compounds (mainly hydroxides of Pb, Cu, and Cr, carbonates of Zn and Cd, organically bonded Cu, Cd, and particles of worn rock that contain Fe, Co, Ni, Cr, and Cu [125], with the composition 34,200 ppm Fe, 333 ppm Pb, 366 ppm Cu, 1096 ppm Zn, 240 ppm Ni, and 284 ppm Cd [130]) were carried out in a closed aerobic shake system with 1 g/liter of zeolite A with 70% Ca loading (Table 6) [122,131].

Before the zeolite was added, the heavy metal ions were brought into relevant, that is, incomplete, equilibria by shaking the sediment suspension for 24 h in river water containing 3 ml/liter of oxygen at pH 7.4–7.9. During this time (system A) higher concentrations of some heavy metals (Cu, Cd, and Ni) were established (partial dissolution). Upon adding the zeolite (system B), comparable trends were observed in both water samples (within 24 h), with reductions in the

TABLE 6 Change in Heavy Metal Concentrations Δc in Rhine and Neckar Water When Sediment Is Added (System A, 24 h) and Then Zeolite, 1 g/liter of 70% Ca-Loaded Zeolite A (System B, 24 h) to a Simulated Water Course in a Closed System

Metal ion Me^{2+}	Δc (A) (%) average value		Δc (B) (%) average value	
	Rhine water pH 7.4–7.9	Neckar water pH 7.4	Rhine water pH 7.4–7.8	Neckar water pH 7.5
Pb	$+^a$	-14	$-^a$	-10
Cu	$+600$	$+220$	$+7$	$+8$
Zn	$+8$	-48	-21	-26
Ni	$+164$	—	-33	—
Cd	$+200$	$+76$	-33	-53

[a]Analytically uncertain value.
Source: From Refs. 122 and 131.

concentrations of the tested heavy metals Cd, Ni, Zn, Pb, and Cu as a result of ion exchange. The analysis values are subject to strong fluctuations in some metals. Because the extent of the temporary reductions in the concentration of certain heavy metals caused by exchange at zeolite A does not correlate strictly with selectivity as the determinant factor for the exchange of heavy metals at zeolite A, the solubility products of the various heavy metal compounds in the sediments and the kinetic effects of the ion exchange equilibrium and solubility equilibrium exercise a strong influence on the exchange.

Tap water containing mixtures of heavy metals but free of complexing agents has been subjected to laboratory tests with combinations of Na-zeolite A or aquatic sediment and the complexing agent NTA [128]. These tests demonstrated that even very low concentrations of the complexing agent considerably inhibit the immobilization of heavy metal ions by zeolite A or by aquatic sediments. The addition of NTA to contaminated waters can result in a partial transfer of heavy metals from poorly soluble compounds to soluble heavy metal complexes. Zeolite A and complexing agents exhibit opposing behavior with heavy metals; zeolite A immobilizes them, whereas complexing agents mobilize them [128].

C. Heavy Metal Ion Exchange During Sewage Treatment

When detergent formulations were such that natural sewage was free of zeolite but contained complexing agents, the different heavy metals were present mainly in soluble form as free heavy metal ions and heavy metal complexes. When such sewage is treated in conventional clarification plants by primary settling, aerobic biological treatment, and subsequent secondary settling, powerful heavy metal removal effects are achieved with Pb, Cu, Zn, Cr, and Cd, but not all heavy metals can be fully removed, for example, Ni. Removal of heavy metals by the activated sludge process, which predominantly involves the adsorption and absorption of heavy metal ions in a flocculated sediment-forming biomass, is more efficient than primary settling, which primarily involves adsorption of heavy metal ions at sediment-forming solid particles in the sewage. This applies on average to Pb, Cu, and Cr, whereas Zn is removed to the same extent in both treatment stages [132–134].

Although the heavy metal removal rates are high, considerable relative standard deviations are registered in laboratory studies and pilot tests with such natural sewage. Natural sewage is of such complexity that the sewage matrix is critical with regard to the quantitative analysis of heavy metal elimination [72,135]. The uncertainties associated with the analysis have a disturbing effect on tests aimed at determining the influence of heavy metal ion exchange at zeolite A on the removal of heavy metals by sewage treatment, so that simulation tests in the laboratory with synthetic sewage without disturbing complexing agents and with constant heavy metal dosage still provide the most reliable information.

The zeolite content in modern phosphate-free detergents results in average zeolite concentrations of 15–30 mg/liter in sewage. Because zeolite, like other suspended solids, is eliminated by about 50% during the primary settling stage in clarification plants, it is realistic to assume an influent concentration of zeolite A of around 10 mg/liter in the biological treatment stage (Sec. II.C.1).

1. Model Tests of Heavy Metal Transfer During Primary Settling

Because insoluble zeolite A, like other suspended solids in sewage, is partially removed during primary settling, it can be fundamentally expected to enhance the removal of heavy metals from sewage with zeolite A. It is appropriate here to assume that two mechanisms are responsible: heavy metal ion exchange and reinforcement of flocculation by the formation of heteroagglomerates in which poorly soluble heavy metal compounds and solids containing adsorbed heavy metals are removed [72]. Reinforced flocculation is less marked, because zeolite A suspended in sewage in environmentally relevant low concentrations does not significantly influence the sedimentation of other solids [72]. The decisive parameter with regard to sewage is probably the mutual competition between the various heavy metals present to enter into ion exchange at the zeolite A, although there is also competition from calcium ion and sodium ion exchange at zeolite A in hard, salt-containing sewage and from heavy metal ion complexing by synthetic and natural complexing agents.

In preliminary laboratory tests of pure water and tap water containing a mixture of added heavy metals (Pb, Cu, Zn, Ni, and Cd, each in a concentration of 1 mg/liter) and a high concentration (3 g/liter) of zeolite A with various loadings of calcium, all these heavy metals, with the exception of Ni, were extensively exchanged, that is, removed from the solution, after 30 minutes in an agitator [136]. This confirmed the rule about the ion exchange of heavy metal mixtures even at Ca-loaded zeolite A in hard water, according to which the more strongly selective heavy metals of a heavy metal mixture, Pb, Cu, and Zn, are preferentially exchanged and the less selective Cd may also be more strongly exchanged, in percentage terms, when it is present in low molar concentration and when there is also a low loading because of the additional high concentration of zeolite A (Sec. IV.B.2). In synthetic hard sewage, however, the percentage exchange of Cu and Ni is clearly less than in pure and hard water so that only Pb, Zn, and Cd were extensively exchanged within 30 minutes [136]. The reversible heavy metal binding protein decomposition products (peptones) in the synthetic sewage are probably also responsible for the exchange reduction in the sewage [136,137].

Static sedimentation tests in the laboratory, carried out with two sedimentation columns (test column with zeolite and control column without zeolite) and stagnant natural sewage containing a mixture of heavy metals, provided information about the influence of zeolite on heavy metal transfer by sedimentation during primary settling [61,136,138]. The tests showed that Ca-loaded zeolite A (30 and 64 mg/liter) had no adverse effect on the removal of heavy metals by the solids that settle out from municipal sewage, but neither did it promote any significant improvement in the removal of heavy metals. However, a higher Na-zeolite A concentration of 100 mg/liter gave a slightly improved degree of heavy metal removal of 5–10% (Cu, Zn, Cd, and Fe) [61].

The failure of zeolite, at the concentrations at which it is present in the environment, to bring about any significant improvement in the level of elimination of the tested heavy metals is apparently attributable to strong kinetically controlled ion exchange at the relatively low zeolite concentrations in stagnant sewage during the static sedimentation test. This is indicated by the fact that a concentration of 22 mg/liter of Ca-loaded zeolite in almost the same sewage produced a significant although weak improvement in natural sewage after being agitated for 30 minutes to accelerate exchange; the heavy metal series in decreasing order of average percentage effect is [136] Pb (8.5) > Cd (5) > Cu (2.75), Zn (2.25) > Ni (1). The evident exchange of the more weakly selective Cd at zeolite A in synthetic sewage thus also takes place in natural sewage.

In comparison to the laboratory tests with natural sewage, a pilot study of primary settling was carried out with two identical sedimentation tanks arranged in parallel (test and control). Tests were carried out on the sewage influent to a clarification plant, with constant and variable retention times [72]. In concentrations of 30 and 60 mg/liter, zeolite A with a 25% loading of Ca brought about a significant but weak improvement in the removal of the heavy metals Pb, Cu,

Zn, and Cr, supported by intensive mixing of the zeolite with the sewage containing the heavy metal mixture before it passed into the sedimentation tank and by the flow conditions in the tank during a period of 3 days. To obtain a differentiated identification of the improvement in the removal of heavy metals by zeolite, therefore, the formula for calculating the percentage metal removal by sewage or sewage and zeolite is not applied (average metal influent concentration minus average metal effluent concentration divided by average metal influent concentration); instead, the percentage so-called metal immobilization by zeolite is calculated [average metal effluent concentration (control) minus average metal effluent concentration (test) divided by average metal influent concentration [72]]. In this way, the influence of the strongly fluctuating metal influent concentrations is eliminated and the percentage immobilization by zeolite is a criterion for percentage heavy metal exchange. The resulting heavy metal series, in descending order of percentage exchange at Ca-loaded zeolite A at average metal influent concentrations in natural sewage of 0.05 mg/liter of Pb, 0.1 mg/liter of Cu, 0.2 mg/liter of Zn, 0.01 mg/liter of Cr, 0.014 mg/liter of Ni, 0.0013 mg/liter of Cd, is as follows [72]: Pb > Zn > Cu > Cr > Cd, Ni. This means that there is a stronger percentage removal of Pb (9%) by zeolite A than of Zn, Cu, and Cr, but Cd and Ni removal is comparatively very low.

2. Model Tests of Heavy Metal Transfer During Biological Treatment

When primary settling is complete, sewage is treated biologically under aerobic conditions in an aerator, where growing populations of flocculated microorganisms (biomass) help to convert the organic matter of the sewage into CO_2, water, and cell material. Preferential bonding of free heavy metal ions at the biomass takes place at bacterially produced extracellular polymers with high molecular weight (polysaccharides, proteins, and nucleic acids). The heavy metal ions combine with the functional groups of the polymers to form poorly soluble, stable organic complexes (mainly heavy metal polysaccharide complexes) [132,139]. The calculated stabilities of the poorly soluble heavy metal-polymer complexes, however, are lower than those of the soluble heavy metal complexes of the synthetic complexing agents EDTA and NTA that used to be present in sewage [132]. Heavy metal ions are also adsorbed at the cell wall of the bacteria and absorbed in the cytoplasm of the bacterial cells [132]. The level of metal uptake by the biomass depends on the concentration of heavy metals and biomass and on the pH, increasing as they increase [139].

Studies of the bonds between individual heavy metals and the biomass indicate that the greater proportion of the heavy metals form bonds of differing strengths within a relatively short time, that is, heavy metal-polymer complexes of varied stability are formed selectively at the biomass [139]. The heavy metal series in descending order of average efficiency of percentage heavy metal uptake is Pb (95–97) > Cu (81–84) > Cd (74–86) > Ni (47–59) as demonstrated by simulated biological sewage treatment with the addition of individual heavy metals in initial concentrations of 2–25 mg/liter at pH 5.8–7.2 [139]. The relative affinity of a specific heavy metal to the biomass is also dependent on the heavy metal mixture, and this leads to metal uptake competition between the various heavy metals. The heavy metal series obtained from mixtures of heavy metals differ to a greater or lesser degree from series of individual heavy metals [139].

If zeolite A is introduced into an aerator used for the biological treatment of sewage that contains hard water and a mixture of heavy metals, competition may be expected between heavy metal ion transfer at the biomass and heavy metal ion exchange at zeolite A, as well as between the various heavy metals participating in both processes and also from calcium. Both processes, which involve the binding of heavy metals, take place at sediment-forming, poorly soluble substances (biomass and zeolite) [77] and can therefore basically bring about the removal of heavy metal ions in sewage by immobilizing heavy metal ions. Because the usual concentration of zeolite in the influent is relatively low (approximately 10 mg/liter) in comparison with the much larger concentration of biomass in aerators used for biological sewage treatment [135], heavy metal removal by zeolite can be expected to be relatively weak. As a poorly soluble builder, therefore, zeolite has an ecological advantage over soluble builders, which form soluble complexes with heavy metals and are able to mobilize heavy metals in sewage, thereby reducing the amounts of heavy metals removed from the sewage by biomass [140,141].

Laboratory tests were carried out to determine the level of heavy metal removal by zeolite A during biological sewage treatment. A model sewage treatment plant [142–145] was used. Simulation of the biological sewage treatment, with synthetic sewage that contained hard water but was free of complexing agents, resulted in a biomass concentration of some 3 g/liter in the aerator, to which a zeolite suspension and a solution containing a mixture of heavy metals (Pb, Cu, Cr, Zn, Ni, and Cd) in concentrations typical of those present in mixtures of domestic and industrial sewage were continuously added [145]. The influent and effluent concentrations of heavy metal ions were used to calculate the percentage removal of all metals.

Zeolite A in the concentration range 15–60 mg/liter, with Ca loadings of 25 and 75%, exhibited no adverse effects on the removal of heavy metals during simulated biological sewage treatment under the conditions of long retention time associated with the apparatus and vigorous mixing [142,143]. The Ca-loaded zeolite A caused a significant but weak improvement in heavy metal removal as its concentration increased; this applied especially to Pb and, to a lesser extent, Zn and Cd, but Cr and Cu were only weakly removed, and there was no clear evidence that Ni was removed at all by zeolite A [142]. Evaluation

TABLE 7 Heavy Metal Series in Order of Decreasing Percentage Heavy Metal Removal During Simulated Biological Sewage Treatment in the Presence of Zeolite at Different Concentrations (Synthetic Hard Water with Heavy Metal Mixture and 25 and 75% Ca-Loaded Zeolite A)

Zeolite concentration (mg/liter)	Heavy metal removal (%)										
0	Cr	>	Cd	>	Zn	>	Cu	>	Pb	>	Ni
15	Cr	>	Cd	>	Zn	>	Pb	>	Cu	>	Ni
30	Cr	>	Pb	>	Zn	>	Cd	>	Cu	>	Ni
60	Pb	>	Cr	>	Zn	>	Cd	>	Cu	>	Ni

Concentrations of the metals of the heavy metal mixture in the influent to the aerator, mg/liter	Pb	Cr	Zn	Cd	Cu	Ni
	0.068	0.1	0.52	0.0093	0.15	0.11

Source: From Ref. 172.

of the sometimes strongly fluctuating analytical data gives a general heavy metal series in descending order of percentage heavy metal removal during simulated biological sewage treatment in the presence of zeolite at a number of different concentrations (Table 7) [142]. The heavy metal series obtained from simulated biological sewage treatment in the presence of the lowest zeolite concentration corresponds largely to the series concerning the transfer at biomass (without zeolite); as the zeolite concentration increases the series changes, tending to resemble a series obtained from the selectivity of the exchange at zeolite A (with the exception of Cu). This indicates that the ion exchange (zeolite) is gaining ground at the expense of the competing ion transfer (biomass) as the zeolite concentration increases.

This simulation of the biological treatment of synthetic sewage that contains a mixture of heavy metals but no complexing agents (Table 7) [142,143] should be compared with a pilot study [135] with natural sewage. The natural sewage contained complexing agents and a mixture of heavy metals and had undergone primary settling. After settling, the average concentrations of heavy metals in the influent to the biological treatment were comparatively low (0.047 mg/liter of Pb, 0.061 mg/liter of Cu, 0.097 mg/liter of Zn, 0.023 mg/liter of Ni, 0.0026 mg/liter of Cd, and 0.01 mg/liter of Cr). The hydraulic retention time was comparatively short. Because the heavy metal concentrations in the sewage fluctuate more or less strongly, a statistical evaluation of the analytical data was performed, taking account of the heavy metal volume balance, from which the improvement in heavy metal removal by zeolite was calculated (test and control pilot apparatus).

In this pilot study no evidence was found that the zeolite concentrations (15 and 30 mg/liter) in sewage that has undergone primary settling have an adverse effect on heavy metal removal during biological treatment. For zeolite A with a 25% Ca loading, Cu was the only one of the heavy metals tested that was removed to a statistically significant extent, although the improvement was only slight, whereas no statistically significant improvement in heavy metal removal was observed with zeolite A with a 75% Ca loading. It is assumed that, when Ca-loaded zeolite is added directly to the aerator of the pilot apparatus, heavy metal exchange at the zeolite, which is present in relatively low concentrations, is reduced because synthetic, nonbiodegradable complexing agents in natural sewage compete to bind heavy metals [135,139,142,143]. It is also suspected that the analytically observed increase in water hardness in the aerator could have reduced heavy metal removal by zeolite, because the water hardness reduces the heavy metal loading of zeolite and reduces the exchange rate of heavy metal mixtures at the zeolite (Secs. IV.B.1 and 2) [135].

Thus, this pilot study proved the expected very poor heavy metal removal by zeolite A in the biological treatment stage. This is a result of the significant reduction in zeolite and heavy metal ion concentrations in the primary stage and the resulting competition of heavy metal complexation and removal mechanisms (e.g., adsorption onto biomass and increase in the water hardness in the aerator) in the biological stage.

REFERENCES

1. D. W. Breck, in *Zeolite Molecular Sieves*, Wiley, New York, 1974.
2. L. V. C. Rees, Proc. Int. Conf. Zeolites 5th, Heyden, London, 1980.
3. R. von Ballmoos and J. B. Higgins, Zeolites *10*:5 (1990).
4. W. M. Meier and D. H. Olson, in *Atlas of Zeolite Structure Types*, Butterworths, London, 1987.
5. A. Araya and B. L. Lowe, Zeolites *6*:111 (1986).
6. O. Grubner, P. Jira, and M. Ralek, *Molekularsiebe*, Verlag der Wissenschaften, Berlin, 1968.
7. E. M. Flanigen, J. M. Bennett, R. W. Grose, J. P. Cohen, R. L. Patton, R. M. Kirchner, and J. V. Smithl, Nature *271*:512 (1978).
8. R. F. Gould, *Molecular Sieve Zeolites*, Part 1 and 2, Adv. Chem. Ser. 101 and 102, Washington, DC, 1971.
9. R. P. Townsend, *The Properties and Application of Zeolites*, Chemical Society Special Publication No. 33, London, 1980.
10. A. Tissler, U. Müller, and K. K. Unger, Nachr. Chem. Tech. Lab. *36*:624 (1988).
11. L. Puppe, Chemie in unserer Zeit *20*:117 (1986).
12. G. A. Ozin, Angewandte Chemie *101*:373 (1989).
13. Society of Chemical Industry, Molecular Sieves (Conference 1967), London, 1968.
14. B. Lutz and S. Schäfer, Pädagogik in der Naturwissenschaft-Chemie *5*:29 (1988).
15. W. Leonhardt, Swiss Chem. *8*:29 (1986).

16. D. W. Breck and R. A. Anderson, in *Kirk-Othmer Encyclopedia of Chemical Technology*, 3rd edition, Vol. 15, Wiley, New York, pp. 639–669.
17. R. P. Townsend. Chem. Ind. 1246, (1984).
18. V. C. Mole and L. V. C. Rees, in *Recent Developments in Ion Exchange* (P. A. Williams, ed.), Heyden, London, (1987), pp. 264–276.
19. L. Puppe, in *Ullmanns Encyklopädie der technischen Chemie*, 4th edition, Vol. 17, VCH, Weinheim, New York, 1979.
20. K. Klier, Langmuir 4:13 (1988).
21. J. Dwyer, Chem. Ind. 258, (1984).
22. N. Y. Chen and W. E. Garwood, Catal. Rev.-Sci. Eng. 28:185 (1986).
23. W. Hölderich, M. Hesse, and F. Näumann, Angewandte Chemie 100:232 (1988).
24. W. Hölderich and E. Gallei, Chem-Ing.-Tech. 56:908 (1984).
25. M. J. Schwuger and H. G. Smolka, Colloid Polymer Sci. 256:1014 (1978).
26. M. J. Schwuger and H. G. Smolka, Colloid Polymer Sci. 254:1062 (1976).
27. C. P. Kurzendörfer, M. Liphard, W. von Rybinski, and M. J. Schwuger, Colloid Polymer Sci. 265:542 (1987).
28. C. P. Kurzendörfer and W. von Rybinski, Progress in Colloid Polymer Sci. 76:303 (1988).
29. I. Yamane and T. Nakazawa, Pure Appl. Chem. 58:1397 (1987).
30. R. M. Barrer, in *Hydrothermal Chemistry of Zeolites*, Academic Press, London, 1978.
31. K. H. Bergk, M. Porsch, and F. Wolf, Chem. Techn. 39:251 (1987).
32. U.S. Patent 4.041.135 to Huber Corp (1977).
33. U.S. Patent 4.278.649 to Henkel KGaA (1981).
34. U.S. Patent 4.267.158 to Henkel KGaA (1981).
35. Eur. Patent 37.018 to Henkel KGaA (1981)
36. German Patent 2.527.388 to Henkel KGaA (1975).
37. J. R. Katzer, Molecular Sieves, Part 2, ACS Symp. Ser. 40, Washington, D.C., 1977.
38. F. Wolf, K. H. Bergk, and K. Pilchowski, Sprechsaal 112:917 (1979).
39. D. E. W. Vaughan, Chem. Eng. Prog. 84:25 (1988).
40. F. Wolf and K. D. Seidig, Tonindustrie Zeitung 97:281 (1973).
41. N. Giordono, V. Recupero, J. C. J. Bart, and L. Pino, Ind. Miner. 83 (1987).
42. O. Koch, Seifen, Ole, Fette, Wachse 106:321 (1980).
43. P. Colombo. EP 0 522 365, Enichem Augusta Industriale, 01.07.1991.
44. DE-OS 2.910.147 to Degussa AG and Henkel KGaA (1971).
45. DE-OS 2.910.152 to Degussa AG and Henkel KGaA (1971).
46. DE-OS 2.941.636 to Degussa AG and Henkel KGaA (1971).
47. DE-OS 2.633.304 to J. M. Huber Corp. (1976).
48. J. Scott (editor), in *Zeolite Technologies and Applications*, Noyes Data, Park Ridge, NJ, 1980.
49. DE 24 12 837 C3 to Henkel KGaA (1974).
50. U.S. Patent 4.605.509 to Procter & Gamble Company (1986).
51. H. G. Smolka and M. J. Schwuger, Tenside Detergents 14:222 (1977).
52. M. J. Schwuger, H. G. Smolka, and C. P. Kurzendörfer, Tenside Detergents 13:305 (1976).

53. A. C. Savitzky, Soap-Cosmetics-Chemical Specialties *53*:29 (1977).
54. C. P. Kurzendörfer and M. Schlag, unpublished results from the laboratories of Henkel KGaA, Düsseldorf.
55. S. Kiesskalt, *Verfahrenstechnik*, Carl Hanser Verlag, München, 1958.
56. A. G. Kassatkin, *Chemische Verfahrenstechnik*, VEB Verlag Technik, Berlin, 1953.
57. T. Allen, *Particle Size Measurement*, Chapman and Hall, London, 1975.
58. M. J. T. Carrondo, R. Perry, and J. N. Lester, Can. J. Civ. Eng. *8*:206 (1981).
59. D. Balzer, unpublished results from the laboratories of Henkel KGaA, Düsseldorf.
60. Y. Yoneyama and K. Ogino, Tenside Detergents *19*:197 (1982).
61. J. E. King, W. D. Hopping, and W. F. Holman, J. Water Poll. Control Fed. *52*:2875 (1980).
62. B. W. Gould, Civil Engineering Transactions of the Institution of Engineers, Australia *11*:55 (1969).
63. F. Hjulström, Bull. Geol. Inst. Univ. Uppsala *25*:221 (1934)
64. W. A. Roland, W. Graupner, and W. Holtmann, GWF Wasser/Abwasser *120*:55 (1979).
65. G. M. Fair and J. C. Geyer, *Water Supply and Wastewater Disposal*, John Wiley & Sons, New York, 1954.
66. Metcalf & Eddy, Inc., *Wastewater Engineering*, McGraw-Hill, New York, 1972.
67. G. Graffmann, W. A. Roland, R. D. Schmid, and H. G. Smolka, Chemiker-Zeitung *103*:123 (1979)
68. R. Wagner, in *Materialien 4/79*: Die Prüfung des Umweltverhaltens von Natrium-Alu-minium-Silikat Zeolith A als Ersatzstoff in Wasch- und Reinigungsmitteln (Umweltbundesamt, ed.), Erich Schmidt Verlag, Berlin, 1979, pp. 58–71.
69. G. Graffmann, W. Hörig, and P. Sladek, Tenside Detergents *14*:194 (1977).
70. P. Christophliemk, P. Gerike, and M. Potokar, in *The Handbook of Environmental Chemistry*, Vol. 3 Part F (O. Hutzinger, ed.), Springer-Verlag, Berlin-Heidelberg, 1992, pp. 205–228.
71. A. C. Savitzky, 50th Ann. Ind. Conv. Soap Deterg. Ass. *29*:1 (1977).
72. A. C. Rossin, R. Perry, and J. N. Lester, Water Res. *16*:1223 (1982).
73. F. Malz and H. Jendreyko, in *Materialien 4/79*, Erich Schmidt Verlag, Berlin, 1979 pp. 49–51.
74. E. R. Baumann, W. D. Hopping, and F. D. Warner, Water Res. *15*:889 (1981).
75. K. Scherb, in *Materialien 4/79*, Erich Schmidt Verlag, Berlin, 1979, pp. 52–57.
76. M. J. T. Carrondo, J. N. Lester, and R. Perry, J. Water Poll. Contr. Fed. *52*:2796 (1980).
77. W. D. Hopping, J. Water Poll. Contr. Fed. *50*:433 (1978).
78. W. F. Holman and W. D. Hopping, J. Water Poll. Contr. Fed. *52*:2887 (1980).
79. K. Scherb, in *Münchener Beiträge zur Abwasser-, Fischerei- und Flußbiologie*, Pfeil, Oldenburg, 1979, pp. 179–193.
80. W.-A. Roland and R. D. Schmid, Vom Wasser *50*:121 (1978).
81. W.-A. Roland and R. D. Schmid, Vom Wasser *50*:177 (1978).
82. A. W. Maki and K. J. Macek, Environ. Sci. Technol. *12*:573 (1978).
83. W. K. Fischer, P. Gerike, and G. Kurzyca, Tenside Detergents *15*:60 (1978).

84. W.-A. Roland, Umwelt *10*:237 (1980).
85. Degussa AG, Hanau, Germany, 1993, personal communication
86. W. K. Fischer and P. Gode, Vom Wasser *49*:11 (1977).
87. Henkel KGaA, Dept. of Ecology Internal report, 1992.
88. D. Mackay, S. Paterson, and W. Y. Shiu, Chemosphere *24*:695 (1992).
89. K. Imhoff and K. R. Imhoff, Taschenbuch der Stadtentwässerung, 23, Aufl., R. Oldenbourg Verlag, München-Wien, 1976.
90. B. Böhnke, Gutachten zur kostenmässigen Beurteilung des Verhaltens von Waschmitteln in kommunalen Kläranlagen, RWTH Aachen, 1990.
91. F. Malz and J. Bortlisz, Abwassertechnik *1*:11 (1988).
92. T. E. Cook, W. A. Cilley, A. C. Savitsky, and B. H. Wiers, Environ. Sci. Technol. *16*:344 (1982).
93. W.-A. Roland, Forum Städte-Hygiene *30*:131 (1979).
94. Association Internationale de la Savonnerie et de la Détergence (AIS), Practical Aspects of Environmental Hazard Assessment of Detergent Chemicals in Europe, Outcome of AIS 2nd workshop, Limelette, June 1992, AIS, Brussels, 1992.
95. H. E. Allen, S. H. Cho, and T. A. Neubecker, Water Res. *17*:1871 (1983).
96. W. Stumm and J. J. Morgan, *Aquatic Chemistry*, Wiley-Interscience, New York, 1970.
97. Internationale Kommission zum Schutze des Rheins gegen Verunreinigung, Zahlentafeln der physikalisch-chemischen Untersuchungen des Rheinwassers, 1984.
98. C. T. Driscoll, in *The Environmental Chemistry of Aluminum* (G. Sposito, ed.), CRC Press, Boca Raton, FL, 1989, pp. 241–277.
99. H. J. M. Bowen, *Trace Elements in Biochemistry*, Academic Press, New York, 1966.
100. G. Graffmann, W. A. Roland, R. D. Schmidt, H. G. Smolka, J. Schneider, and H. Vogg, in *Materialien 4/79*, Erich Schmidt Verlag, Berlin, 1979, pp. 39–44.
101. G. Müller, in Reference *Materialien 4/79*, Erich Schmidt Verlag, Berlin, 1979, pp. 95–104.
102. W. Lortz and P. Kuhm, unpublished data, lecture to be held at the 10th International Zeolite Conference, Garmisch-Partenkirchen, Germany, July 17–22, 1994.
103. L. Kintrup, R. Schürgers, P. Kuhm, and G. Preuhs, Electron Microscopy, 2, EU-REM 92, Granada, Spain, 1992.
104. P. A. Belinskaya, S. P. Zhdanov, E. A. Materova, and M. A. Shubaeva, Theoretica Ionnogo Obmena Khromatografica, 37 (1965).
105. D. P. Roelofsen, E. R. Wils, and H. Van Bekkum, J. Inorg. Nucl. Chem. *34*:1437 (1972).
106. C. V. McDaniel and P. K. Mahler, ACS Monogr No. 171, 296 (1976).
107. P. D. Hopkins, in *Synthesis of Microporous Materials* (M. L. Occelli and H. E. Robson, eds.), Van Nostrand-Reinhold, New York, 1992.
108. D. Drummond, A. De Jonge, and L. V. C. Rees. J. Phys. Chem. *87*:1967 (1983).
109. R. L. Patton, E. M. Flanigen, L. G. Dowell, and D. E. Passoja, ACS Symposium Series No. 40, 64 (1977).
110. P. Kuhm, unpublished results from the laboratories of Henkel KGaA, Düsseldorf.
111. J. S. Chappel and J. D. Birchall, Inorg. Chim. Acta *153*:1 (1988).

112. J. S. Chappel and J. D. Birchall, Lancet *1*(8644):953 (1989).
113. B. Palmer and R. J. Gilkes, Aust. J. Soil Res *20*:243 (1982).
114. P. A. Moore, K. R. Reddy, and D. A. Graetz, J. Environ. Qual. *20*:869 (1991).
115. M. P. Silverman, in *Biochemical Cycling of Mineral-forming Elements* (P. A. Trudinger and D. J. Swaine, eds.), Elsevier, Amsterdam, 1979.
116. M. J. Schwuger, Ber. Bunsenges. Phys. Chem. *88*:1123 (1984).
117. F. Wolf and H. Fürtig, Kolloid-Zeitschrift und Zeitschrift für Polymere *206*:48 (1965).
118. F. Helfferich, *Ionenaustauscher*, Vol. 1, Verlag Chemie GmbH, Weinheim, 1959.
119. R. M. Barrer, Proc. Chem. Soc. *99*:135 (1958).
120. A. Cremers, ACS-Symposium, Series *40*:179–193 (1977).
121. H. Krumschmidt, Diploma Thesis, Universität Düsseldorf, 1987.
122. C. P. Kurzendörfer, M. J. Schwuger, and H. G. Smolka, Tenside Detergents *16*:123 (1979).
123. B. H. Wiers, R. J. Grosse, and W. A. Cilley, Environ. Sci. Technol. *16*:617 (1982).
124. W. A. Roland and R. D. Schmid, Tenside Detergents *15*:281 (1978).
125. U. Förstner and S. R. Patchineelam, Chemiker Zeitung *100*:49 (1976).
126. F. Dietz, Z. f. Wasser- und Abwasserforschung, *10*:20 (1977).
127. U. Förstner and G. Müller, *Schwermetalle in Flüssen und Seen*, Springer Verlag, Berlin, 1974.
128. G. Müller, Chemiker-Zeitung *103*:131 (1979).
129. F. Frimmel, Z. f. Wasser- und Abwasserforschung *9*:170 (1976).
130. S. R. Patchineelam, Thesis, Universität Heidelberg, 1975.
131. G. Müller, Institut für Sedimentforschung der Universität Heidelberg, unpublished results, 1976.
132. M. J. Brown and J. N. Lester, Water Res. *13*:817 (1979).
133. K. Y. Chen, C. S. Young, and N. Rohatgi, Journal Water Poll. Control Fed. *46*:2663 (1974).
134. H. G. Brown, C. P. Hensley, G. L. McKinney, and J. L. Robinson, Envir. Lett. *5*:103 (1973).
135. M. J. T. Carrondo, J. N. Lester, and R. Perry, Journal Water Poll. Control Fed. *53*:344 (1981).
136. L. A. Obeng, M. J. T. Carrondo, R. Perry, and J. N. Lester, J. Amer. Oil Chem. Soc. *58*:81 (1981).
137. H. Beyer, *Lehrbuch der Organischen Chemie*, Hirzel Verlag, Leipzig, 1955, pp. 628–629.
138. W. W. Eckenfelder, *Industrial Water Pollution Control*, McGraw-Hill, New York, 1966.
139. M. H. Cheng, J. W. Patterson, and R. A. Minear, Journal Water Poll. Control Fed. *47*:362 (1975).
140. J. Barica, M. P. Stainton, and A. L. Hamilton, Water Res. *7*:1791 (1973).
141. S. E. Manahan and M. J. Smith, Water Sewage Wks. *120*:102 (1973).
142. L. A. Obeng, J. N. Lester, and R. Perry, Environment International *3*:225 (1980).
143. S. Stoveland and J. N. Lester, Environmental International *3*:49 (1980).

144. S. Stoveland, M. Astruc, J. N. Lester, and R. Perry, Sci. Total Environment 9:263 (1978).
145. S. Stoveland, J. N. Lester, and R. Perry, Water Res. *13*:949 (1979).
146. M. J. Schwuger, W. von Rybinski, and P. Krings, Progress in Colloid & Polymer Sci. *69*:167 (1984).

III
Polycarboxylates

5

Significance of Polycarboxylates in Detergents

DIETER KIESSLING   Specialty Chemicals, BASF AG, Ludwigshafen, Germany

I. PROPERTIES

Since early in the 1980s, ecological reasons have led in many countries to a reduction in or complete elimination of pentasodium triphosphate (STP) in detergents, replacing it with other builder systems. In several countries, such as Germany, Austria, Switzerland, and Italy, the STP content of detergents is regulated by law; in other countries there are voluntary agreements between the relevant authorities and the detergent industry.

The properties of detergents are determined, apart from the builder system, by surfactants, bleaching agents, and additives. Each component fulfills specific tasks. Synergistic effects can be achieved by suitable selection and appropriate quantity ratios, so that the detergent formulation can be optimized with respect to the washing result.

A reduction or even elimination of the STP content mitigating the environmental load has more or less grave consequences for the detergent effect, depending on water hardness, degree of soiling, fiber type, washing time, and temperature. If phosphate reduction or elimination is carried out without accompanying measures, the disadvantages to the consumer are as follows:

Reduction in detergency (soil removal)
Hardening of the wash by depositions (encrustation)
Graying of the wash (soil redeposition)
Deposits on washing machine parts (reduction in the service life of these parts or
    the washing machine)

The cleaning properties are of significance for the evaluation of the efficiency of
a detergent.

Primary detergency is defined as the whitening effect, the bleaching effect,
and the soil- and fat-removing effect of the detergent after one washing cycle.

Secondary detergency comprises fabric deposits (encrustation), graying, and
color transfer after several washing cycles.

STP largely fulfills the requirements to be met by a detergent builder:

Sequestering of the magnesium and calcium ions
Dispersion of sparingly soluble precipitates
Support of the active detergent surfactants in soil removal
Stabilization of the colloidally distributed soil in the washing liquor
Sufficient buffering effect (keeping the pH value constant)
Ensuring good powder structure

The STP substitute zeolite A is not soluble in water and primarily acts as an ion
exchanger, giving off sodium ions and chiefly taking up calcium ions from the
hardness constituents. The other requirements, especially the dispersion of sparingly soluble precipitates and the stabilization of colloidally distributed soil, are
only insufficiently fulfilled by zeolite A. Consequently, dispersive and precipi-
tation-preventing additives are important to prevent deposition during the washing and rinsing process.

Besides zeolite A, Na citrate, layered silicate, and amorphous silicate have
been proposed as further phosphate substitutes. However, these builders are not
in a position to fulfill all the preceding requirements.

In-depth studies have shown that phosphate-free detergents with single or
combined phosphate substitutes as the builders require auxiliary agents to produce an optimized builder system. Polymers containing carboxyl groups (polycarboxylates) have proved very suitable for this purpose [1–5]. Their properties
are as follows:

Good to very good binding power for magnesium and calcium ions (or salts, such
    as carbonates)
Pronounced dispersive effect on magnesium and calcium carbonate
Good compatibility with detergent constituents, such as surfactants and auxiliary
    agents
Good gray-inhibiting effect (antiredeposition effect)
Favorable toxicologically and ecological properties

Copolymers of acrylic and maleic acid are particularly effective, the weight ratio of the monomers varying between 4:1 and 1:1 and the optimum molar mass (MM) ranging between 40,000 and 70,000.

The input concentrations of polycarboxylates are in the range 3–7% in phosphate-free detergents. The consumption quantities (1993 calculated as 100%) are as follows: for Europe, approximately 80,000 t/a, for the United States, approximately 40,000 t/a, for Japan, approximately 10,000 t/a, and for the rest, approximately 10,000 t/a. This gives a total consumption of approximately 140,000 t/a for 1993.

As already mentioned, polycarboxylates are particularly advantageous for the formulation of phosphate-free detergents. Their good dispersive effect intensifies primary detergency and reduces the occurrence of disturbing deposits on textile material and washing machine parts.

The composition of the phosphate-free detergent used for testing primary detergency and its effect as an encrustation inhibitor is (mass fractions, %):

Zeolite A, 20
Soda, 0–20
Polymer (100%), 0–5
Na-$C_{10}C_{13}$ alkylbenzene sulfonate, 50%, 12.5
$C_{13}C_{15}$ fatty alcohol + 7EO, 4.7
Soap, 2.8
Na metasilicate $\times$ $5H_2O$, 6
Mg silicate, 1.25
Na perborate $\times$ $4H_2O$, 20
Bleaching activator, 3
Carboxymethylcellulose (CMC), 1
Na Sulfate, balance

The conditions for testing the primary cleaning effect of the phosphate-free detergent are as follows:

Washing device, Launder-O-Meter
Washing temperature, 60°C
Water hardness, 2.5 mMol Ca^{2+} (approximately 14°d), Ca/Mg, 3:2
Washing time, 30 minutes
Washing cycles, one
Detergent concentration, 8 g/liter
Liquor ratio, 1:25
Test fabric, WFK fabric with standard soiling

The following polymers are tested:

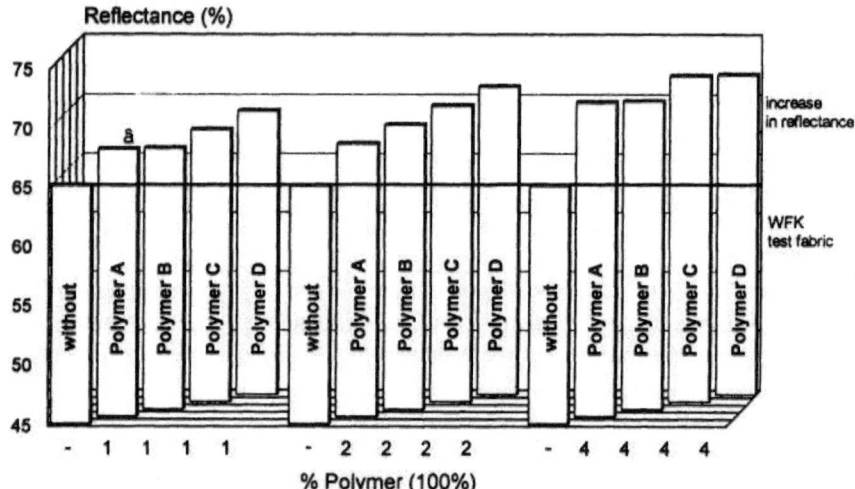

FIG. 1 Phosphate-free detergent, influence of polymers on the primary detergency (polymers A–D; see also text).

Polyacrylic acid, MM approximately 100,000 (A)
Polyacrylic acid, MM approximately 250,000 (B)
Copolymer from acrylic acid and maleic acid,
MM approximately 50,000 (C)
Copolymer from acrylic acid and maleic acid,
MM approximately 70,000 (D)

The acrylic acid/maleic acid weight ratio is different for the copolymers.

When a combination of zeolite A and polymer is used, an optimum is achieved with 4% polymer. Higher concentrations cause only a slight additional improvement. The MM 70,000 copolymer shows the best results (Fig. 1).

If the phosphate-free detergent also contains soda ash in addition to zeolite A, the copolymers are in most cases superior to the homopolymers. Particularly effective is the combination of soda and the copolymer with MM 50,000 (Fig. 2).

To determine the effect of polycarboxylates as an encrustation inhibitor, the ash content of cotton fabric is determined after 20 washing cycles.

Washing conditions for the determination of the effect as an encrustation inhibitor are as follows:

Washing device, Launder-O-Meter
Washing temperature, 30–95°C
Water hardness, 4 mMol Ca^{2+} (approximately 22.4°d), Ca/Mg, 3:2

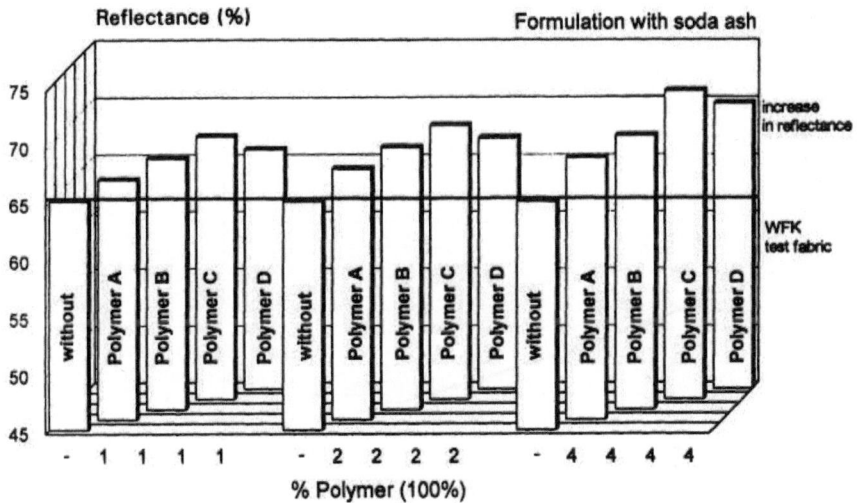

FIG. 2 Phosphate-free detergent: influence of polymers on the primary detergency of a soda ash-containing formulation (polymers A–D; see also text).

Washing time, 45 minutes
Washing cycles, 20
Detergent concentration, 8 g/liter
Liquor ratio, 1:12.5
Test fabric, cotton cloth
Determination method, ash content after burning the fabric in a muffle furnace at 700°C

Soda ash added to phosphate-free detergents to achieve optimum alkalinity of the washing liquor causes significantly greater encrustation. The best values are found for the MM 50,000 copolymer (Figs. 3 and 4).

II. COMPLEXING PROPERTIES, DISPERSIVE EFFECT, AND INTERACTION WITH WATER HARDNESS

Basically, polycarboxylates are not phosphate substitutes. In fact, they are not suitable as builder substances because they have no skeleton-forming or water-softening effect. They do not have complexing properties, either. Their efficiency is based exclusively on their dispersive properties and their threshold effect (inhibiting crystal growth). They are therefore added to detergents only in the preceding low concentrations and are thus auxiliary substances.

The number and density of the carboxyl groups has an influence on the charge

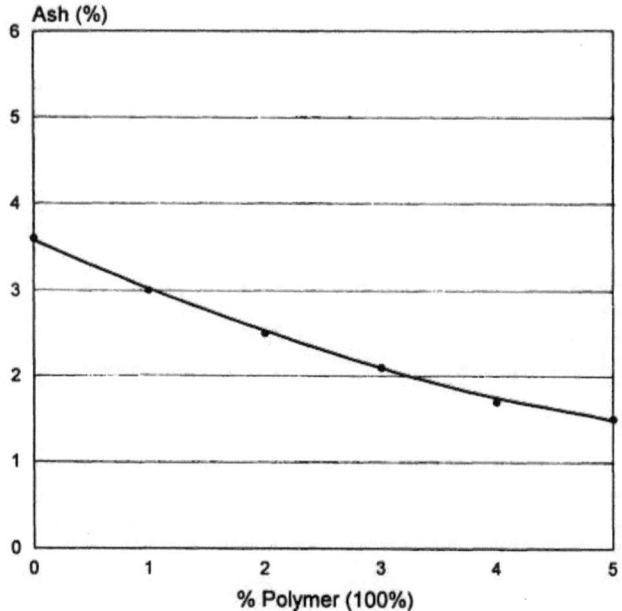

FIG. 3 Phosphate-free detergent: encrustation inhibition (polymer D; see also text).

density and thus on the mutual repulsion of negatively charged groups. Polycarboxylates are either adsorbed directly on solid surfaces with positive charge or deposited on negatively charged surfaces by intermediate cations, such as calcium ions.

Apart from the distribution of the carboxyl groups, the behavior of polycarboxylates is influenced by the chain length, that is, by the degree of polymerization. This means that copolymers based on maleic acid are generally more efficient in phosphate-free detergents.

The interaction of polycarboxylates with the ions of water hardness is possible in two ways:

1. Binding magnesium and calcium ions (or their salts, such as carbonates)
2. Dispersing magnesium and calcium carbonate

Both mechanisms play an important role in application. The bonding constants of polycarboxylates for magnesium and calcium ions can be determined by titration. The ions are presented in a certain concentration and titrated with a solution of the polycarboxylate.

The concentration of calcium ions, for example, in the presence of different quantities of polycarboxylate can be measured with ion-sensitive electrodes or

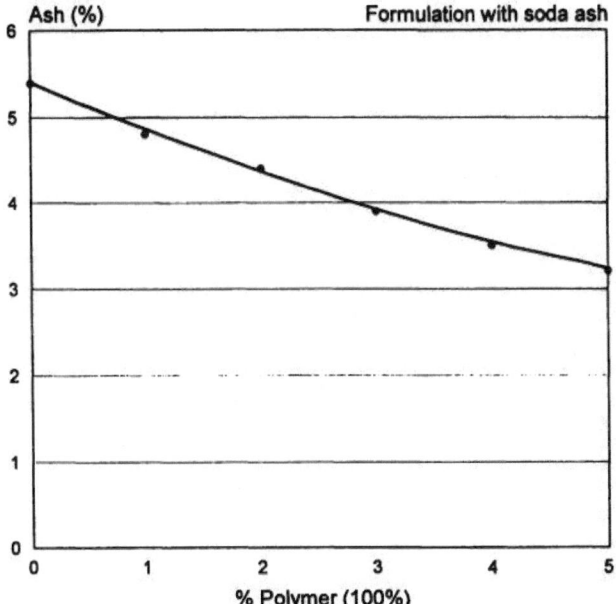

FIG. 4   Phosphate-free detergent: encrustation inhibition of a soda ash-containing formulation (polymer C; see also text).

color indicators. This permits the simultaneous determination of the bonding constants and the number of carboxylate groups necessary for bonding the metal ion. There are different calculation methods for evaluation [6–13].

The binding constants thus determined significantly depend, however, on the type and concentration of further, especially ionic components in the solution. An additional feature of polycarboxylates is that the constants are also subjected to such influences as occupancy by calcium ions. For this reason, calcium binding constants between 10^2 and 2×10^5 are specified for polyacrylic acid alone.

These constants are below those of the usual complexing agents by several orders of magnitude; they are so low that equilibrium is always on the side of the sparingly soluble calcium carbonate over the entire candidate concentration range of the different ions (polycarboxylate, calcium, and carbonate ions) for polycarboxylate application. Hence it follows that the encrustation- and deposit-inhibiting properties of the polycarboxylates must primarily be attributed to their effect as dispersive agents.

Figure 5 shows the binding capacity of some polycarboxylates for calcium ions as a function of the polymer concentration.

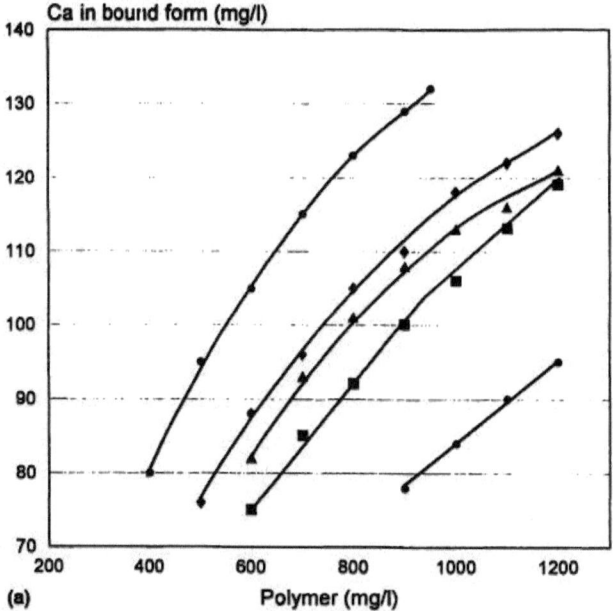

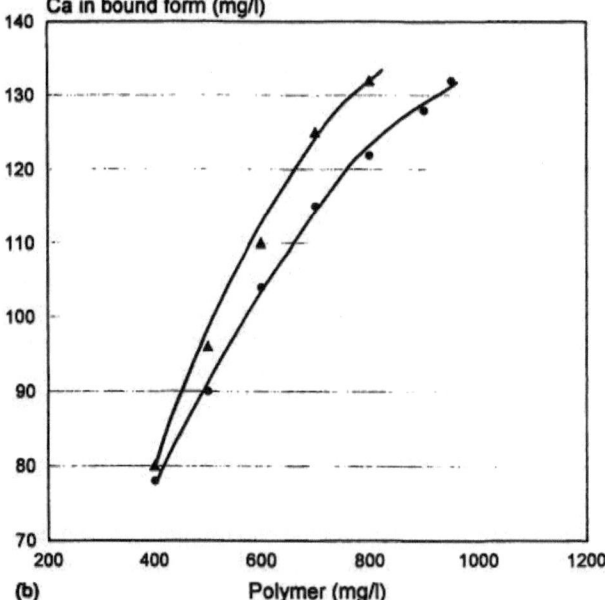

FIG. 5 Binding capacity for calcium ions as a function of polymer concentration. (a) polymers: A (left circles); B (triangles); C (squares); D (diamonds); E (right circles); (b) polymers: F (circles); G (triangles; polymers A–G; see also text).

Polymer A: polyacrylic acid, MM approximately 1200
Polymer B: polyacrylic acid, MM approximately 4000
Polymer C: polyacrylic acid, MM approximately 15,000
Polymer D: polyacrylic acid, MM approximately 70,000
Polymer E: polyacrylic acid, MM approximately 250,000
Polymer F: maleic acid/acrylic acid copolymer MM approximately 70,000
Polymer G: maleic acid/acrylic acid copolymer MM approximately 50,000

Encrustation inhibitors are above all those polycarboxylates based on copolymers with higher molar mass. Their effect can be tested only by washing in formulations, if standardization is also possible.

Simplified methods have therefore been developed to reduce the working effort. They permit rough classification of the polymers independently of the formulation and fabric. One method is the determination of the calcium carbonate dispersing capacity (CCDC) [14].

The calcium carbonate dispersing capacity is determined by titration of the polymer with the solution of a calcium salt. This corresponds in principle to the well-known Hampshire test. Carbonate is generally used to determine the end point. If the calcium complexing constant of a polymer is known, the processes taking place in a turbidimetric titration like the Hampshire test can be completely described by calculation. The solubility product of $CaCO_3$ (10^{-8}) is known in the same way as the protolysis constant of HCO_3^- (4.8×10^{-11}).

With a complexing agent, such as nitrilotriacetic acid (NTA, pK = 6.4), the calcium ions are sequestered at the beginning of titration; the solubility product of $CaCO_3$ is not achieved, as shown by Fig. 6. Exhaustion of the complexing capacity is indicated by precipitation of the calcium carbonate. It follows from the calculations that the theoretical complexing capacity of nitrilotriacetic acid is not fully utilized.

The relatively low complexing constant means that the solubility product of $CaCO_3$ is already surpassed because the theoretical equivalence point (Fig. 6) has been reached. This is different for polycarboxylates. Assuming a complexing constant of 3×10^4, it can be calculated that nearly all calcium ions are present as carbonate in the first phase of titration, whereas appreciable binding of calcium ions to the polymer takes place only when the carbonate ions are largely consumed. Because of the dispersion of calcium carbonate by polycarboxylate, however, the solution remains clear until all polymer molecules are saturated with calcium ions and precipitate as calcium salt (Fig. 7). The region between complete solubility and complete precipitation is very narrow.

This simplified model can explain the comparatively high values found for polycarboxylates in the Hampshire test and related turbidimetric titrations. The anion serving as the indicator is nearly always added in quantities sufficient for an indication of excess calcium ions but in most cases do not exploit the polymer's dispersion capacity. Thus, for example, a maximum of 190 mg calcium

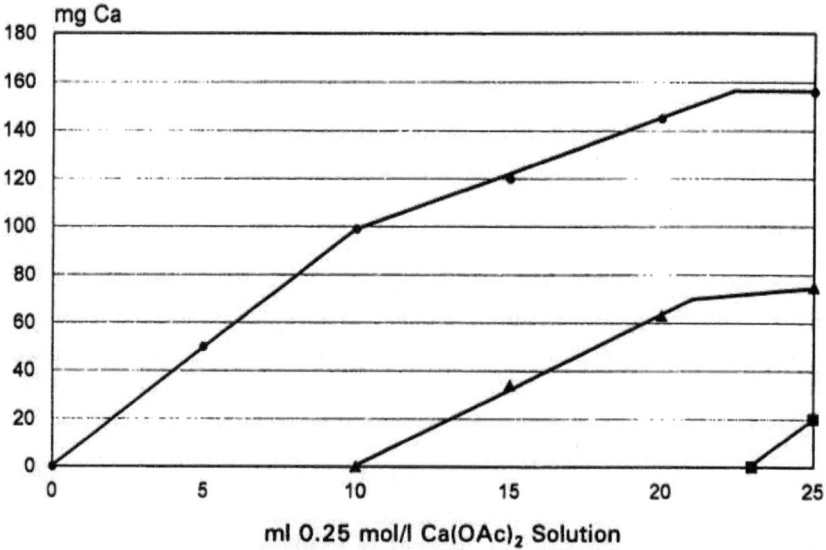

FIG. 6 Calculated titration process in testing Na_3NTA according to the Hampshire test: Ca as Ca^{2+} (squares); Ca as $CaCO_3$ (triangles); Ca as $CaNTA^-$ (circles).

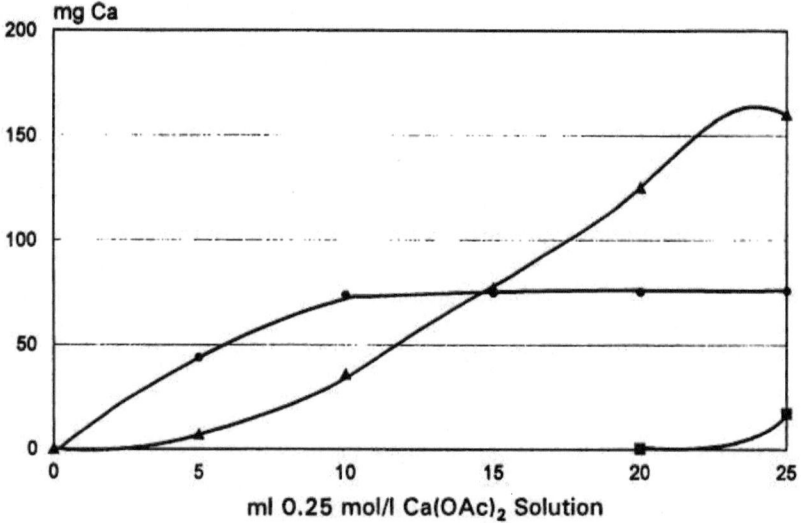

FIG. 7 Calculated titration curve in testing polycarboxylates according to the Hampshire test: Ca as Ca^{2+} (squares); Ca as $CaCO_3$ (circles); Ca as Ca-polycarboxylate (triangles).

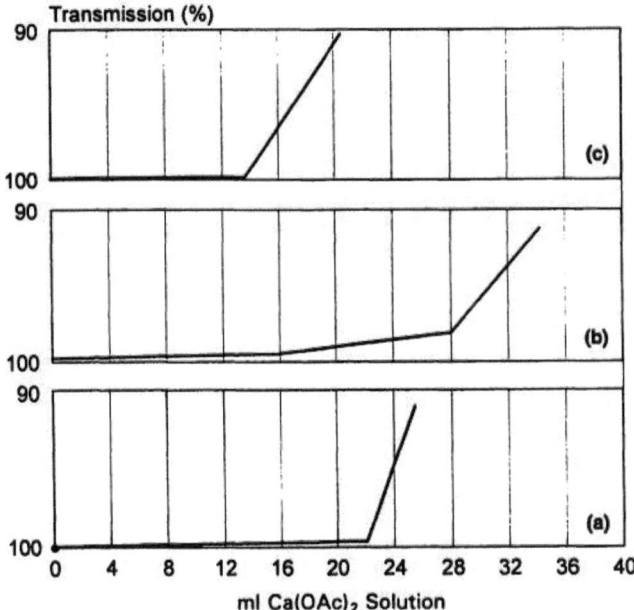

FIG. 8 Turbidimetric titration of a copolymer as a function of the carbonate content: (a) 200 mg Na_2CO_3; (b) 300 mg Na_2CO_3; (c) 500 mg Na_2CO_3.

carbonate/liter can be produced in the Hampshire test, but the measured binding power values reported in the literature are generally much higher.

The individual processes taking place during turbidimetric titration can be experimentally visualized by titration in the presence of higher carbonate quantities. An example is again the copolymer with a molar mass of approximately 70,000.

Titration is carried out analogously to the Hampshire test: 1 g polymer is dissolved in 100 cm^3 water, neutralized, mixed with 200, 300, and 500 mg Na_2CO_3, and adjusted to pH 11. Titration is effected with 0.25 M $Ca(OAc)_2$, keeping the pH value automatically constant.

Figure 8 shows transmission of the solution as a function of the amount of calcium acetate. With 200 mg Na_2CO_3 (corresponding to the Hampshire test), the polymer is in a position to disperse all the calcium carbonate produced. The turbidity at the end of titration is caused by precipitation of the calcium polymer salt.

With 300 mg Na_2CO_3 the dispersion capacity of the polymer is slightly exceeded; the excess calcium carbonate can be identified by a weak turbidity, which initially hardly increases with further titration until the end point of titration is reached, as before.

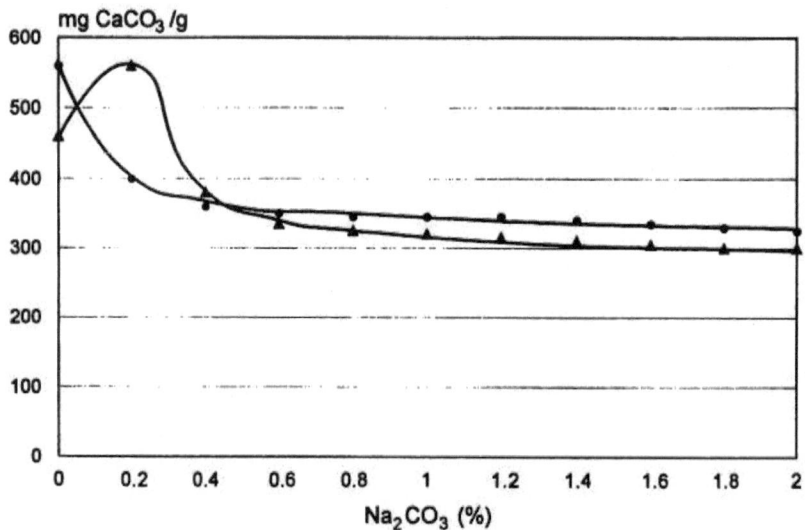

FIG. 9 Titration of copolymer (MM 70,000) in the presence of different amounts of carbonate: Na_3NTA (circles); copolymer (triangles).

With 500 mg Na_2CO_3 the dispersion capacity is clearly exceeded. In this case, the end point of titration is the formation of calcium carbonate.

The result of turbidimetric titration for the polycarboxylate with a molar mass of approximately 70,000 and NTA as a function of the Na_2CO_3 content is shown in Fig. 9.

NTA does not produce any turbidity in the absence of carbonate even after the addition of large amounts of calcium ions. The consumption of calcium acetate solution (as a result of the competition between carbonate and complexing agents for the calcium ions) steadily decreases with increasing carbonate quantity. On the other hand, the curve for the polycarboxylate passes through a maximum.

In the absence of sodium carbonate the titration end point is reached with the precipitation of the calcium polymer salt. It is initially postponed because of small amounts of carbonate, because part of the calcium ions are bound as calcium carbonate, without the possibility for the calcium carbonate to precipitate as a result of the dispersive effect of the polymer.

Finally, at high carbonate concentrations, titration furnishes the calcium carbonate dispersion capacity of the polycarboxylate. This proves the synergism between carbonate and polymer; each individually leads to the precipitation of calcium carbonate and calcium polycarboxylate, whereas both together have a precipitation-inhibiting effect. In practice (as in the washing process), the remaining carbonate content of the water is sufficient to achieve this effect.

REFERENCES

1. J. Perner and W. Trieselt, Tenside Detergents *18*, 5 (1981).
2. A. Hettche, W. Trieselt, and P. Diessel, Tenside Detergents *23*, 12 (1986).
3. W.J. Wirth, Proc. Second World Conference on Detergents, 138 (1987).
4. J. Perner and H.-W. Neumann, Tenside Detergents *24*, 334 (1987).
5. P. Diessel, J. Stabenow, and W. Trieselt, Tenside Detergents *25*, 268 (1988).
6. E.A. Matzner, M.M. Gutchfield, R.P. Langguth, and R.D. Swisher, Tenside Detergents *10*, 119 (1973).
7. I.F. Schaffer and R.T. Woodhams, Ind. Eng. Chem. Prod. Res. Dev., *16*, 3 (1977); Tenside Detergents *16*, 82 (1979)
8. I.A. Blay and I.H. Ryland, Anal. Lett. *4*, 653 (1971)
9. K.H. Nagarajan and H.L. Paine, J. Am. Oil Chemists Soc. *61*, 1475 (1984).
10. D.M. Chang, J. Am. Oil Chemists Soc. *60*, 618 (1983).
11. G.A. Rechnitz and Z.F. Lin, Anal. Chem. *40*, 696 (1968).
12. A. Craggs, G.I. Moody, and I.O.R. Thomas, Analyst *104*, 961 (1979).
13. N. Schönfeldt, J. Am. Oil Chemists Soc. *45*, 80 (1968).
14. F. Richter and E.W. Winkler, Tenside Detergents *24*, 213 (1987).

6

Trace Analysis of Polycarboxylates in Water

ANGELIKA BARTELT Development OEM Spray Coatings, BASF L + F AG, Münster, Germany

ULRICH SCHRÖDER Specialty Chemicals, BASF AG, Ludwigshafen, Germany

DIETER HORN Polymer Physics, BASF AG, Ludwigshafen, Germany

I. INTRODUCTION

Since the beginning of the 1980s, polycarboxylates (PCAs) have been used in detergents and cleaning agents. As additives in low-phosphate and phosphate-free detergent formulations, they prevent incrustation and soil redeposition [1]. In 1990, the market share for phosphate-free detergents had increased to about 95%, with an annual consumption of 19,000 t [2]. An average polycarboxylate amount of about 2% in German detergents corresponds to a PCA concentration of about 2–3 mg/liter in untreated sewage [3]. As a result of the high degree of elimination (91–98%) during sewage treatment [4–7], however, the possible amount of polycarboxylates in surface water is dramatically lower (ppb concentration range).

As a result, an analytical method with a high sensitivity is needed that allows the detection of PCA concentrations of a few micrograms per liter. Turbidity methods are normally used for determination of PCAs, but these methods are successful only in the ppm concentration range [8]. For a.c. tensammetry and voltammetry, the detection limit for low molecular weight polyacrylic acids (2000–5000 g/mol) is 1 ppm [9]. Another method for the ppm range developed especially for analysis of polycarboxylates with average molecular weights of between 800 and 10,000 g/mol, is based on a colorimetric detection method after removal of water-soluble impurities by adsorption/desorption [10]. For the quantitative determination of PCAs in concentrations below 1 mg/liter, polyelectrolyte titration is suitable [11–13]. To our knowledge, polyelectrolyte titration is the only method for trace analysis of PCAs in drinking and surface water. Therefore, a detailed description of this method and results are given in the following sections.

II. POLYELECTROLYTE TITRATION

Polyelectrolyte titration or colloid titration for the characterization of polyelectrolytes was already presented by Terayama in 1952 [14]. The method is based on the stoichiometric reaction between a polycation and polyanion. For the end point determination, Terayama used the color change in a special indicator dye (o-toluidine blue). In principle, turbidimetric, potentiometric, conductometric, or streaming potential measurements can also be used instead of optical measurements for end-point determination, but the optical measurement has the advantage of being simpler to carry out and very sensitive.

An automated method for the investigation of polyelectrolytes, which is based on the principle described by Terayama, was developed and presented by Horn [15–17]. By application of a modified and optimized version of the original polyelectrolyte titration, it is possible to determine polycarboxylates down into the trace analysis range, that is, in concentrations <100 ppb [11–13].

A. Experimental Section

1. Reagents

3,6-Ionene bromide (poly[(dimethyliminio)-1,3-propanediyl(dimethyliminio)-1,6-hexanediyldibromide]; Polybren; Aldrich Chemie, Steinheim, purity at least 97%) with a molecular weight $M_w = 7800$ (relative to the standard Pullulan) as determined by gel permeation Chromatography (GPC) was stored in a desiccator and used without pretreatment.

The purity of Eriochrome black T (ECBT; 3-hydroxy-4-[(1-hydroxy-2-naphthalenyl)azo]-7-nitro-1-naphthalenesulfonic acid, monosodium salt; Merck, Darmstadt) was determined titrimetrically to be 55%. The indicator solution was prepared fresh daily using double-distilled water saturated with N_2 (99.995%, 100 mg/liter), because the dye slowly decomposes owing to its sensitivity to oxidation. The concentration of the indicator in the titration was 1×10^{-6} M.

Na polyacrylate [poly(2-propenoic acid); BASF AG, Ludwigshafen] was polymerized by free radical polymerization. It has a wide molecular weight distribution (M_w = 125,000 g/mol, M_w/M_n = 13, determined by GPC).

The sodium salt of acrylic acid/maleic acid copolymer [poly(2-propenoic acid-co-2-butenedioic acid, Z); weight ratio 70:30; BASF AG, Ludwigshafen] was synthesized by free radical polymerization. It has a wide molecular weight distribution (M_w = 85,000 g/mol, M_w/M_n = 10, determined by GPC).

The other reagents were employed in p.a. purity. All solutions were prepared in double-distilled water.

2. Measuring Device

For the trace analysis a special phototitrator was developed; the setup of the equipment is seen in Fig. 1. Two light-emitting diodes (LED, maxima at 630 and 565 nm) are used as light sources and a photodiode serves as detector. Lock-in amplification results in very high sensitivity. The titrant is added using a commercially available servo burette (Metrohm). The titrations are controlled by a microprocessor. The signal emitted—the difference between the light intensities of the two diodes after being transmitted through the solution—is transferred to a PC (personal computer), where it is processed. For the determination of trace amounts of polyanions, it proved to be best to use cuvettes made of plexiglass to reduce the adsorption of polyions at the walls of the cuvette.

3. Method

The basis of the determination of polyelectrolytes is the formation of a polysalt between two oppositely charged polyions. In most cases, the polysalt reaction occurs quantitatively according to a 1:1 charge saturation [11–21].

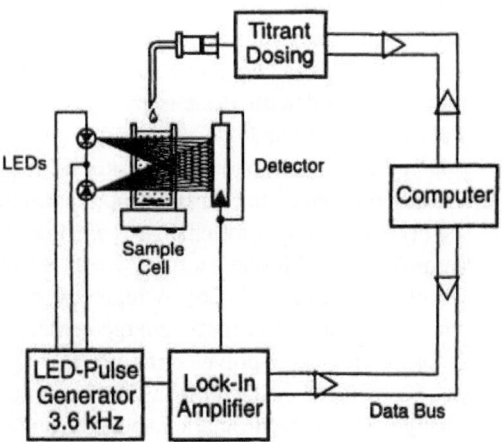

FIG. 1 The phototitrator.

Polyelectrolyte Complex Formation

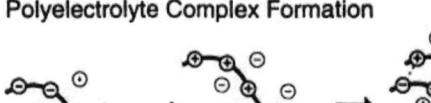

Indicator Reaction

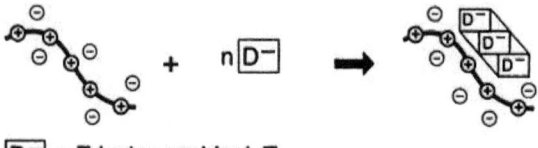

$\boxed{D^-}$ = Eriochrome black T

FIG. 2   Direct determination of anionic polyelectrolytes using chromotropic polycations and an anionic metachromatic dye for end-point detection.

In its original version, polyelectrolyte titration was described for the determination of polycations [14]. These are titrated with appropriate polyanions, such as potassium polyvinyl sulfate. A backtitration was successfully employed in the investigation of polyanions, that is, in the charge determination of proteins [17]. The charge densities found agreed well with those calculated from the composition of the proteins. This result confirms the stoichiometric 1:1 reaction between the oppositely charged polymer molecules.

To analyze polyanions at very low concentrations, a titration method had to be developed employing a negatively charged indicator dye and a suitable polycation as titrant [11]. A suitable system was found with Eriochrome black T as indicator and 3,6-ionene bromide (Polybren) as titrant.

The 1:1 reaction between polyanion and polycation and the indicator reaction is shown schematically in Fig. 2. At the end point, the chromotropic cationic titrant reacts with an anionic dye. The corresponding color change is ascribed to a polymer-induced interaction of dye molecules with one another, which occurs when the distances between the molecules are small [22–26]. A high structural charge density and thus a small distance (<1 nm) between the charge centers is a prerequisite for the chromotropic properties of a polyelectrolyte. The dye molecules may then interact to produce a metachromatic shift in the absorption bands and hence a detectable color change. The metachromatic band shift on the formation of the ECBT/ionene bromide complex is shown in Fig. 3, which shows

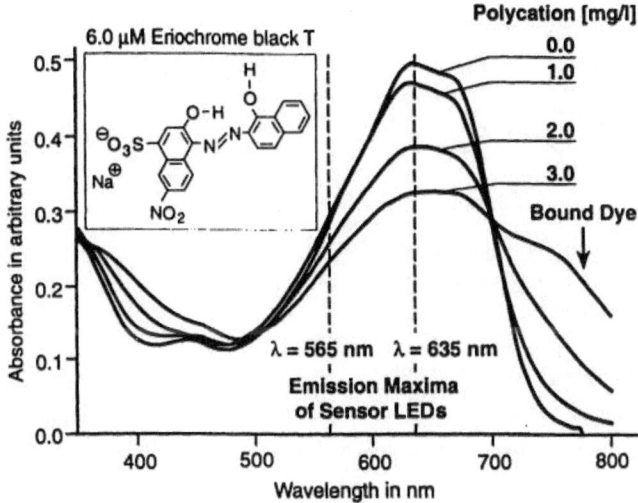

FIG. 3 Metachromatic band shifts caused by the formation of the Eriochrome black T/ionene bromide.

the absorption spectra for various ECBT/ionene bromide concentration ratios. The dashed lines mark the emission maxima of the LEDs.

The overall reaction can be described by the following two equilibria:

$$P^- + P^+ \rightleftharpoons P^-P^+ \qquad K_1$$

$$D^- + P^+ \rightleftharpoons D^-P^+ \qquad K_2$$

K_1 and K_2 are the two equilibrium constants. Because the interaction of opposite charges forms the basis of these equilibria, the concentrations of the charges must be used as concentrations. If each monomer unit carries a charge, these are identical to the concentrations of the monomer units.

Both the polysalt formation and the indicator reaction are based on electrostatic and cooperative interaction [16,27,28]. The equilibrium constants K_1 and K_2 differ owing to differences in the degree of cooperativity of the polymer/polymer versus the polymer/dye interaction [27]. With a sufficiently large ratio of the two constants ($K_1/K_2 > 100$), the dye binds with excess polycations only after complete polyanion/polycation reaction. This is expressed by a sharp color change and a sharp breakpoint in the titration curve.

The aim of polyelectrolyte titration measurements can be either to determine the charge densities of polyions or to analyze solutions with respect to the concentration of a known polyion. For both purposes, the equivalence amount of

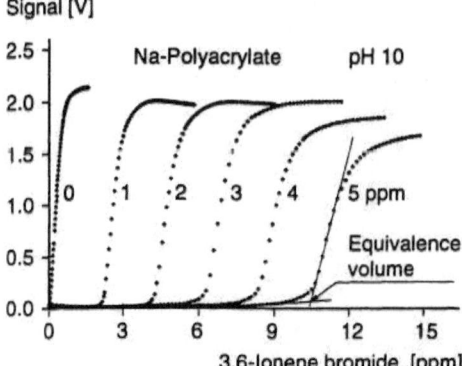

Signal [V]

FIG. 4 Photometric titration curves of sodium polyacrylate against 3,6-ionene bromide at pH 10.0 (cNa-PAA = 0–5 mg/liter). Titrant: ionene bromide at 0.1 g/liter.

titrant must first be determined for a set of solutions containing different concentrations of the polyelectrolyte under test.

Figure 4 shows such a set of curves for the titration of different concentrations (0–5 ppm) of sodium polyacrylate (Na-PAA) with a ionene bromide solution at pH 10. The curve at 0 mg/liter represents the blank test. The curves are displaced to higher values along the x axis with increasing concentrations of sodium polyacrylate in the cuvette, but the shape of the curve does not change substantially. This result indicates that the titration proceeds according to the proposed mechanism and that the necessary condition for the ratio of the two equilibrium constants, $K_1 \gg K_2$, is fulfilled.

B. Titration in the Trace Analysis Range

Figure 5 shows that polyanions may even be titrated in the concentration range 10–100 ppb. The curves presented result from the titration of solutions containing the sodium salt of an acrylic acid/maleic acid copolymer (AA/MA copolymer). The curves correspond to those of Fig. 4. The titration curve at 10 ppb (0.01 mg/liter, corresponding to 0.11×10^{-6} M monomer units) is clearly separated from the blank test curve of a solution containing no polyanion (0 ppb). By plotting the equivalent-point volumes against the initial sample concentration, a linear relationship results (Fig. 6). From the slope of the regression line, the structure-related charge density of the polyion investigated can be determined, taking into account the 1:1 stoichiometry and the charge density of the titrant. On the other side, the regression line is the calibration line for the determination of concentrations in the trace-analysis range. The reproducibility of two following

Signal [V]

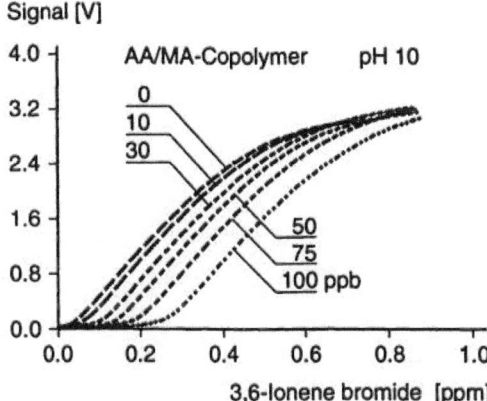

FIG. 5 Photometric titration curves of AA/MA-copolymer against 3,6-ionene bromide at pH 10.0 (cAA/MA-copolymer = 0–0.1 mg/liter). Titrant: ionene bromide at 0.03 g/liter.

measurements in general is better than ±5 ppb. This indicates the precision of the method.

1. Effect of Impurities

(a) Salt Ions. In the determination of polyanions in water samples, other substances are often present that may have an effect on the titration curves. Monovalent ions (e.g., NaCl) in concentrations of more than 10 mM are often responsible for a flattening of the titration curves. The reason for this is to be sought in a change of the equilibrium constants caused by screening of the electrostatic interactions [29]. At low salt concentrations, the equivalence volume is independent of the ionic strength. These concentrations are not exceeded at the ionic strengths usually present.

Divalent metal ions that would interfere with the color reaction of ECBT can be masked by adding EDTA (ethylenediaminetetraacetic acid) without this having any effect on the calibration curve. In the titrations shown in Fig. 5 (10–100 ppb), 3 mM $MgCl_2$ had been added as an example of a divalent metal ion to simulate water hardness.

(b) Natural Polyanions. In addition to salt ions, real water samples may also contain natural polyanions, such as humic acids, which stem from the degradation of organic materials. They also react in the polyelectrolyte titration and therefore consume titrant. Such natural polyanions may be distinguished from synthetic polycarboxylates by selective oxidation with $KMnO_4$. This is demonstrated in Fig. 7. The points give the amount of titrant used by various concen-

3,6-Ionene bromide [ppb]

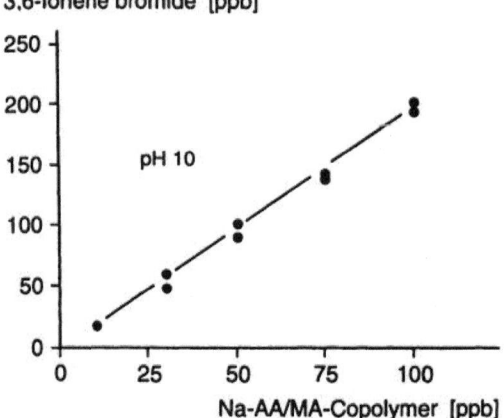

FIG. 6 Equivalence point amount of titrant as a function of the concentration of AA/MA-copolymer at trace amounts of the copolymer.

trations of AA/MA copolymer and humic acid solutions before and after oxidizing the samples. The regression lines correspond to the calibration curves for these two substances. For humic acid, practically no polyanions are detectable after oxidation, regardless of the initial concentration used. For AA/MA copolymer, on the other hand, up to 80% of the amount of titrant used for the unoxidized sample is required to neutralize the negative charges on the polymer after oxidation. The slope of the equivalence lines for the oxidized samples is 70% of that for the unoxidized samples. Treating samples with $KMnO_4$ therefore makes it possible to oxidize naturally occurring polyanions selectively and thus distinguish between these substances and synthetic polycarboxylates.

2. Analytical Procedure

The results shown earlier lead to a series of measurements that must be carried out when a real water sample should be analyzed with respect to the presence of polyanions and, in particular, synthetic polycarboxylates. The following samples should be investigated:

1. Model drinking water
2. Real water sample
3. Sample with standard addition (e.g., 50 ppb AA/MA copolymer)
4. Model drinking water, oxidized
5. Real water sample, oxidized
6. Sample with standard addition, oxidized

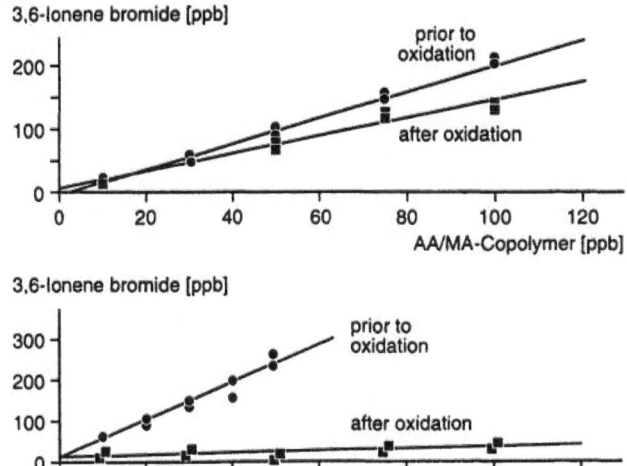

FIG. 7 Calibration curves for AA/MA-copolymer and humic acid before and after oxidation with KMnO₄.

"Model drinking water" means a solution of a well-defined concentration of MgCl₂ or MgCl₂/CaCl₂ mixtures as a model for water hardness in distilled water. A real water sample might be a tap water sample or a surface water sample. Measurements 1 and 4 serve to determine the blank values of the oxidized and unoxidized samples. From 1 and 2, the total polyanion concentration is calculated; from 4 and 5, the concentration of unoxidized polyanions. The oxidation is monitored by 3 and 6, and the recovery rate after oxidation of the standard addition of polyanions should be at least 30%. Otherwise, one cannot be sure that any polycarboxylates that may be present are not changed by the oxidation in such a way that they are no longer detectable.

C. Results and Discussion

1. Tap Water

Figure 8 shows the results of measurements on a drinking water sample. A standard addition of 50 ppb AA/MA-copolymer was added, 53% of which was detectable after the oxidation. The polyanions present in the original sample (34 ppb; all concentrations are given in equivalents of AA/MA-copolymer) were reduced by the oxidation to below the detection limit. These could therefore not have been synthetic, nonoxidizable polyanions. They were presumably substances of the humic acid type.

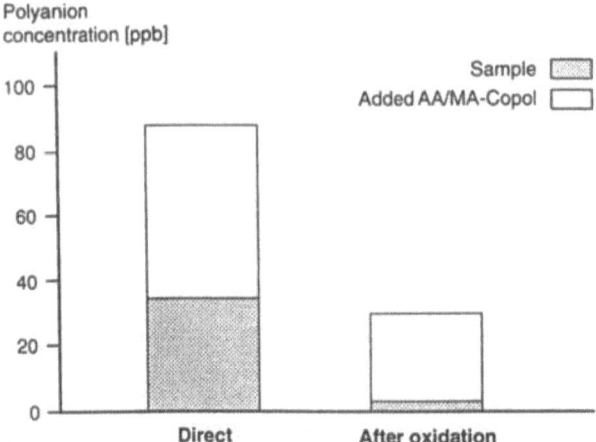

FIG. 8 Concentrations of polyanions in a drinking water sample determined by poly-electrolyte titration, with and without standard addition, before and after oxidation (standard addition: 50 ppb AA/MA-copolymer).

Table 1 shows the results of testing the method on several drinking water samples. In every case, direct titration revealed the presence of polyanions; after oxidation, polyanions were no longer detectable. Before oxidation, the recovery rate of a standard addition of 50 ppb AA/MA-copolymer is with one exception $100 \pm 30\%$; after oxidation it is between 30 and 60%. In all the drinking water samples analysed, it was possible to differentiate between natural, oxidizable polyanions and synthetic, only partially oxidizable polyanions.

These results were supported by an analysis in which a standard addition of 300 ppb humic acid was made instead of 50 ppb AA/MA-copolymer. After oxidation, no polyanions, including the added humic acid, were detectable.

2. Surface Water

Surface water samples normally contain more impurities than tap water samples. With the method just described, the total concentration of polyanions can be determined. A distinction between naturally occuring oxidizable polyanions and synthetic polyanions is often not possible because the recovery rate of added AA/MA copolymer becomes too small. The reason for this is not yet known. Catalytic effects may play a role, or the impurities might interfere with the titration. Therefore, a sample treatment before the titration should be developed, which leads to purification and/or possibly concentration of the polyanions. By adsorption/desorption at inert substrates, this could not be achieved because of the irreversibility of the polymer adsorption.

TABLE 1 Determination of Polyanions in Drinking Water Samples for Testing the Method (Concentrations in ppb AA/MA-Copolymer)

	Polyanion concentration (blank value subtracted)		Recovery rate of 50 ppb AA/MA-copolymer addition (%)	
Sample	Direct	After oxidation	Direct	After oxidation
A	34	<10	108	53
B	21	<10	(184)	43
C	90	<10	83	49
D	33	<10	94	57
E	33	<10	90	42
F	19	<10	64	32
G	24	<10	80	32
H	17	<10	(20)	38
A[a]	27	<10	59	<10
F[a]	19	<10	63	<10
H[a]	17	<10	49	<10

[a]Addition of 300 ppb humic acid instead of AA/MA-copolymer.
Source: From Ref. 13.

The results of measurements using purification of the samples by dialysis, however, look very promising. A dialysis membrane with a cutoff of 1000 was used to minimize leaking of polymer molecules. Table 2 shows the results of the determination of polyanions in tap water samples. No polyanions are detectable after oxidation without or with dialysis. The recovery rate of added AA/MA copolymer is clearly higher with dialysis than without. This gives higher reliability to the conclusion that the polyanions originally present in the samples cannot be of the AA/MA copolymer type.

Application of the method to surface water samples shows that dialysis improves the analysis, which can be clearly seen by comparing titration curves (Fig. 9): curves resulting after dialysis of the sample have a significant greater steepness and therefore satisfactory evaluation of the titration curves becomes possible. After optimization and standardization of experimental conditions for sample pretreatment, dialysis, and oxidation, reproducible and dependable data can be achieved even for surface water. Investigations of various surface water samples (from the Rhein) were performed: in water samples under test, 50–110 ppb nonoxidizable polyanions can be detected with polyelectrolyte titration. Sources for polyanions in surface water can be washing agents, although contamination by dispersing agents from industrial processes or by polymeric flocculants, which are used for waste water treatment, is possible.

TABLE 2   Determination of Polyanions in Drinking Water Samples Without and With Pretreatment by Dialysis Before Titration (Concentrations in ppb AA/MA-Copolymer)

| | Polyanion concentration (blank value subtracted) after oxidation | | | Recovery rate of AA/MA-copolymer addition after oxidation | | |
| | | | | Added concentration (ppb) | Undialyzed (%) | Dialyzed (%) |
Sample	Direct	Undialyzed	Dialyzed			
F	19	<10	<10	50	32	92
H	17	<10	<10	50	38	82
I	47	<10	<10	50	0	114
J	13	<10	<10	100	70	68
K	13	<10	<10	100	0	75
L	57	<10	<10	250	43	80

III. CONCLUSION

The polyelectrolyte titration described here—the direct titration of polyanions using ionene bromide as a titrant and ECBT as the indicator, in combination with a specially developed phototitrator—is eminently suitable for the investigation of polyanions. In particular, in the trace analysis range, previously unachievable detection sensitivities have been obtained. Important features are its relatively low sensitivity against the presence of monovalent and divalent ions. Monovalent ions do not interfere at the concentrations usually present, and divalent ions can be masked by EDTA.

The method can therefore be used for determining polyanions in drinking and surface water samples even in concentrations below 0.1 ppm. In pure water samples, the precision of the polyanion determination is about ±5 ppb. It is lower in real water samples, however, because of the impurities. In the concentration range below 0.1 ppm, the accuracy of the total polyanion concentration in these samples can therefore be estimated as ±20 ppb. By selective oxidation, it is possible to distinguish in such samples between natural polyanions, such as humic acids, and synthetic polycarboxylates of the acrylic acid type. For surface water purification of the sample by dialysis is required. An extension of the method to analysis of the discharge of waste water treatment plants may be possible.

In all investigated drinking water samples no polycarboxylates could be detected. In the surface water samples under test, 50–110 ppb nonoxidizable polyanions could be detected by polyelectrolyte titration.

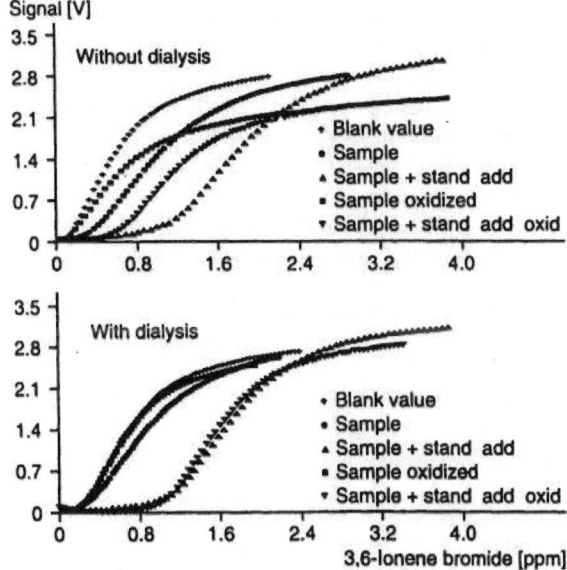

FIG. 9 Titration curves of a surface water sample, without and with standard addition, before and after oxidation (standard addition: 300 ppb AA/MA-copolymer) without and with pretreatment by dialysis.

ACKNOWLEDGMENTS

The authors thank Dr. K.-H. Wassmer, who developed the basis for the direct titration of polyanions. The technical assistance of Dipl.-Ing. (FH) K. Becker in the development of the phototitrator is gratefully acknowledged. All experiments were skillfully assisted by Mrs. A. Scherrer, Mrs. H. Debus, and Mrs. B. Gussner.

REFERENCES

1. J. Perner and H.-W. Neumann, Tenside Surfactants Detergents 24, 6 (1987).
2. Stellungnahme des Hauptausschusses "Phosphate und Wasser" der Fachgruppe Wasserchemie in der Gesellschaft Deutscher Chemiker zur Beurteilung der Umweltverträglichkeit von Polycarboxylaten aus Waschmitteln, Schreiben der Fachgruppe Wasserchemie vom 02.02.1990 an das Bundesministerium für Umwelt, Naturschutz und Reaktorsicherheit.
3. H.-J. Opgenorth, Tenside Surfactants Detergents 24, 366 (1987).
4. H.-J. Opgenorth, in The Handbook of Environmental Chemistry (O. Hutzinger, ed.), Springer-Verlag, Berlin, 1992, Vol. 3, Part F.

5. H. Schumann and S. Kunst, Wasser Abwasser *132*, 376 (1991).
6. P. Schöberl and L. Huber, Tenside Surfactants Detergents *25*, 99 (1988).
7. H.G. Hauthal, Chemie in unserer Zeit *26*, 293 (1992).
8. J.W. Wimberley and D.E. Jordan, Anal. Chim. Acta *56*, 308 (1971); F.J. Angenend and U. Schulte-Wieschen, VGB Kraftwerkstechnik *59*, 995 (1979).
9. P.M. Bersier, W. Neagle, and D. Clark, Analyst *117*, 863 (1992).
10. Hatch Company, *Technical Information Manual*, 1992.
11. K.-H. Wassmer, U. Schröder, and D. Horn, Makromol. Chem. *192*, 553 (1991).
12. U. Schröder, D. Horn, and K.-H. Wassmer, Seifen, Ole, Fette, Wachse *117*, 311 (1991).
13. U. Schröder and D. Horn, in Studies in Polymer Science *11*, Polymer Solutions, Blends, and Interfaces (I. Noda and D.N. Rubingh, eds.), Elsevier, Amsterdam, 1992.
14. H. Terayama, J. Polym. Sci. *8*, 243 (1952).
15. D. Horn, Progr. Colloid Polym. Sci. *65*, 251 (1978).
16. D. Horn, in *Polymeric Amines and Ammonium Salts* (E.J. Goethals, ed.), Pergamon Press, Oxford, 1980, pp. 333–355.
17. D. Horn and C.-C. Heuck, J. Biol. Chem. *258*, 1665 (1983).
18. A. Domard and M. Rinaudo, Macromolecules *13*, 898 (1980).
19. A. Domard and M. Rinaudo, Macromolecules *14*, 620 (1981).
20. B. Philipp, H. Dautzenberg, K.-J. Linow, J. Kötz, and W. Dawydof, Progr. Polym. Sci. *14*, 91 (1989).
21. E. Tsuchida and K. Abe, in *Interactions Between Macromolecules in Solution and Intermolecular Complexes*, Adv. Polymer Sci. Vol. 45, Springer Verlag, Berlin, Heidelberg, New York, 1982.
22. G. Schwarz, S. Klose, and W. Balthasar, Eur. J. Biochem. *12*, 454 (1970).
23. G. Schwarz and W. Balthasar, Eur. J. Biochem. *12*, 461 (1970).
24. B. D. Gummow and G. A. F. Roberts, Makromol. Chem. *186*, 1239 (1985).
25. B. D. Gummow and G. A. F. Roberts, Makromol. Chem. *186*, 1245 (1985).
26. K. Toei, Anal. Sci. *3*, 479 (1987).
27. E. Tsuchida, Y. Osada, and K. Abe, Makromol. Chem. *175*, 583 (1974).
28. E. Tsuchida and Y. Osada, Makromol. Chem. *175*, 593 (1974).
29. H. Dautzenberg, G. Rother, K. Linow, and B. Philipp, Acta Polym. *39*, 157 (1988).

7

Interactions of Polycarboxylates with Major Inorganic Soil Components

JEAN-MARIE SÉQUARIS   Institute of Applied Physical Chemistry, Research Center Jülich GmbH, Jülich, Germany

I. INTRODUCTION

The quantity of the synthetic polymer, polycarboxylate (PCA), used in phosphate-free detergents is growing. Although a large part of this polymeric material can be discarded in modern water treatment plants [1,2], the persistence of the polymer in the environment because of its low biodegradability requires our attention. In this context, the behavior of PCA in soils or water when introduced or inadvertently discharged must be investigated.

General pathways for PCA transport into soils and sediments are governed by the equilibria between its dissolved state and other flocculated or adsorbed states on natural surfaces that immobilize or possibly retard further transport. In these processes, not only the ionic composition, pH, and temperature of the surrounding water medium play important roles but the electrostatic and chemical properties of particle surfaces should also be considered.

With regard to its composition, the water medium exerts variable solvation properties on the solubility and conformational structures of the PCA. Thus a highly soluble, totally ionized polyelectrolyte with a stretched structure is obtained at alkaline pH and low ionic strength, but either a lower solubility neutral polymer in a coiled structure or an insoluble flocculated PCA in the presence of an excess of divalent cations is formed at acidic or neutral pH, respectively. It follows that these multiple properties of the PCA in solution complicate the description of its behavior under environmental conditions.

A comprehensive review requires systematic studies in which the simultaneous variations in the parameters, such as pH and ionic strength at constant temperature, are considered. This parameter list is not exhaustive because the chemical composition of PCA (homopolymer and copolymer), as well as the effects of the average molecular weight average and distribution should also be emphasized. The resulting variable hydrodynamic and interfacial properties thus affect the transport of PCA.

The results of studies of this behavior in soil are scarce and generally concern the use of PCAs as soil conditioners. Polyacrylates have been introduced to keep soils in proper flocculated states that encourage drainage and root growth [3]. Thus this application has shown that the colloidal properties of soil particles as clay surfaces in particular are modified by the strong adsorption of small amounts of PCA.

The sorption properties of clay minerals for synthetic and natural organic polymers, such as humic substances, are well known and have been reviewed [4–6]. These results show the importance of interactions of the soil clay fraction with polymers. For PCA, the interactions with other metal oxide particles must also be considered. Indeed, specific adsorption processes as a result of interactions of carboxylate groups with metal active centers from the soil particle surface are of prime importance [7].

II. POLYCARBOXYLATES IN PHOSPHATE-FREE DETERGENTS

Typical PCAs used as detergent ingredients are generally either copolymers of the acrylic-maleic type (MW 70,000) or low molecular weight acrylic acid homopolymers [1,2].

PCA is a macromolecule bearing ionizable carboxyl groups whose charge density is controlled by varying the pH [8].

The polyelectrolyte features of PCA are characterized by a locally strong electrical field that exerts a long-range influence on charged species, such as colloid surfaces. However, the increase in the concentration of small mobile electrolyte ions limits the efficacy of the electrical field by screening the structural charges. Thus, depending on the pH and electrolyte ion concentration, a variable electrostatic repulsion between ionized carboxyl groups can induce conformational changes in the polymer in solutions (Fig. 1).

With a high screening effect of the polymer charge by proton or salt neutralization, a coiled structure of high density in polymer segments (random coil conformation) can be induced. In contrast, total ionization of the carboxyl groups at alkaline pH and low ionic strength stretches the polymer chain in solution through electrostatic repulsions (rod-like conformation).

These solution-dependent structures of PCA may also be reflected in the adsorbed state at the metal oxide surface.

Another important property of PCA containing oxygen donor ligands in carboxyl groups is the ability to complex hard A-type metal cations, such as calcium and aluminum. This effect is enhanced by the polyelectrolyte character of the polyanionic form of PCA at neutral and alkaline pH. By raising the Ca^{2+} cation concentration, intra- and intermetal bridges between segments of the polymers

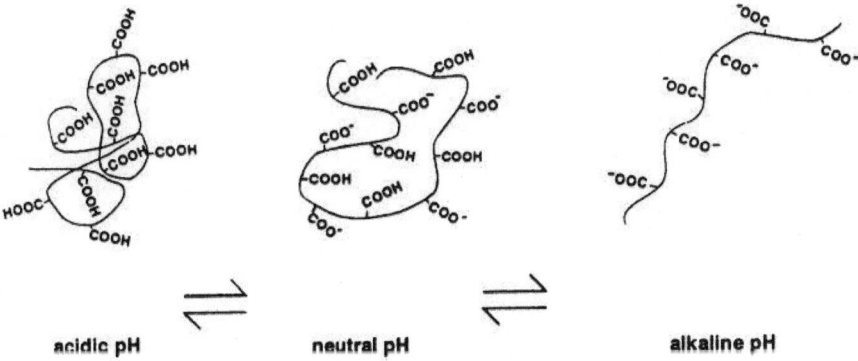

acidic pH   neutral pH   alkaline pH

FIG. 1 pH effect on the conformational structure of polycarboxylic acids.

induce a shrinking of the PCA structure and lead to aggregation [9,10]. A decrease in the solubility of the thus neutralized polymer follows, and it precipitates out of the aqueous phase.

III. MAJOR METAL OXIDE PARTICLES IN SOILS

The most abundant metal oxides in soils are SiO_2, Al_2O_3, and Fe_2O_3. It must be recalled that the respective weight percentages of these metal oxides in the earth's crust are 55, 14, and 7%, respectively, in total oxide components, whereas H_2O is only 8% [11]. Most "active" metal oxides in soils are clay minerals that principally consist of an assembly of dimensional sheet structures containing SiO_2 and Al_2O_3 layers. Kaolinite, montmorillonite, illite, and mixed-layer minerals are the principal clay minerals found in soil and have been widely used for experimental studies of particle/polymer interactions [4–6].

In Fig. 2, for example, the profile depth of a typical Parabraun soil shows the relative importance of the clay fractions. Especially in the B horizon level, where

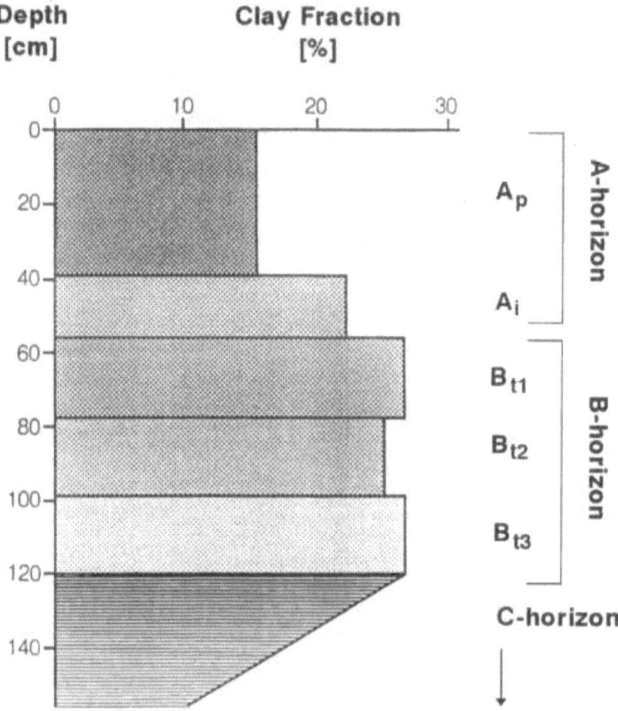

FIG. 2 Clay fraction in a profile depth of a typical Parabraun soil.

there are fewer biological materials, a clay mineral fraction of up to 30% can be found.

This "activity" of the clay minerals is related not only to the large surface area of the small particles <2 μm but also to the ionic properties of their surfaces [6]. In contrast, sand and silt particles in soils are relatively large and have surfaces that are fairly inert.

As already noted, PCA shows a polyanionic character in neutral and alkaline media. The electrostatic properties of the metal oxide particles must thus play an important role in their sorption capacities. Indeed, in the presence of water, the surface of these oxides is generally covered with surface hydroxyl groups that show pH-dependent ionization [12]. Thus, electrostatic attraction or repulsion forces determine parameters in the interactions of both charged surfaces.

The acid base properties of the hydroxide form of the metal center (>M–OH) at the particle surfaces can be described as following. In two pK model system, it can be generalized that at acidic pH the metal hydroxide group is protonated (>MOH_2^+); at more alkaline pH the deprotonation leaves a negative charge at the surface (>MO^-).

The pH of the point of zero charge pH_{pzc} of the surface can thus be defined from the intrinsic acid and basic pKs of the amphoteric metal hydroxide group. At $pH_{pzc} = (pK_a + pK_b)/2$, the net electrical charge of the surface is zero. The surface consists of neutral hydroxyl sites plus an equal number of positively and negatively charged binding sites. At $pH < pH_{pzc}$, positive binding sites are dominant; at $pH > pH_{pzc}$, negative binding sites are dominant.

The pH ranges over which the electrical surface charges occur are related not only to the specific or intrinsic proton affinity of the oxide type but also to the ionic strength of the surrounding solution. Indeed, in view of the polyelectrolyte character of the surface, the charge developed at the surface can be modeled [13,14], where free H^+ and electrolyte concentrations as well as the valence of ions are considered.

In the clay minerals, this pH-dependent charge is mainly located at the edge face of structural basal layers. For kaolinite, a representative 1:1 crystalline mixed-oxide phyllosilicate, the fundamental units of its structure consist of an extended sheet of two basal layers that contain Si and Al in the siloxane and gibbsite layers, respectively.

The edge face exposes most active singly coordinated OH groups to Si and Al, silanol and aluminol groups, which have acid base functions. The aluminol groups are associated with Lewis acid sites, a combination of the metal cation and water at the interface Al (III) · H_2O [15]. Under mild pH conditions, from pH 4 to 10, aluminol groups are protonated in acidic pH and are dissociated in alkaline pH [16]. Silanol groups tend to be dissociated at only extremely alkaline pH. The pH_{zpc} of the edge surface is about 7.5.

For clay minerals, the presence of a permanent negative structural electrical

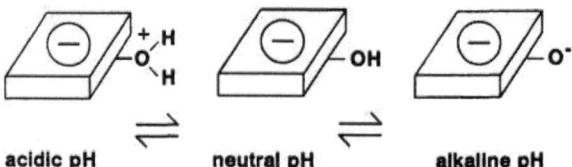

acidic pH   neutral pH   alkaline pH

FIG. 3 Location of the electrical charge at a clay mineral surface: pH-dependent charge at edge faces and permanent negative charge at basal faces.

charge must be added to this pH-dependent charge. This charge is mainly localized on the SiO_2 or siloxane layer in kaolinite and montmorillonite and results from the isomorphic substitution of tetravalent silicon atom by a trivalent aluminum atom [6]. This charge difference is compensated by the interaction of exchangeable cations.

The presence of this permanent negative charge at the clay surface dominates the electrostatic interaction with charged molecules, such as PCA. Thus polyanions tend to be repelled from the surface and little adsorption occurs in comparison with neutral and cationic polymers [4,5].

A schematic representation of the charge localization at the clay mineral surface is shown in Fig. 3.

IV. INTERFACIAL BEHAVIOR OF POLYCARBOXYLATES

The general parameters that play an important role in the adsorption of polyelectrolyte are presented here in brief. For more ample information, readers are referred to recent reviews about this subject [17–20].

The theoretical approach to polyelectrolyte adsorption behavior is generally treated using the results obtained with uncharged polymers; in addition, specific parameters, such as the charge density of the polymer and the ionic strength of the solution, are considered [17,18].

Thus, the favorable adsorption of a polymer at a surface increases with the polymer concentration and the length of the polymer chain or the polymer molecular weight. Indeed, a minimum of binding segments per macromolecule may be required to overcome the weak adsorption energy of the individual segment. In the same way, in the adsorbed layer, a high fraction of segments directly interacting with the surface in trains promotes the high-affinity character of the polymer adsorption. With polyelectrolytes, it must be remarked that the repulsive interaction between charged segments limits the effect of the increased molecular weight on the building of loops in the adsorbed layer.

The solubility of the polymer is also an important parameter adjustable by controlling the pH and the ionic strength in PCA solutions. Indeed, the decrease in the net surface charge density of PCA by lowering the pH and its screening by increasing the salt concentration in solution diminish polymer molecule solubility, which promotes their adsorption at interfaces. In the same way, the charge screening of the lateral electrostatic repulsions between chains in an adsorbed polymer layer leads to an increase in adsorbed amounts. The complexation of multivalent cations by PCA also contributes to lower solubility in the aqueous phase [9,10].

Until now these parameters have mainly influenced the solution properties of polymers, but as with pH and ionic strength, they also contribute to the electrical properties of the metal oxide surfaces. Thus, in the presence of PCA, the simultaneous variations in the electrical properties from both surfaces make it difficult to obtain a comprehensive overview of the adsorption processes.

According to DLVO theory, long-range electrostatic interactions (attraction/repulsion) operate under low ionic strength conditions. Other physical forces—dipole interactions and London-van der Waals and hydrophobic attraction forces—require shorter distances as well as hydrogen bonding; chemical forces that result in coordination linkage or chelation directly modify the surface structure of the adsorbent.

Bonding of PCA to the metal oxide surface results in a combination of these solution and surface properties: however, important PCA/metal oxide surface interactions must be briefly defined before reviewing experimental results:

Hydrogen Bonding: direct bonding through neutral carboxyl groups from PCA with hydroxyl groups of the metal oxide surface >M–OH..O=COHR or indirect bonding through the water bridge associated with cations lying at the surface. The oxygen of the carboxyl groups serves as the hydrogen acceptor.

Electrostatic interaction: electrostatic attraction between the negatively ionized carboxylate groups from PCA and protonated hydroxyl groups of the metal oxide surface >M–OH$_2^+$..$^-$OOCR. at pH < pH$_{pzc}$ of the surface. This reaction corresponds to an anion exchange mechanism in which electrolyte anions are displaced by the polyanion. This outer surface complex type involves a water molecule interposed between the positive surface and the complexed anions. Displacement of counterions at the surface and from the polyelectrolyte greatly increases the entropy, which favors adsorption. For pH > pH$_{pzc}$, electrostatic repulsion forces between negatively ionized carboxylate groups and binding sites >M–O$^-$ decrease the adsorption.

Ligand exchange reaction: direct bond formation between carboxylate groups and the metal oxide surface involves the displacement of the hydroxyl group >M–OOCR from Lewis acid sites [15]. This reaction is thus enhanced at acidic pH. The anionic groups enter the inner coordination layer of the surface metal ions to form an inner sphere surface complex.

Cation bridging mechanism: multivalent cations are involved in ternary complexes between structural permanent or negatively ionized charge surfaces and carboxylate groups $>M-O^-$. M^{n+}. ^-OOCR. Depending on the nature of the cation, inner or outer sphere complexations can occur.

Nature and concentration of competitive anions: As already noted, the effect of electrolyte concentration on the adsorption of PCA is a crucial parameter. However, possible antagonistic effects related to the nature and concentration of the anions interacting with the surface must be indicated here. Thus the increase in salt concentration primarily diminishes electrostatic repulsions in PCA or at the metal oxide surface by a cationic screening that favors adsorption. However, anions can also electrostatically compete with carboxylate groups from PCA in an anion exchange mechanism for the same positively charged binding sites at the surface, which limits adsorption. In oxyanion molecules, such as phosphate compounds, a specific chemical interaction through a ligand exchange mechanism can block Lewis acid metallic binding sites [21] or displace previously adsorbed molecules from them, such as carboxylate groups from PCA. It is obvious that this desorption effect depends on the available concentration of oxyanions, such as borate, phosphate, or other carboxylate compounds, in soil. Thus the presence of naturally occurring low molecular weight weak organic acids and highly polymerized fulvic and humic acids may affect the mobility of PCA by competitive processes.

V. ADSORPTION OF POLYACRYLIC ACID ON METAL OXIDE SURFACES

Most experimental results on the adsorption behavior of PCA at metal oxide surfaces concern polyacrylic acids (PAA): However, recent results obtained with acrylic/maleic acid copolymers, typical ingredients of phosphate-free detergents, are also reported in terms of their interactions with clay minerals.

A. Adsorption of PAA on SiO_2

It is shown that PAA adsorption on a SiO_2 Aerosil and clean quartz surface is low or negligible [22–24]. Indeed, the silica surface becomes increasingly negatively charged from pH_{pzc} 2.5 [24] or 3.5 [22] to pH 10 through ionization of the silanol surface groups. No adsorption in this pH range 3–9 > pH_{pzc} was observed in 10^{-3} M $NaClO_4$ [24] for PAA. In 10^{-2} M NaCl, no adsorption can be found for PAA at neutral or alkaline pHs either because of electrostatic repulsion effects with the increasingly dissociated carboxyl groups from PAA. At low pH, the formation of hydrogen bonds between silanol groups and carboxylic groups has been proposed [25].

However, an enhanced adsorption of PCA on quartz surfaces is found in the presence of multivalent cations. Thus, hydrolyzed ferric ions strongly bound to the quartz surface in the pH range from 3 to 6 facilitate PAA adsorption through a bridging mechanism [23].

B. Adsorption of PAA on Al$_2$O$_3$

Adsorption of PAA onto alumina has been reported [26–28] in a pH range from pH 2 to 11, which can be correlated with the presence of positively protonated binding sites interacting with carboxylate groups. Indeed, the pH$_{pzc}$ of Linde alumina [27] is 8.3; Pechiney alumina has an isoelectric point (pH$_{ise}$) at about pH 8 [28]. Thus a large decrease in the adsorbed amount of PAA can be shown at pH > pH$_{pzc}$ [26–28] under electrostatic repulsion conditions at low ionic strength.

The variations in the adsorbed polymer conformation under changing pH conditions have been investigated by a fluorescence spectroscopic technique using pyrene-labeled PAA [27]. The variations in the excimer-monomer ratio I_e/I_m allow following the interfacial transformation of a coiled form of preadsorbed PAA at low pH into a stretched structure, schematized in Fig. 4a–c, when the pH is raised. It has been shown that this transformation is not reversible when the pH is decreased. Thus, the original stretched form for the preadsorbed PAA structure at pH 10 is adsorbed in a flatter configuration at low pH (Fig. 4d–f). These pH-monitoring polymer interfacial conformations are correlated with the flocculation process of alumina in the presence of PAA.

The nature of the interaction of the carboxyl group of PAA with the alumina surface has also been investigated by diffusion reflectance Fourier transform infrared spectroscopy [26]. A shift in the antisymmetric C-O stretching vibration from 1570 to 1600 cm^{-1} for the adsorbed PAA shows that carboxylic groups are dissociated and react with the surface aluminum atom as proposed in a ligand exchange reaction.

A tentative and schematic description of the adsorption of PAA, taking into account the interaction of solubilized aluminum from the surface, has been proposed [29]. The complexity of the interfacial reactions has been followed by the potentiometric estimation of aluminum-hydrogen exchange kinetics during PAA adsorption at pH 5 in the presence of added 3×10^{-4} N AlCl and 1 mM KCl. It appears that under these conditions, which simulate a dissolution equilibrium state for alumina, only the complexed form of PAA with aluminum ions can be adsorbed.

The effect of the calcium ions also shows marked enhancement of PAA

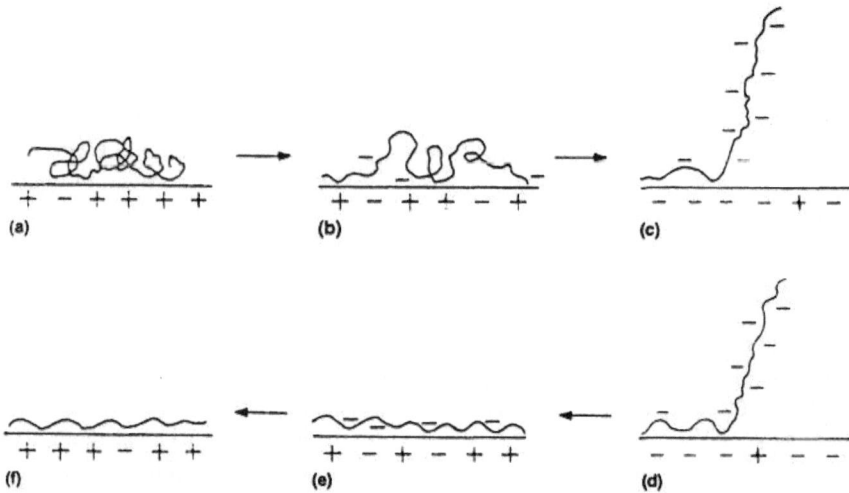

FIG. 4 Variation in polymer conformation at the alumina/liquid interface ($pH_{zpc} = 8.3$) under changing pH conditions. (a) pH 4; (b) pH 5–7; (c–d) pH 10; (e–f) acidic pH region (Fig. 7 in Ref. 27).

adsorption at pH 9 [28]. The presence of a maximum in adsorption isotherms of PAA at a constant calcium concentration has been explained by a speciation model of calcium ions. It was shown that the main parameter controlling the adsorption is a critical ratio of the total concentration of calcium ions to the PAA carboxylate group of close to 0.4. As with aluminum ions [29], a maximum adsorption is observed for the less soluble complexed form of PAA with calcium ions.

C. Adsorption of PAA on Fe$_2$O$_3$

The pH_{pzc} and $pH_{iše}$ of hematite (Fe$_2$O$_3$) occur at pH 8.1, which indicates that surface positively charged sites are available over a large pH range of environmental interest [24]. Thus, the adsorption results for PAA at low ionic strength (10^{-3} M) show a pH dependence adsorption that decreases to near zero values close to pH_{pzc}. At pH < pH_{pzc}, it is postulated that the adsorption of PAA may be a result of electrostatic attraction between positively charged binding sites and negatively ionized carboxylate groups from PCA, as well as through hydrogen bonding between the metal hydroxyl groups and the double-bonded oxygens of the carboxyl groups.

Electrokinetic measurements (Fig. 5) also show a large decrease in the electrophoretic mobilities of the oxide particles as a result of the overcompensation

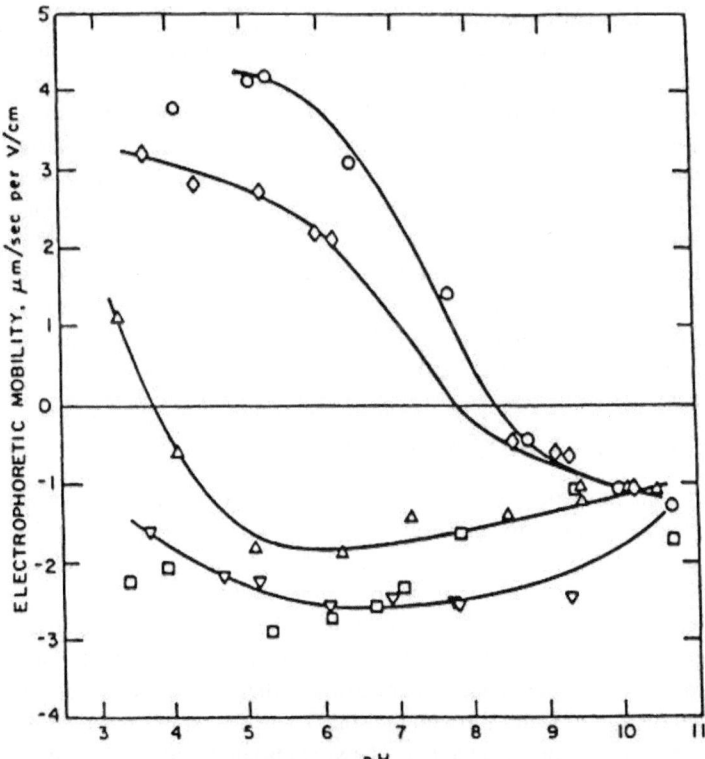

FIG. 5 Effects of pH and polyacrylic acid (PAA, MW 2×10^6) concentration on the electrophoretic mobility of Fe_2O_3 (pH_{zpc} 8.1): 0 mg/liter of PAA; (circles); 0.01 mg/liter of PAA (diamonds); 0.1 mg/liter of PAA (triangles); 1 mg/liter of PAA (squares); 10 mg/liter of PAA (upside-down triangles); and 1 mM $NaClO_4$ (Fig. 8 in Ref. 24).

of original positive charges by the adsorbed PAA. A reversal charge effect is observed under saturation conditions, which indicates that the mobilities of the modified particles depend on the electrical properties of the negatively ionized PAA. An effective electrostatic repulsion that prevents the adsorption of PAA is observed only when the pH is increased several units above the pH_{pzc}. At pH = pH_{pzc}, sensitive electrophoretic mobility measurements indicate that some specific adsorption occurs.

The dispersing properties of low molecular weight PAA [30] can also be related to this net negative surface charge.

Recently the colloidal stability of hematite in the presence of polyelectrolytes and natural organic matter was correlated with the ionization of the adsorbed structure [31]. The aggregation kinetics of hematite particles in the presence of PAA was monitored using a light-scattering technique. The experimental results and model systems show that the colloidal stability is affected by the net electrostatic repulsion forces originating from modified particle surfaces at low ionic strength or from the adsorbed PAA at high ionic strength ($I > 0.1$ M). Indeed, the importance of the relative thickness of the adsorbed PAA layer in comparison with the decreasing ionic diffuse layer (Debye length) along the ionic strength has been clarified.

The effect of calcium ions on colloidal stability has also been reported. In the presence of PAA, it was shown that not only does a screening effect by calcium ions lead to destabilization but that this flocculation process is also dominated by a cation bridging mechanism of a more chemical nature.

D. Adsorption of PAA on Clay Minerals

Clay minerals as a major component of most soils of interest may focus research on their interactions with PCA for different reasons. Thus the transport of PCA concerns adsorption/desorption processes at the solution/clay interface. Other processes, such as weathering of the clays or aggregation/deflocculation phenomena, can also be considered. In the early 1950s, the latter process aroused considerable interest in the use of polyelectrolytes as aggregating agents for soil [3]. Indeed, these so-called soil conditioners can stabilize soil aggregates, encouraging the percolation of water through the soil. These first studies also chose clays to obtain insight into the mechanism of the stabilizing action of PCA in soils. Thus, the reported results pointed out different controversial aspects of the nature of PCA interactions with clay surfaces, as well as the effects of the solution concentration and molecular weight on their dispersing and flocculating properties [3].

Before describing some recent results obtained with detergent PCA copolymers, important conclusions from adsorption studies of PAA should be emphasized with regard to the general adsorption behavior of polyanions at clay surfaces [4].

The adsorption of polyanions at the clay surface is limited by the electrostatic repulsion forces arising from the permanent negative charge of the clay surface. Adsorption is thus favored under charge screening conditions of PAA by protonation or in the presence of a high electrolyte concentration [3,32–34]. In swelling montmorillonite clays, a nonintercalation of PAA between flat surfaces as a result of electrostatic repulsions can be demonstrated [3,35].

At acidic pH, the adsorption of PAA takes place at the crystal edges of clay surfaces through the electrostatic interactions of carboxylate groups with positively charged exposed aluminum atoms or aluminol groups. This exclusively anion exchange mechanism at the edge face can explain the limited surface area

accessible to the PAA adsorption [3,32–34,36]. It must be added that the interactions of the positively charged edge faces with negatively charged plate faces
stabilize a "house of cards" structure typical of pH-aggregated clay particles.
The addition of low molecular weight PAA has been shown to disperse this
structure; higher molecular weights promote a flocculated state by the establishment of polymer bridges between particles [3,37,38]. Deflocculation occurs as a
result of a superequivalent adsorption of carboxylate groups of PAA. Thus,
neutralization of the edge face alone is not enough to obtain a dispersed state for
clay particles. An increase in the negative overall charge caused by the ionized
adsorbed PAA layer leads to the peptidization of the clay suspension [39–40].

The adsorption competition of PAA with phosphate compounds (metaphos-
phate [3] at montmorillonite and pyrophosphate and polyphosphate at kaolinite
[33, 34, 39]) also confirms the importance of the positively charged edges at the
gibbsite sheet plane. In the same way, an adsorption competition between PAA
and an anionic surfactant can be shown [33]. The sharing of the binding sites at
the kaolinite surface in acidic medium for PAA and sodium dodecyl benzene
sulfonate (SDBS) is demonstrated in Fig. 6. The relative decrease in the plateau
value for adsorbed PAA corresponds to a parallel extent of binding by competing
SDBS molecules.

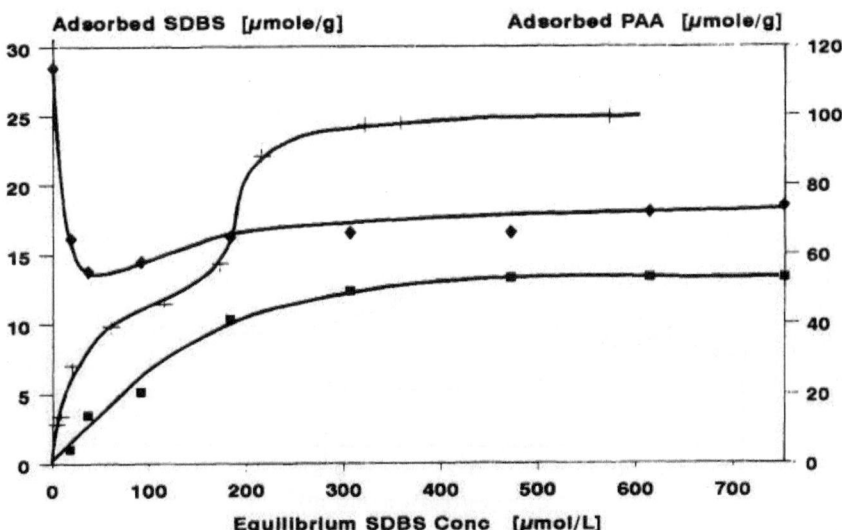

FIG. 6 Variations in the simultaneous adsorption of sodium dodecyl benzene sulfonate
(SDBS) and 300 mg/liter of polyacrylic acid (PAA, M_w 250,000) on kaolinite (5 g/liter)
at pH 4.5 and 0.01 M NaCl: SDBS alone (plus); SDBS in the presence of PAA (squares);
PAA in the presence of SDBS (diamonds) [34].

A ligand exchange reaction by which the carboxylate group complexes the aluminum surface atom has not been directly demonstrated for adsorbed PAA at the clay mineral surface. Indeed, the low amount of PAA adsorbed per clay material, depending on the rather low surface area of edge sites, renders a direct investigation with spectroscopic methods difficult.

The adsorption of PAA can be promoted by the presence of multivalent cations, calcium ions for example, which may act as cation bridges between the carboxylate groups of PAA and negatively charged binding sites at the surface clays. However, recent studies show that primarily the ratio of the total concentration of Ca^{2+} to the total PAA carboxyl group concentration in solution is a crucial parameter for the dependence of the PAA adsorption [41]. An adsorption maximum for PAA on kaolinite occurs when the binding ratio of Ca^{2+} per carboxyl group reaches a saturation value of 0.31, which corresponds to an equivalent ratio of total concentration of 0.45 in solution at alkaline pH. As also reported in the adsorption of PAA on alumina [28], "humped" adsorption isotherms (Fig. 7) can thus be obtained in the presence of calcium ions, depending on the ratio of calcium ions to PAA monomer concentration. It has been proposed

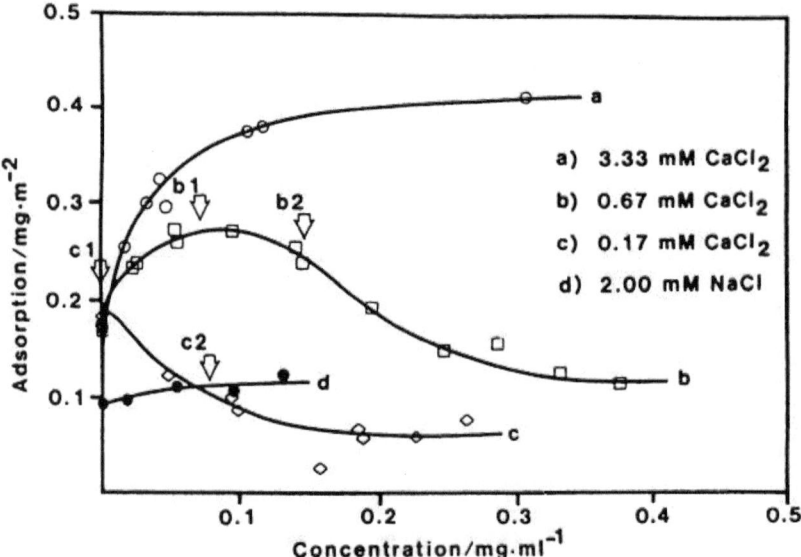

FIG. 7 Adsorption of polyacrylic acid (PAA, M_w 2600) on Ca-kaolinite from solution of $CaCl_2$ or NaCl at pH 8 (25 g/liter of kaolinite). The equilibrium concentration of PAA when the ratio $[Ca^{2+}]_{tot}/C_{PAA} = 0.45$ is indicated by arrows (arrow b refers to curve b and arrow c refers to curve c). Index 1: the Ca^{2+} ions originating from the kaolinite are not included in $[Ca^{2+}]_{tot}$. Index 2: these Ca^{2+} ions are indicated in $[Ca^{2+}]_{tot}$ (Fig. 6 in Ref. 41).

that this enhanced adsorption in the presence of calcium ions (compare results in Fig. 7b and d for an equivalent ionic strength) may take place on energetically less favorable adsorption sites at the negatively charged basal planes of kaolinite. This adsorption of PAA saturated with calcium ions also leads to aggregation in concentrated kaolin solution. It must be noted that the low molecular weight PAA (MW 2600) has dispersing properties on kaolinite particles. In the presence of Ca^{2+}, however, it can induce aggregated structures of kaolin different from the pH-aggregated structures, as has been characterized by rheological measurements and scanning electron microscopy [42].

The flocculation of PAA-stabilized kaolinite with calcium ions leads to a reduced repulsion between the faces of the kaolinite particles. This effect on interparticle interactions has been investigated by surface force measurements in the adsorption of PAA onto large negatively charged sheets of muscovite mica [43]. In the presence of Ca^{2+}, the position of the attractive minimum for layer-layer contact is located 1–3 nm away from mica-mica contact. It can thus be argued that attraction forces are caused by cation bridges between opposite surfaces through adsorbed low molecular weight PAA in a very flat conformation. These results also give some evidence that the adsorption of PAA at mica surfaces involves a cation bridge mechanism.

VI. ADSORPTION OF ACRYLIC ACID/MALEIC ACID COPOLYMERS ON CLAY MINERALS

In this part, recent results for the adsorption behavior of PCA copolymer (acrylic acid/maleic acid) on clay minerals are reviewed. Evidence is presented about the location and nature of the binding sites at clay minerals, the effects of pH and ionic strength, and the binding mechanism of PCA [44–46].

A. Binding Sites of Clay Minerals for PCA Copolymers

As described earlier, the heterogeneity of the clay minerals can be roughly related to the chemical composition in terms of siloxane and gibbsite layers and also to the structure in terms of basal face and edge face sites.

Information about the location of these binding sites can thus be obtained by comparing the adsorption behavior of PCA on well-defined clay mineral microstructures and under identical solution conditions (Fig. 8).

For montmorillonite, kaolinite, and illite, it can be shown [45,46] that the maximum surface area occupied by PCA molecules matches the estimated area of the edge surface of the different clay minerals.

An analysis of the PCA adsorption isotherms also provides insight into the nature of the binding site. With the help of labeled ^{14}C-PCA, it has been possible to cover a large domain of the free concentration of PCA. Thus, a semilogarith-

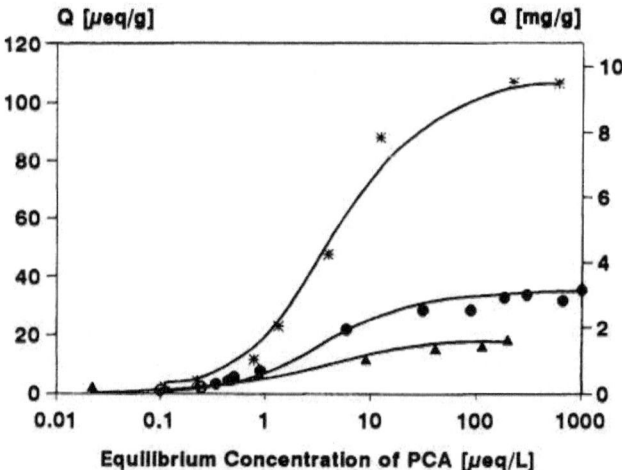

FIG. 8 Semilogarithmic representation of the adsorption isotherms of [^{14}C] acrylic-maleic acid copolymers on clay minerals (5 g/liter) at pH 4.5 and 0.01 M NaCl. Curves are calculated with a Langmuir equation [45,46]: Na-montmorillonite (asterisks); Na-illite (triangles); Na-kaolinite (circles).

mic representation in Fig. 8 reveals that the adsorption isotherm can be fitted by a Langmuir equation. The derived affinity constants are very close for the three clay mineral types [45,46], which indicates that the nature of the binding site at the edge faces is independent of the Si/Al composition. It can thus be argued that a specific adsorption of carboxyl groups from PCA copolymer takes place at the aluminol binding sites. Thus, these results quantitatively confirm previous assumptions.

B. Effects of pH and Ionic Strength

The adsorption behavior of PCA copolymers at clay minerals is governed by forces that have been roughly divided into long-range ionic interactions of an electrostatic nature and other short-range interactions of a more chemical nature.

The ionic interactions depend on the polyelectrolyte properties of both surfaces, which is controlled by a pH-dependent surface ionization variably screened by the effect of the ionic strength. The influence of these parameters has been investigated in the adsorption study of PCA copolymer on kaolinite.

In Fig. 9, the pH dependence of the maximum adsorption values of PCA copolymer are shown for three NaCl salt concentrations corresponding to ionic strengths 0.01, 0.1, and 1 M. A rapid overview of the experimental results

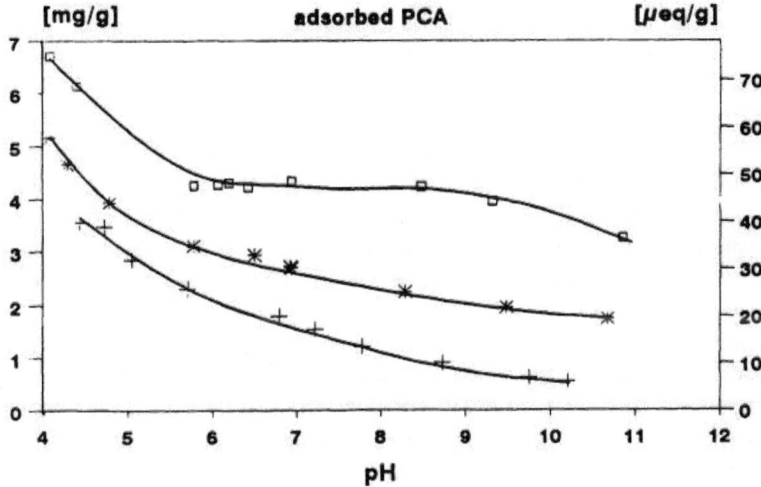

FIG. 9 Effects of pH and ionic strength (NaCl concentration) on the maximum adsorption values of [^{14}C] acrylic-maleic acid copolymers on Na-kaolinite (5 g/liter) [45, 46]: 0.01 M (plus); 0.1 M (asterisks); 1 M (squares).

indicates that the adsorption of PCA copolymer decreases by increasing the pH or lowering the salt concentration.

Taking into account a global negative charge of the kaolinite particles caused by a predominant effect of permanent negative charge, as well as the growing importance of the ionization of the carboxyl groups up to alkaline pH, these results show that long-range electrostatic repulsion forces, according to the DLVO theory, operate at low ionic strength. Charge neutralization by protonation at low pH and nonspecific charge screening for higher ionic strengths decrease the electrostatic repulsion between PCA copolymer and kaolinite surface and thus promote adsorption through other nonelectrostatic forces.

For high-energy binding sites located at the edge sites, it can also be added that the increase in the ionic strength decreases the internal repulsion between adsorbed PCA copolymer segments and thus favors a close packing of coiled structures of higher segment density.

C. Ligand Exchange Mechanism

As already noted, nonelectrostatic forces have been involved in the adsorption processes. With regard to the weak acid properties of PCA and the aluminol nature of the binding site, a ligand exchange mechanism can be generally considered [21] in which a hydroxyl group from aluminol >AlOH is exchanged with a carboxylate group from PCA. A pH increase as a result of the release of OH⁻ and PCA adsorption found under unfavorable electrostatic conditions at pH > pH_{pzc} are reported as indirect evidence of this mechanism.

Furthermore, an exclusive displacement of PCA from the kaolinite surface by specifically adsorbed anions, such as phosphate compounds, which form inner sphere complexes with surface aluminum, is also indirect confirmation. Indeed, the adsorption competition between simple organic acids with phosphates on soil components has been widely investigated [47,48]. It has been explained that adsorption occurs at the same binding sites of soil oxide surfaces and results in mutual blocking effects.

In a similar way, it is thought that the mobility of PCA copolymer in soils also depends on the relative displacement efficiency of phosphate compounds in which oligomeric triphosphates (STP) are more efficient than di- or monophosphate compounds. The configuration of the adsorbed layer of polymers also seems to be of crucial importance in the displacement mechanism [49].

This observation is also demonstrated by desorbing the copolymer in the entire region of an adsorption isotherm with a fixed STP concentration of 0.15 M phosphate monomer in Fig. 10. The observed percentages of desorption are plotted against the amount of copolymer initially adsorbed at the kaolinite surface. The percentage of desorption is found to be lowest for surface concentrations lower than 0.7 mg/g of kaolinite (or 8 µEq/g) of adsorbed copolymer. This

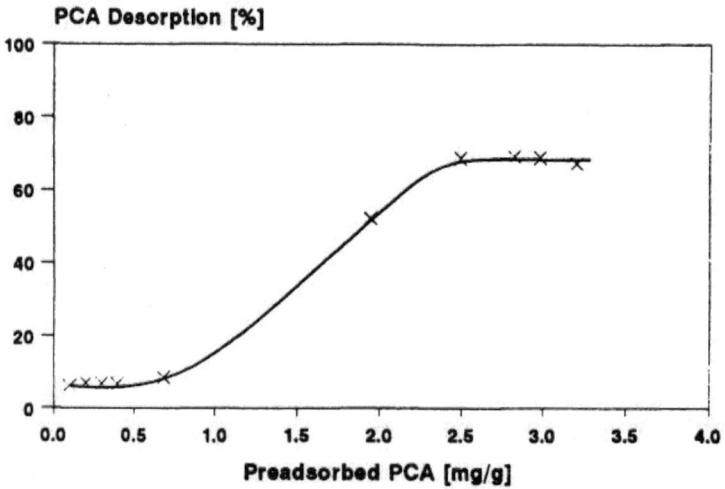

FIG. 10 Desorption percentage of [¹⁴C] acrylic-maleic acid copolymers as a function of preadsorbed PCA copolymer on Na-kaolinite with a fixed concentration of triphosphate (0.15 Eq/liter of STP) at pH 4.5 (see other conditions in Fig. 8) [45,46].

is followed by a rise in the percentage of desorption up to a constant value of about 70% under surface saturation conditions, as shown in Fig. 8.

The initial, almost nondesorbing region indicates that the PCA copolymer is strongly bound to the edge surfaces and is only desorbable under very drastic conditions. It can be assumed that the adsorbed PCA copolymer adopts a completely flat configuration exclusively formed by train segments. The observed desorption in the intermediate and final regions of the adsorption isotherm confirms that adsorbed polymer layers containing more loops and tails are formed at higher equilibrium concentrations. The relatively fewer adsorbed segments per copolymer molecule in this region could thus be displaced more easily by strong competing displacers than in the flat configuration state. A simple ligand exchange mechanism has been used to measure the affinity of the PCA copolymer at the low and high concentration extremes of the adsorption isotherm [45,46].

VII. CONCLUSION

Under environmental conditions, it can be concluded that the active carboxyl groups from PCA in the aqueous phase or at the surface of metal oxide surfaces have a high tendency to interact with multivalent cations, which promote immobilization through precipitation and adsorption. The relative ratio of carboxyl groups from PCA to metallic ions in the free state or partially engaged in a crystalline structure at surface binding sites is thus a crucial parameter in these mechanisms.

With respect to the expected trace amounts of synthetic PCA, a favorable concentration excess of "active" metallic ions and sites for the immobilization processes is always encountered in various environmental matrices [1,2].

Other naturally occurring anions, such as small organic acids or humic and fulvic acids of higher molecular weights, may also compete in the transport of PCA by blocking surface binding sites and lowering the free metal concentration through complexation. However, lysimeter studies [1] show that PCA is strongly immobilized in the upper soil layers, which restricts the potential influence of these competing anions.

REFERENCES

1. H.-J. Opgenorth, in *The Handbook of Environmental Chemistry* (O. Hutzinger, ed.), Springer-Verlag, Berlin, 1992, pp. 337–350.
2. M.B. Freeman, and T.M. Bender, Environ. Technol. *14*; 101–112 (1993).
3. B.P. Warkentin and R.D. Miller, Soil Sci. *85*; 14–18 (1958).
4. B.K.G. Theng, Clays Clay Min. *30*; 1–10 (1982).
5. B.K.G. Theng, in *Formation and Properties of Clay-Polymer Complexes*, Elsevier, Amsterdam, 1979.

6.   G. Lagaly, in *Surfactant Sciences Series* (B. Dobiás, ed.), Vol. 47, Dekker, New York, 1993, pp. 427–494.
7.   W. Stumm, Colloids Surfaces A: Physicochem. Eng. Aspects *73*; 1–18 (1993).
8.   C.F. Anderson and H. Morawetz, in *Encyclopedia of Chemical Technology* (Kirk-Othmer, ed.), John Wiley & Sons, New York, 1982, pp. 495–530.
9.   F.T. Wall and J.W. Drenan, J. Polym. Sci. *7*; 83–88 (1951).
10. K. Huber, J. Phys. Chem. *97*; 9825–9830 (1993).
11. A.A. Denio, J. Chem. Educ. *57*; 272–275 (1980).
12. P.W. Schindler and W. Stumm, in *Aquatic Surface Chemistry* (W. Stumm, ed.), Wiley-Interscience, New York, 1987, pp. 83–110.
13. J. Westall and H. Hohl, Adv. Colloid Interface Sci. *12*; 265–294 (1980).
14. L.K. Koopal, in *Surfactant Sciences Series*, (B. Dobiás, ed.), Vol. 47, Dekker, New York, 1993, pp. 101–207.
15. G. Sposito, in *The Chemistry of Soils*, Oxford University Press, New York, 1989, pp. 127–147.
16. E. Wieland and W. Stumm, Geochim. Cosmochim. Acta *56*; 3339–3355 (1992).
17. M.A. Cohen Stuart, G.J. Fleer, J. Lyklema, W. Norde, and J.M.H.M. Scheutjens, Adv. Colloid Interface Sci. *34*; 477–535 (1991).
18. G.J. Fleer, in *Surfactant Sciences Series* (P. Somasundaran and B.M. Moudgil, eds.), Vol. 27, Dekker, New York, 1988, pp. 105–158.
19. S.V. Krishnan and Y.A. Attia, in *Surfactant Sciences Series*, (P. Somasundaran and B.M. Moudgil, eds.), Vol. 27, Dekker, New York, 1988, pp. 485–518.
20. M.E. Lewellyn and P.V. Avotins, in *Surfactant Sciences Series* (P. Somasundaran and B.M. Moudgil, eds.), Vol. 27, Dekker, New York, 1988, pp. 559–578.
21. G. Sposito, in *The Chemistry of Soils*, Oxford University Press, New York, 1989, pp. 42–65.
22. G.R. Joppien, J. Phys. Chem. *82*; 2210–2215 (1978).
23. J. Drzymala and D.W. Fuerstenau, in *Flocculation in Biotechnology and Separation Systems* (Y.A. Attia, ed.), Elsevier, Amsterdam, 1987, pp. 45–60.
24. J.E. Gebhardt and D.W. Fuerstenau, Colloids Surfaces *7*; 221–231 (1983).
25. K.B. Musabekov, K.I. Omarova, and A.I. Izimov, Acta Phys. Chim. Szeged *29*; 89–101 (1982).
26. N. Shashidhar, J.R. Varner, and R.A. Condrate, Sr., Cer. Transact. Westerville *17*; 443–450 (1990).
27. K.F. Tjipangandjara, Y.-B. Huang, P. Somasundaran, and J. Turro, Colloids Surfaces *44*; 229–236 (1990).
28. L. Dupont, A. Foissy, R. Mercier, and B. Mottet, J. Colloid Interface Sci. *161*; 455–464 (1993).
29. E. Ringenbach, G. Chauveteau, and E. Pefferkorn, J. Colloid Interface Sci. *161*; 223–231 (1993).
30. M.K. Nagarajan, JAOCS *62*; 949–955 (1985).
31. C.L. Tiller and C.R. O'Melia, Colloids Surfaces A: Physicochem. Eng. Aspects *73*; 89–102 (1993).
32. A.S. Michaels and O. Morelos, Ind. Eng. Chem. *47*; 1801–1809 (1955).
33. J.I. Bidwell, W.B. Jepson, and G.L. Toms, Clay Minerals *8*; 445–459 (1970).

34. N.V. Sastry, J.-M. Séquaris, and M.J. Schwuger, J. Colloid Interface Sci. *171*; 224–233 (1995).

35. W.W. Emerson, Nature *176*; 461 (1955).

36. B. Siffert and P. Espinasse, Clays Clay Minerals *28*; 381–387 (1980).

37. W. Flaig and H. Söchtig, Z. Pflanzenernahr. Dung. Bodenkunde *87*; 44–57 (1959).

38. M.K. Nagarajan, Tenside Surf. Det. *28*; 230–234 (1991).

39. I.H. Joyce and W.E. Worrall, Trans. Br. Ceram. Soc. *69*; 211–216 (1970).

40. F. Miano and M.R. Rabaioli, Colloids Surfaces A: Physicochem. Eng. Aspects *84*; 229–237 (1994).

41. L. Järnström and P. Stenius, Colloids Surfaces *50*; 47–73 (1990).

42. P. Stenius, L. Järnström, and M. Rigdahl, Colloids Surfaces *51*; 219–238 (1990).

43. J.M. Berg, P.M. Claesson, and R.D. Neuman, J. Colloid Interface Sci. *161*; 182–189 (1993).

44. F. Blockhaus, J.-M. Séquaris, and M.J. Schwuger, Tenside Surf. Det. *28*; 447–451 (1991).

45. F. Blockhaus, Ph.D. Thesis, Univ. Düsseldorf (1996).

46. (a) F. Blockhaus, J.-M. Séquaris, H.D. Narres, and M.J. Schwuger, Prog. Colloid Polymer Sci. *100*; (1996). (b) F. Blockhaus, J.-M. Séquaris, H.D. Narres, and M.J. Schwuger, submitted manuscript.

47. S. Nagarajah, A.M. Posner, and J.P. Quirk, Nature *225*; 83–85 (1970).

48. U. Kafkafi, B. Bar-Yosef, R. Rosenberg, and G. Sposito, Soil Sci. Soc. Am. J. *52*; 1585–1589 (1988).

49. P.J. Dodson and P. Somasundaran, J. Colloid Interface Sci. *97*; 481–487 (1984).

8

Biological and Physicochemical Aspects of Polycarboxylate Behavior in the Environment

INGE LANGBEIN BASF AG, Ludwigshafen, Germany

I. PRACTICAL RESULTS

The use of polycarboxylates in detergents only began to play a major role in the early 1980s. Despite the relatively short application period, there are already a number of publications and reviews on environmental data for polycarboxylates [1–5]. Publications suggesting that the environmental behavior of polycarboxy-lates has not been sufficiently studied [6,7] can be refuted today in many points [2,3,8]. Extensive studies were recently carried out by manufacturers and users documenting the environmental behavior of polycarboxylates.

The polycarboxylates exclusively used in detergents today are linear, water-soluble homopolymers of acrylic acid and copolymers of acrylic and maleic acid of different molar mass.

Studies with homopolymers (4500 M_w) show that they are not adsorbed on the material in the washing liquor under normal conditions, such as high pH range and short exposure time [9], so that they are almost completely passed into the waste water after use. It therefore appears meaningful also to discuss the results for environmental behavior according to input and possible propagation paths:

Waste water
Surface waters
Soil
Drinking water treatment
The environment

II. BEHAVIOR OF POLYCARBOXYLATES IN WASTE WATER

In highly industrialized countries, waste water, from households, industry, or agriculture, is generally purified in biological sewage plants. When this is not the case, solids and, in part, also suspended matter are removed in settling pits or mechanical sewage plants and the effluents are then discharged into waters. Polycarboxylates can be eliminated from waste water by precipitation of an insoluble calcium polymer salt. Studies of a copolymer with an average molar mass of 70,000 have shown, for example, that an elimination rate of more than 80% was achieved by sedimentation of the sparingly soluble calcium polymer salt after about 4 days in raw waste water [10]. These results show that good elimination can also be expected for polycarboxylates in settling pits as a result of long residence times. In contrast, the elimination rates in mechanical sewage plants are dependent on both the residence time and the extent to which oppor-tunities for sedimentation were available in the sewage system.

Conditions similar to those in mechanical sewage plants are given in preclar-ification for biological waste water purification in which settlability also plays an important role, in addition to molecular weight and solubility. This was shown by results in a model plant for the simulation of pre-clarification. With a residence time of 2 h, 8–29% of the polycarboxylates (homopolymer, 4500, copolymers, 12,000 and 70,000 M_w) was eliminated in a settling tank [4]. These results illustrate the dependence of the sedimentation rate on the residence time and also show the differences expected between mechanical sewage plants and settling pits.

A. Sewage Plant Conditions

In biological waste water treatment, copolymers (70,000 M_w) are mineralized up to a maximum of 10% [11,12] and investigations of homopolymers (4500 M_w) yielded maximum values around 15% [4,9]. These values vary within specified bounds and are dependent on the test system used. The degradation values found in respirometric tests, such as determination of the $BOD_5/ThOD$ [13], the BOD_{30}/COD [11,14], the modified MITI test (I) [15,16], and the "closed bottle test" [17,18] are just as low as, for example, in SCAS tests in which higher bacterial cell densities are applied. On the whole, the investigations available for biodegradation show that the values determined are generally below 20%, irrespective of the test procedure. An exception are homopolymers of low molar mass (1000 M_w), which show biodegradation rates of up to 43% [4]. This supports the assumption that only short-chain, low molecular weight fractions are biodegraded.

Studies of the elimination of polycarboxylates under sewage plant conditions were carried out in various test systems [4,11,12,19–21]. In batch tests, similar to the Zahn-Wellens test, up to 90% and more [11] of the copolymer (70,000 M_w) was bound on activated sludge after about 2–3 of contact in a molar excess of calcium, and the balance was biodegraded. Homopolymers show different elimination behaviors under these test conditions. Some are not eliminated, others almost completely with activated sludge [15,16,22,23]. The elimination behavior continuously increases with growing chain length. Similar results were also obtained in semicontinuous testing (SCAS test; Table 1). The high elimination rate of the homopolymer of $M_w = 1000$ mainly results from biodegradation.

In continuous testing [4,9,24] (see also Ref. 43), that is, in sewage plant simulation tests, higher elimination rates are generally also achieved with increasing molecular weight (Table 2). For the 70,000 M_w copolymer it was possible to balance the behavior under sewage plant conditions completely by ^{14}C radiolabeling [21]. The copolymer was added as a constant batch and as a pulsed batch (Table 3). In the outlet, 2–3% of the radioactivity was found, so that a total elimination of 97–98% may be assumed here. Similar results are obtained for low molecular weight homopolymers (4500 M_w) if iron salts are added to the waste water, as in phosphate precipitation [4].

TABLE 1 Elimination of Polycarboxylates as the Sum of Adsorption/Precipitation and Biodegradation

Polycarboxylates	Elimination (%)	Reference
Homopolymers (MW)	45	Hennes, 1991 [4]
1000		
2000	21	
4500	40	
10,000	58	
60,000	93	
Copolymers (MW)		
12,000	83	
Homopolymers (MW)	40	Unilever, 1989 [40]
9400		
23,000	48	
111,000	81	
152,000	95	
215,000	95	
Copolymers (MW)		
50,000	98	
Copolymers (MW)	94–99	Opgenorth, 1987 [19]
70,000		Unilever, 1989 [40]
		Hennes, 1991 [4]

In summary, the results from screening and simulation tests suggest that polycarboxylates in biological waste water treatment are mainly eliminated from the waste water by adsorption/precipitation with the sewage sludge. Nonadsorbed, low molecular weight polycarboxylates are eliminated from the waste water by biodegradation, forming biomass, carbon dioxide, and water.

B. Anaerobic Conditions

To examine the biodegradability under anaerobic conditions, radiolabeled polycarboxylate was added to a mixture of digesting sludge and nutrient solution. The results for biodegradability were between 11 and 16% [11]. Studies in a laboratory digester did not show further biodegradation of the adsorbed polycarboxylate during sludge digestion. In other experiments under anaerobic conditions, about 80% of the radiolabeled carbon was adsorbed on the sludge phase and about 3% was mineralized, identified as radiolabeled CO_2 [12]; that is, polycarboxylates are not more easily biodegradable under anaerobic conditions than under aerobic conditions.

TABLE 2 Elimination of Polycarboxylates Under Continuous Test Conditions as a Function of Molecular Weight

Polycarboxylates	Elimination (%)	Reference
Homopolymers (MW)		
1000	9–24	Hennes, 1991 [4]
2000	13–18	Hennes, 1991 [4]
4500[a]	16–27	Hennes, 1991 [4]
	75	Rohm & Haas, 1991 [43]
	>80	Freemann and Bender, 1992 [9]
Copolymers (MW)		
12,000	70–80	Hennes, 1991 [4]
70,000	82–93	Hennes, 1991 [4]
	94	Opgenorth, 1987 [19]

[a]Differences for the 4500 MW homopolymer result from different initial concentrations.

TABLE 3 Fate of the 70,000 (MW) Copolymer in Biological Waste Water Treatment

Fate	Pulsed dosage (%)	Continuous dosage (%)
Adsorption	~96	>90
CO_2 formation	<2	~5
Outlet	<3	2

C. Influence of Ultraviolet Pretreatment

Investigations in which polycarboxylates were subjected to ultraviolet (UV) pretreatment before their introduction into the laboratory sewage plant showed formation rates of radiolabeled CO_2 above 67%. Further investigations revealed, however, that the major fraction of about 60% was attributable to direct oxidation of polycarboxylate in the UV irradiation facility, not to biological oxidation [12,21]. The required UV energies are many times higher (factor of 1800) than those of the UV treatment applied for bacterial reduction in the outlet of sewage plants.

D. Influence of Polycarboxylates on Microorganisms

At the concentrations at which polycarboxylates are present in practice, they do not disturb the activity of bacteria in activated sludges [4,11]. Additional studies have shown that neither the settling behavior of sludges during pre-clarification and final clarification nor the dehydration of excess sludges nor simultaneous phosphate precipitation is influenced by the presence of polycarboxylates [4,11,19]. However, even outside sewage plants they influence only the growth of microorganisms at high concentrations many times above environmentally relevant concentrations [4,19,24].

E. Behavior of Polycarboxylates in the Presence of Heavy Metals

The issue of heavy metal remobilization has been studied both with activated sludge and sediment-containing river water and also with soil.

In shaking experiments with activated sludge, various polycarboxylates (homopolymers 1000, 2000, and 4500 M_w; copolymers 12,000 and 70,000 M_w) were added in batch tests at concentrations of 0.3–100 mg/liter, and the influence on Cd, Cu, Ni, Pb, and Zn was studied [4]. The results show that the copolymer with a molar mass of 70,000 does not influence the elimination behavior of these heavy metals at any concentration. An addition of 10 mg/liter each of the homopolymers of 1000 and 2000 M_w increased the Cu concentration. The addition of 100 mg/liter of these two homopolymers and of the homopolymer with 4500 M_w caused an increase in Zn concentration and in that of dissolved Cu. The addition of 100 mg/liter of the copolymer with a molar mass of 12,000 showed an increase in Cu concentration (Table 4).

It was furthermore examined in two semitechnical sewage plants, both operated with the same raw waste water, whether the elimination of heavy metals is impaired by adding 10 mg/liter of a copolymer (70,000 M_w) [19]. The copolymer was alternately fed into the two sewage plants over a period of several weeks. No influence on the elimination behavior of Cr, Cu, Ni, Pb, and Zn with the activated sludge was found when adding the copolymer to the waste water.

Other investigations [4] concerning the remobilization behavior of heavy metals in river sediments under the influence of polycarboxylates were carried

TABLE 4 Polycarboxylate Concentration (mg/liter) Without Significant Effects on the Remobilization of Heavy Metals from Activated Sludge

Heavy metal	Homopolymer (MW)			Copolymer (MW)	
	1000	2000	4500	12,000	70,000
Cd	100	100	100	100	100
Cu	10	10	30	30	100
Ni	100	100	100	100	100
Pb	100	100	100	100	100
Zn	30	30	30	100	100

TABLE 5 Polycarboxylate Concentration (mg/liter) Without
Significant Effects on the Remobilization of Heavy Metals
from Sediments

Heavy metal	Homopolymer (MW)			Copolymer (MW)	
	1000	2000	4500	12,000	70,000
Cd	30	30	30	30	30
Cu	30	30	30	30	30
Ni	30	30	30	30	30
Pb	30	30	30	30	30
Zn	30	10	10	30	30

out similarly to the shaking tests with activated sludge, adding various polycarboxylates at concentrations from 0.1 to 30 mg/liter. No remobilization of heavy
metals from sediments took place up to a concentration of 10 mg/liter. An
increase in the Zn concentration was found when adding 30 mg/liter of the
homopolymers of 2000 and 4500 M_w (Table 5).

In another test series [19], soil material was suspended, loaded with the salts
of the metals Cd, Cr, Cu, Hg, Ni, and Pb, and mixed with 10 and 100 mg/liter
of copolymer (70,000 M_w) each. The addition of 10 mg/liter of copolymer did not
show differences from the control test. Slightly increased concentrations compared with the control batch were found for Cu and Cr when adding 100 mg/liter
of copolymer.

All investigations show that polycarboxylates in environmentally relevant
concentrations (see point 5) neither influence the heavy metal elimination in
sewage plants nor produce an effect on the remobilization of heavy metals from
sewage sludge, soil, or river sediments.

III. BEHAVIOR OF POLYCARBOXYLATES IN SURFACE WATERS

The results obtained for the behavior of polycarboxylates from detergents under
sewage treatment conditions show that the higher molecular fractions are eliminated by adsorption on the sewage sludge and the low molecular weight balance
is biodegraded. For this reason, only traces of synthetic polycarboxylates can be
present in surface waters. This is confirmed by examination of surface water
samples using highly sensitive polyelectrolyte titration [25], which reveals a
fraction of <0.010 mg/liter for synthetic polymers. A selective separation of
synthetic from natural polyanions is currently possible only with great effort (see
preceding chapters), so only isolated values are available at present.

Should any polycarboxylates be passed directly into surface waters, they would be subjected to elimination processes by adsorption/precipitation and biodegradation similar to those in sewage treatment plants.

A number of studies are available on the acute fish and daphnia toxicity of polycarboxylates. A survey of the aquatic toxicity of polycarboxylates is given in Table 6. In most cases, no adverse effect on aquatic organisms was found even for the highest concentrations tested.

Investigations of the chronic and subchronic toxicity of algae showed growth inhibition at concentrations of 180 mg/liter for homopolymers (4500 M_w) [4]. Very different results were obtained for copolymers (70,000 M_w) for which the EC_{10} values ranged between 32 mg/liter [21] and 200 mg/liter [19].

Similar situations are observed in daphnia reproduction tests. For example, Rohm and Haas [27] found an NOEC value of 12 mg/liter, Freeman and Bender [9] an NOEC of 5.6 mg/liter, and Hennes [4] an NOEC value of 450 mg/liter for a homopolymer (4500 M_w). Studies with higher molecular weight polycarboxylates (70,000 M_w) also showed a very wide range of NOEC values from 350 mg/liter [4] through 6.2 mg/liter [19] to 1.3 mg/liter [21]. This is because the copolymer (70,000 M_w) is precipitated as an insoluble calcium polymer salt in excessive water hardness in the test systems, that is, at low polycarboxylate concentrations. The assumption that the result is definitively based on the precipitation of calcium polymer salt, not on typically toxic properties of the polycarboxylate, was confirmed by the fact that no negative effects were found over the entire concentration range up to 400 mg/liter when the test solutions, after allowing them to stand for 24 h, were filtered before introducing the daphniae [19]. Under environmentally relevant conditions these test results are therefore of restricted significance, because polycarboxylate is preferentially adsorbed on solid particles and is thus not freely available in waters.

Despite the variations, it can be stated that polycarboxylates in environmentally relevant concentrations, that is, concentrations applicable in practice, do not show any acute or chronically toxic effects on algae, daphniae, or fish. Because polycarboxylates consist exclusively of carbon, hydrogen and oxygen, any growth-stimulating effect toward a eutrophication reaction is also highly improbable. Moreover, according to Zitko [29], only those substances can pass through biological membranes whose molar mass is <600, so that the bioaccumulation of polycarboxylates from detergents through algae is not expected.

IV. BEHAVIOR OF POLYCARBOXYLATES IN SOIL

In biological wastewater treatment, polycarboxylates are eliminated up to 95% by adsorption on sewage sludge. They even remain bound to the sludge after digestion. Sewage sludge disposal is normally as follows:

TABLE 6 Aquatic Toxicity of Polycarboxylates

Test system/test organism	Test substance					
	Homopolymer (MW)				Copolymer (MW)	
	1000	2000	4500	78,000	12,000	70,000
Acute toxicity						
Fish LC$_{50}$, 96 h zebra fish	>200 [4]	>200 [4]	>200 [4]	>1000 [24]	>200 [4]	>100 [4]
Golden orfe				1590 [39]		>1000 [40] >200^a [19]
Daphnia EC$_{50}$, 48 h	>200 [4] ≥1000 [44]	>200 [4]	>200 [4] >1000 [45] 21,9	750 (24 h) [24]	>200 [4]	>100 [4] >200 [4] 908 [21]
Subchronic/chronic toxicity						
Fish NOEC, 14 days zebra fish						40^a [19]
Eggs and larvae of zebra fish NOEC, 28 days			1000^a [5] 450 [4]			
Young zebra fish NOEC, 21 days						40^a [19]
Daphnia NOEC, 21 days (reproduction test)			5.6 [9] 12.0 [27] 450.0 [4]	>400 [24]		1.3 [21] 6.2 [19] 350.0 [4]
Algae EC$_{10}$, 96 h			180 [4]	82 (4–14 days) >1000 (30 days) [24]		32 [21] ≥200 [4]

^{a}Maximum tested concentration.

Thermal disposal in incinerator plants
Dumping on a suitable landfill
Use in agriculture

Both the thermal disposal and the dumping of polycarboxylate-containing sewage sludge pose no relevant environmental problem. A direct introduction of
polycarboxylates into the environment, that is, into soil, is caused only by the
agricultural use of sewage sludges. The biodegradation results from radiolabeled
investigations are similar to those for biological wastewater purification [4]. The
highest mineralization rates (35%) were found for polymers with a molecular
weight of about 1000. The mineralization rates decrease with increasing molecular weight for homopolymers (4–11%). Radiolabeled long-term tests with
higher molecular copolymers (70,000 M_w) revealed low mineralization values of
4–7% [30]. On the whole, polycarboxylates adsorbed on solids are biodegraded
to only a minor extent by soil organisms within reasonable periods of time.

This high adsorption tendency (see preceding section) also has the effect that
the migration potential of polycarboxylates in soil is very low despite rain or
artificial irrigation. This was demonstrated in lysimeter experiments with radiolabeled polycarboxylates. For this purpose, soils of different compositions were
mixed with homopolymers (4500 M_w) and copolymers (12,000 and 70,000 M_w),
and 84–93% of the polymers examined was retained in the first 15 mm of the
100 mm thick soil layer [4]. The remaining 7–14% was found as a mobile
fraction in the lysimeter eluate and represents the low molecular weight biodegradable balance. This mobile fraction was not present in the lysimeter eluate
from sewage sludge containing adsorbed copolymer (70,000 M_w), because it had
already been eliminated by biodegradation in the sewage plant [11]. The higher
molecular weight fraction of the copolymer remained completely in the soil layer
in which it was applied.

In connection with the agricultural use of sewage sludge containing strongly
bound polycarboxylates, any rinsing of polycarboxylates by rain or irrigation
into deeper layers or groundwater is not expected.

A. Effect on Plant Growth

To study the effect on higher plants, homopolymers (4500 M_w) were added to soil
at concentrations of up to 225 mg/kg. Up to this concentration, no adverse effect
on germination, growth, or crop yield was observed for maize, soybeans, wheat,
or grass [4]. Similar results were obtained for a homopolymer of 78,000 M_w, for
which an NOEC of 1000 mg/kg was determined for the growth of beet seed [24].
No adverse effect by copolymers (70,000 M_w) was observed in a growth test with
oat up to the highest concentration of 400 mg/kg soil tested [19].

B. Effect on Soil Organisms

Behavior with respect to soil organisms was studied for the earthworm (*Eisenia foetida foetida*) with homopolymers (4500 and 78,000 M_w) and copolymers (70,000 M_w). The values for acute toxicity ranged between 1000 and >1600 mg/kg soil. In toxicity tests of homopolymers (4500 M_w) with bloodworms (*Chironomus riparius*), no adverse effects were observed up to the highest concentrations of 10,000 mg/kg soil tested [4].

V. BEHAVIOR OF POLYCARBOXYLATES IN DRINKING WATER TREATMENT

Drinking water treatment processes, such as bank filtration, groundwater replenishing, and filtration over sand and activated carbon, are likely to eliminate small residual quantities of polycarboxylates from raw water because polycarboxylates are adsorbed on solids. Studies with homopolymers (4500 M_w) and copolymers (12,000 and 70,000 M_w) have shown that coagulation with alum and ferric salts can also lead to elimination values above 90% [4].

VI. ASSESSMENT OF THE ENVIRONMENTAL CONCENTRATIONS OF POLYCARBOXYLATES

Investigation results for concentrations of synthetic polycarboxylates in the environment are currently available only in isolated cases. This is because of the lack of a selective detection method. As described in the preceding chapters, intensive work is devoted to the establishment of a selective detection method.

In its comment on the environmental compatibility of polycarboxylates [2] the phosphates and water central committee compiled an estimate of the concentrations of copolymers (70,000 M_w) to be expected (Table 7). This was based on the conditions in Germany at the beginning of 1990, and the concentrations were determined purely computationally according to input and propagation paths.

Based on the input quantity of polycarboxylates in detergents and determining a specific sewage flow of about 300 liters per inhabitant and per day, concentrations of about 3 mg/liter are expected in municipal sewage. Considering the wastewater treatment techniques in Germany, an average degree of elimination of about 91% is assumed, so that a calculated polycarboxylate concentration of about 0.3 mg/liter can occur in purified waste water. If the purified waste water is discharged into the waters, a dilution of at least 1:5 will occur, so that a maximum polycarboxylate concentration of 0.05 mg/liter would be expected in surface waters. However, most rivers carry significantly more water, so that much lower concentrations normally result. River water investigations by Schroeder and Horn [25] show polymer contents below the detection limit of

TABLE 7 Estimate of the Environmental Concentration of Polycarboxylates from Detergents for Germany

	Concentration	
Source	mg/liter	g/kg
Raw waste water	2.9	
Digested sewage sludge		14.2
Soil after sewage sludge application[a]		0.013–0.027
Purified waste water	0.26	
Highly polluted water (dilution 1:5)	0.05	
Drinking water	0.005	

[a]After 3 year application of the maximum quantity legally permissible in Germany on agriculturally used soils.

0.010 mg/liter. These results demonstrate that the purely computational determination of polycarboxylate concentrations is very generous and that the concentrations actually present are obviously much lower.

The content of synthetic polycarboxylates in drinking water is estimated on the basis of the concentration of 0.05 mg/liter assumed in surface waters. As already described, synthetic polycarboxylates can probably also be eliminated by drinking water treatment techniques and by ferric iron and alum coagulation. The content of synthetic polycarboxylates would then amount to about 0.005 mg/liter as a very conservative estimate. In contrast to surface waters, a selective determination of synthetic polycarboxylates is possible in drinking water [25]. The investigation of drinking waters of different origin has not revealed the presence of synthetic polycarboxylates to date.

Investigation results for wastewater treatment have shown that copolymers are separated with activated sludge, except for a small biodegradable fraction. Digested sewage sludge therefore contains about 1.4% polycarboxylates relative to the sewage sludge dry weight. Because of the agricultural use of sewage sludge, polycarboxylates are introduced into soil and the environment. In many countries, fertilization with sewage sludge is limited in terms of time and volume as a result of possible heavy metal accumulation in sewage sludges. According to German law, the maximum permissible quantity of sewage sludge may be applied for only 3 years the same soil followed by a pause of several years. Depending on soil conditions, a polycarboxylate concentration between 0.013 and 0.027 g/kg soil can be reached within 3 years.

Estimating the bioconcentration potential, log P_{ow}, is only possible for substances with a molar mass <600 [32] and nonionic compounds [33]. Moreover, as already described, it is only possible for substances with a molar mass below 600 to pass biological membranes, and bioaccumulation is not expected because of the high solubility in water and the capability of precipitating as insoluble calcium salt in water with excess calcium.

VII. STUDIES OF THE TOXICOLOGICAL BEHAVIOR OF POLYCARBOXYLATES

There are a number of summary reports of toxicological behavior [2–4,11, 19,34,35], which on the whole show that the salts of linear polyacrylic acids are toxicologically harmless.

Acute oral toxicity in rats is generally very low [28,38]. Subchronic studies show that high doses are tolerated without problems [5]. In dermal application, neither the results of acute [9,28,38] nor those of subchronic studies [19,28,34] in animal experiments showed irritation or signs of accumulation in the skin. Patch tests in humans did not reveal any irritant effects [34]. Studies of subchronic toxicity involving the inhalation of polycarboxylate aerosols showed slight, reversible lung irritation in animal experiments [9,41]. Numerous results are available from animal experiments on sensitization by polycarboxylates [19,20,37,42]. In no case was a sensitizing potential detected. Patch tests in humans with different sodium polyacrylates did not show sensitizing reactions [34]. Polycarboxylates were tested for mutagenic and teratogenic behavior in various in vitro and in vivo systems [9,19,26,34,36,46], and the results did not indicate any genetic changes or malformations.

REFERENCES

1. P. Berth and P. Krings, Einfluss der PHöchstMengV auf Waschmittel, in Kompendium-Auswirkungen der Phosphathöchstmengenverordnung für Waschmittel auf Kläranlagen und in Gewässern, A. Hamm, (ed.), Academia Verlag, Richarz GmbH, St. Augustin, 1989, pp. 69–79.
2. Comment by the central committee "Phosphates and Water" of the Study Group on Water Chemistry within the Society of German Chemists concerning the evaluation of the environmental compatibility of polycarboxylates from detergents, letter from the Study Group on Water Chemistry dated 2/2/90 to the Federal Ministry for the Environment, Nature Conservation and Nuclear Safety.
3. G. Chiaudani and P. Poltronieri, Study on the environmental compatibility of polycarboxylates used in detergent formulations, Ingegneria Ambientale *11*, 1–43 (1990).
4. E.C. Hennes, Fate and effects of polycarboxylates in the environment, Procter & Gamble, B. Strombeek-Beever, 1991.
5. H.J. Opgenorth, Polymeric materials polycarboxylates, in *Handbook of Environmental Chemistry*, Vol. 3, Part F, *Detergent Chemicals*, (O. Hutzinger, ed.), Springer, Berlin, 1992, pp. 337–350.
6. W. Giger and T. Conrad, Wasser Berlin '85, Wissenschaftsverlag Berlin, 1985, pp. 362–377.
7. M. Hunter, D.M.L. da Motta Marques, J.M. Lester, and R. Perry, Environ. Technol. Lett., *9*, 1–22 (1988).
8. ECETOC Report, final draft, Joint assessment of commodity chemicals No. X, Polycarboxylate polymers as used in detergents, ISSN-07773-6339-xx, Brussels, 1993.

9. M.B. Freeman and T.M. Bender, An environmental fate and safety assessment for a low molecular weight polyacrylate detergent additive, Environ. Technol. *14*, 101–112 (1992).

10. BASF, study report, unpublished.

11. H.-J. Opgenorth, Münchener Beitr. Abwasser-, Fisch- Flussbiologie, *43*, 338–351 (1990).

12. H. Schumann and S. Kunst, GWF Wasser Abwasser, *132*, 376–383 (1991).

13. V. Abe, S. Matsumura, H. Yajima, R. Suzuki, and Y. Masago, Yakugaku, *33*, 228 (1984).

14. G. Jakobi, Angew. Makromol. Chem. *123/124*, 119–145 (1984).

15. W. Schefer, and K. Romanin, Textilverdlung *23*, 340–344 (1988).

16. W. Schefer and K. Romanin, Gas, Wasser, Abwasser *69*, 131–136 (1989).

17. W.K. Fischer, P. Gerike, and R. Schmidt, Wasser-Abwasserf. *7*, 99 (1974).

18. W.K. Fischer, Tenside Detergents *12*, 53 (1975).

19. H.-J. Opgenorth, Tenside Detergents *24*, 366–369 (1987).

20. Henkel, Experimental toxicology of Degapas 4104, personal communication by J. Steber to M. Richold, 25.04.90, Henkel, Düsseldorf, quoted in ECETOC Report, final draft, Joint assessment of commodity chemicals No. X, Polycarboxylate polymers as used in detergents, ISSN-07773-6339-xx, Brussels, 1993.

21. H. Schumann, VDI Verlag, Progress Reports, Series 15, Umwelttechnik No. 81, ISBN 3-18-148115-7, 1990.

22. W. Schefer, Textilveredlung *17*, 541 (1982).

23. W. Schefer, Seifen-Ole-Fette-Wachse *109*, 423 (1983).

24. Henkel Laboratory, Ecological data, polyacrylic acid (PAA), MW 78.000 (revised summary 2/4/92), Degussa, Hanau, 1987.

25. U. Schroeder and D. Horn, Studies in Polymer Science, Proceedings of the P&G UERP Symposium (I. Noda and D.N. Rubingh, eds.), Elsevier, 1992.

26. M. Ishidate, Jr., T. Sofuni, K. Yoshikawa, M. Hayashi, T. Nohmi, M. Sawada, and A. Matsuoka, Food Chem. Toxicol. *22*, 623 (1984).

27. Rohm and Haas, Chronic toxicity of Acusol TM 445N to the daphnid, *Daphnia magna*, Rohm and Haas, Spring House, PA, 1991.

28. BASF, unpublished results.

29. V. Zitko, Uptake and excretion of chemicals by aquatic fauna, in *Exotoxicology and the Aquatic Environment* (P.M. Stokes, ed.), Pergamon, New York, 1981.

30. H.-J. Opgenorth, Umweltverträglichkeit von Wasch- und Reinigungsmitteln, Polycarboxylate in Abwasser und Klärschlamm, 44, Abwasserbiologischer Fortbildungskurs, October 1989, BASF, Ludwigshafen, 1989, pp. 1–15.

31. Rohm and Haas, Acute toxicity of Acusol TM 445N to the earthworm, *Eisenia foetida foetida*, Envirosystems, Rohm and Haas, Spring House, PA, 1991.

32. G.D. Veith and P. Kosian, Estimating bioconcentration potential from octanol/water partition coefficients, in Reference manual for quantitative structure activity relationships (QASR's) and other useful relationships used in PMN assessment (R.G. Clements, ed.), Environmental Effects Branch, U.S. EPA, 1988.

33. OECD, Guideline for testing of chemicals, Partition coefficient (*n*-octanol/water), May 12, 1981, OECD, Paris.

34. BIBRA (British Industrial Biological Research Association), Toxicity profile, poly-acrylic acid and its sodium salt, BIBRA, Carshalton, Surrey, 1987.

35. NN, acrylic acid, methyl acrylate, ethyl acrylate and polyacrylic acid, in IARC Monographs on the evaluation of the carcinogenic risk of chemicals to humans, Vol. 19, February 1979.

36. E.D. Thompson, M.J. Aardema, and R.A. LeBoeuf, Lack of genotoxicity with acrylate polymers in five short-term mutagenicity assays, Environ. Mol. Mutag. *14*, 98–106 (1989).

37. Unilever, Summaries of toxicity studies: the mineral status of rats fed polyanions for 4 weeks, rep. PES 88 1031; absorption and metabolism of polyacrylic acid phosphinate (14C) DKW 125 in the rat, rep. AM 85.04, Unilever, Environmental Safety, Sharnbrook, Bedford, UK, 1990.

38. Rohm and Haas, Toxicity report, Acute oral, dermal, skin and eye, range finding report, Rohm and Haas, Spring House, PA, 1982.

39. Degussa, Degapas 4104 N, Bestimmung der Fischtoxizität an Goldorfen (LC$_{50}$), personal communication by E. Roth, 4/12/85, with appendix (D.M.M. Adema, ed.), 1985. The acute toxicity of Degapas 4104 N (30%) to *Leuciscus idus* (L.) (Gold-orfe), TNO report R85/073. Unpublished report US-IT No. 85-0039-DKO. Degussa, Hanau; quoted in: ECETOC Report, final draft, Joint assessment of commodity chemicals No. X, Polycarboxylate polymers as used in detergents, ISSN-07773-6339-xx, Brussels, 1993.

40. Unilever, Personal communication by P.A. Gilbert to E.C. Hennes concerning ASSOCASA review on polymers, Unilever, LDC, Port Sunlight, 1989; quoted in ECETOC Report, final draft, Joint assessment of commodity chemicals No. X, Polycarboxylate polymers as used in detergents, ISSN-07773-6339-xx, Brussels, 1993.

41. Procter and Gamble, Summary of 91-day inhalation toxicity (rats), personal communication by J. David Innis, December 16, 1991 based on a unpublished report, P&G, Cincinnati, OH, 1991, quoted in ECETOC Report, final draft, Joint, assessment of commodity chemicals No. X, Polycarboxylate polymers as used in detergents, ISSN-07773-6339-xx, Brussels, 1993.

42. Rohm and Haas, Acrysol SP-02 N, skin sensitization, Magnusson-Kligmann, Bio-Tox, Rohm and Haas, Spring House, PA, 1988.

43. Rohm and Haas, Assessing the removal of Acrysol LMW-45N during secondary wastewater treatment, Roy Weston, Rohm and Haas, Spring House, PA, 1991.

44. Rohm and Haas, Acute toxicity of Acrysol LMW-10NX to *Daphnia magna*, final static bioassay report, ABC (Analytical Bio-Chemistry Laboratories), Rohm and Haas, Spring House, PA, 1983.

45. Rohm and Haas, Acute toxicity of Acrysol LMW-45NX to *Daphnia magna*, final static bioassay report, ABC (Analytical Bio-Chemistry Laboratories), Rohm and Haas, Spring House, PA, 1983.

46. G.A. Nolen, A. Monroe, C.D. Hassall, J. Lavicoli, R.A. Jamieson, and G.P. Daston, Studies of the developmental toxicity of polycarboxylate dispersing agents, Drug Chem. Toxicol. *12*, 95–110 (1989).

IV
Complexing Acids

9

Nitrilotriacetic Acid: Physicochemical and Ecological Properties

DIETER KIESSLING Specialty Chemicals, BASF AG, Ludwigshafen, Germany

ULRICH KALUZA Product Safety and Environmental Protection, BASF AG, Ludwigshafen, Germany

I. INTRODUCTION

In nearly all processes in which water plays a role, disturbances by metal ions dissolved in the water pose a central problem. Drinking water contains different fractions of dissolved metal ions, depending on its origin. The magnesium and calcium ion content governs water hardness: disturbances can occur as a result of the catalytic influence of heavy metal ions and precipitates from relatively insoluble metal salts can impair the different effects of detergents and cleaning agents. Complexing or chelating agents largely prevent these undesirable reactions.

II. PHYSICOCHEMICAL PROPERTIES

Nitrilotriacetic acid is generally referred to as NTA and has the CAS number 139-03-93. *N,N*-bis(carboxymethyl)glycine is also used as a notation.

The trisodium salt of NTA, Na_3NTA, has the CAS number 5064-31-3:

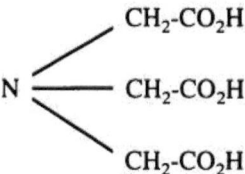

Nitrilotriacetic acid ($C_6H_9NO_6$, NTA) was synthesized from ammonia and chloroacetic acid by Heintz in 1861. Industrial preparation began in Ludwigshafen in 1936. NTA has since maintained its position as a complexing agent and is used in many branches of industry. It serves to bind polyvalent metal ions that otherwise would form low-solubility precipitates or initiate disturbing catalytic reactions (see Ref. 1).

NTA has a molar mass of 191.14 and is a tribasic acid crystallizing from water in colorless needles. The melting point is in the range of 242°C involving decomposition.

A. Solubility

NTA is only sparingly soluble in water at room temperature; the solubility increases with temperature:

Solubility of NTA in water (g/100 g), at	
5°C	0.13
23°C	0.13
80°C	0.95
100°C	3.3

The alkali-metal salts of NTA are readily soluble in water. Approximately 640 g trisodium salt dissolves in 1 liter water at room temperature. The pH value of the 1% aqueous solution of this salt is approximately 11.

B. Dissociation, Complexation Constants, and pH Dependence

Aminocarbonic acids of the NTA type react in solution as polybasic acids [1]. NTA thus dissociates in three steps:

Dissociation constants [2] at
25°C, ionic strength 0.1 M
pK_1	1.80
pK_2	2.48
pK_3	9.65

The pH value of the saturated solution is 2.3.

The acid strength of the individual protons, defined by the dissociation constant K_j ($j = 1-m$), thus decreases significantly with progressing dissociation. In particular, the protons fixed at the nitrogen atom dissociate only at high pH values:

$$K_1 = \frac{[H^+][H_{m-1}Z^-]}{[H_mZ]} \qquad K_2 = \frac{[H^+][H_{m-2}Z^{2-}]}{[H_{m-1}Z^-]}$$

The progressive protonation of the anions at lower pH values is taken into account by the factor α_z, which is defined as the reciprocal fraction of the complex anion Z^{m-} relative to the total amount of complexing agent present in solution, which is not bound to the metal ion:

$$\alpha_z = \frac{[H_mZ] + [H_{m-1}Z^-] + \cdots + [Z^{m-}]}{[Z^{m-}]}$$

The effective complexation constant K_{MeZ}^{eff} at certain pH values is obtained if K_{MeZ} is divided by α_z:

$$K_{MeZ}^{eff} = K_{MeZ}\frac{1}{\alpha_z}$$

Above a pH value of 12, K_{MeZ}^{eff} (pH dependent) corresponds approximately to the value of K_{MeZ} (pH independent).

The following table shows the pH dependence of log α_z for NTA [3]:

pH	log α_{NTA}
1	11.4
3	7.0
5	4.8
7	2.8
9	0.87
10	0.22
11	0.03
12	0.00
13	0.00
14	0.00

The higher stability of the complexes in the alkaline range is opposed to the fact that many metal ions form hydroxo complexes. This effect can be treated like the protonation of the complex anions [3]; analogously, the fraction of free Me^{n+} ions is represented by the factor α_{Me}:

$$\alpha_{Me} = \frac{[Me^{n+}] + [Me(OH)^{(n-1)+}] + \cdots}{[Me^{n+}]}$$

Both effects are taken into account by introducing the conditional complexation constant K_{MeZ}^{cond}, which describes the actual efficiency of the complexing agents:

$$K_{MeZ}^{cond} = K_{Mez} \frac{1}{\alpha_z} \frac{1}{\alpha_{Me}}$$

The following lists log α_{Me} for important metal ions [3]:

pH value	log α_{Mg}	log α_{Ca}	log α_{Cu}	log α_{Zn}	log $\alpha_{Fe(II)}$	log $\alpha_{Fe(III)}$
1	0.00	0.00	0.00	0.00	0.00	0.00
3	0.00	0.00	0.00	0.00	0.00	0.40
5	0.00	0.00	0.00	0.00	0.00	3.71
7	0.00	0.00	0.04	0.00	0.00	7.70
9	0.00	0.00	1.04	0.18	0.12	11.70
10	0.02	0.00	2.00	2.41	0.62	13.70
11	0.15	0.01	3.00	5.41	1.51	15.70
12	0.70	0.08	4.00	8.45	2.50	17.70
13	1.61	0.48	5.00	11.75	3.50	19.70
14	2.60	1.32	6.00	15.53	4.50	21.70

The conditional complexation constants of all metal complexes pass through a maximum as a function of the pH value; Figure 1 shows the pH-dependent stability of the NTA complexes of Ca^{2+} and Fe^{3+}:

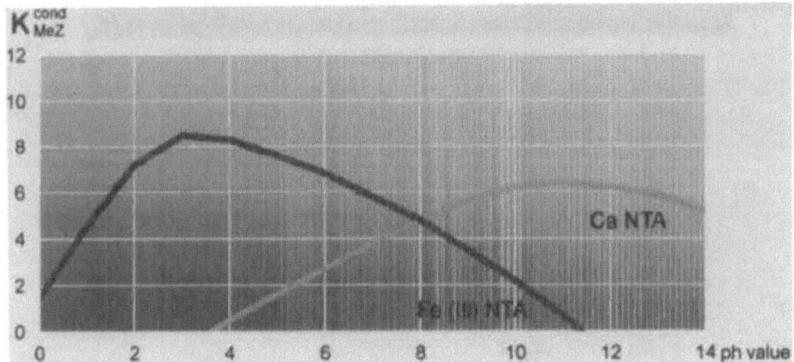

FIG. 1 Stability of NTA-metal complexes and dependence on pH.

The conditional complexation constants make it possible to include side re-actions, such as the formation of amine complexes (Cu^{2+}) or low-solubility salts [3,4].

C. Formation and Structure of Complexes

As mentioned earlier, NTA forms water-soluble complexes with a large number of polyvalent metal ions. The metal ion is almost completely enclosed, so that many of its typical—and sometimes disturbing—reactions are prevented.

Depending on the stoichiometric ratio of NTA to metal ion, 1:1 or 2:1 complexes are formed.

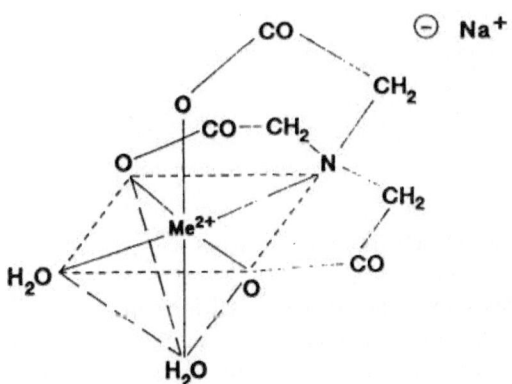

The stabilities of the complexes of various metal ions are different, as can be seen from the following list of stability constants for NTA chelates [2]:

Metal ion	$\log K_1$	$\log K_2$	pK_d
Al^{3+}	11.4		5.09
Ca^{2+}	6.39	8.76	
Cd^{2+}	9.78	14.39	11.25
Co^{2+}	10.38	14.33	10.80
Cu^{2+}	12.94	17.42	9.14
Fe^{2+}	8.33	12.80	10.60
Fe^{3+}	15.90	24.30	4.1/7.8[a]
Hg^{2+}	14.60		
Mg^{2+}	5.47		
Mn^{2+}	7.46	10.94	
Ni^{2+}	11.50	16.32	10.86
Pb^{2+}	11.34		
Zn^{2+}	10.66	14.24	10.06

[a]Reacts as dibasic acid, ionic strength (25°C) 0.1 M.

The stability of the metal complexes is described by the stability constants just listed (K_1 for 1:1 complexes and K_2 for 2:1 complexes). As a result of polarization of the OH bond in the chelate, the 1:1 complexes behave like weak acids and also dissociate.

This effect is expressed by the dissociation constant K_d:

$$K_1 = \frac{[MNTA]}{[M][NTA]}$$

$$K_2 = \frac{[MNTA_2]}{[M][NTA]^2}$$

$$K_d = \frac{[MNTAOH][H^+]}{[MNTA]}$$

M denotes a polyvalent metal ion and NTA is the anion $N(CH_2COO^-)_3$. The charges have been omitted, except H^+.

The water-soluble complex compounds of the metal ions are produced (see Sec. II.B) in competition with the low-solubility precipitates these ions form with such anions as sulfide, hydroxide, carbonate, or oxalate. Hydrogen ions compete with metal ions for the NTA anion. These equilibrium states must therefore be taken into consideration for the overall equation.

The chelates of NTA are stable over differently wide pH ranges:

Ca^{2+}	pH 9–12
Mg^{2+}	pH 7–10
Cu^{2+}	pH 3–12
Fe^{3+}	pH 1.5–3

On the assumption that a 1:1 complex is formed, 1 g Na_3NTA binds the following amounts of ions in the optimum pH range:

156 mg	Ca^{2+}
94 mg	Mg^{2+}
247 mg	Cu^{2+}
217 mg	Fe^{3+}

D. Preparation

The only preparation processes of industrial significance are the single-stage alkaline and the two-stage acid processes. Both are based on the cyanic methylation of ammonia or ammonium sulfate with formaldehyde and sodium cyanide or hydrogen cyanide.

1. Alkaline Process

This process has long been the only method for the preparation of NTA. The reaction can be carried out discontinuously or continuously, the continuous process being more economical:

$$NH_3 + 3HCHO + 3NaCN \rightarrow N(CH_2CN)_3 + 3NaOH$$

$$N(CH_2CN)_3 + 3NaOH + 3H_2O \rightarrow N(CH_2COONa)_3 + 3NH_3$$

A 40% solution is obtained, which can be either directly marketed or transformed into $Na_3NTA \cdot H_2O$ by crystallization.

2. Acid Process

The proportion of secondary products resulting from the alkaline process has led to an elaboration and improvement in the acid process in the past few years, which provides a much purer product.

In the first step, ammonia reacts with formaldehyde to hexamethylene-tetra-mine, which reacts with hydrogen cyanide in sulfuric acid solution, forming tricyanomethylamine. The latter is a solid substance that does not dissolve readily in acid medium, so that it can be easily filtered. Cleaning and saponification with NaOH provides Na_3NTA. The result is a purer product that can be processed in the same way as with the alkaline process.

E. Economic Significance and Commercial Products

The economic significance of NTA is shown by the following production figures (calculated as H_3NTA):

Consumption in Western Europe	
1990	~17,350
1991	~16,190
Consumption in Germany	
1990	~2,160
1991	~2,220

Commercial products are the following:

Dissolvine A brands (Akzo)
Hampshire NTA brands (W.R. Grace)
Masquol NTA brands (Protex)
Nervanaid NTA brands (ABM Chemicals)
Rexene NTA brands (Rexolin Chemicals)
Trilon A brands (BASF)
Versene NTA brands (Dow)

F. Analytical Determination

The following methods can be used for the quantitative determination of NTA and its salts:

Potentiometric titration with Fe^{3+} chloride [1,3]
Ion pair chromatography [4]
Gas chromatography [5]
Polarographic determination [6]

Ion pair chromatography, gas chromatography, and polarographic determination are also suitable for the analysis of trace amounts.

G. Use

NTA and its salts are used in very different applications to eliminate the disturbing effects of metal ions as a result of chelation, to dissolve precipitates, to modify the redox potential of metal ions, or to produce metal ion buffers.

1. Water Softening

The elimination of 1 mMol Ca^{2+} requires 191 mg H_3NTA or 257 mg Na_3NTA. The chelation optimum of the hardness constituents is in the neutral to basic range.

Water softening is important in the paper and textile industry, for soap production, in the cosmetics and toilet article industry, for the production of detergents and cleaning agents, in water treatment, and in many other chemical processes.

2. Replacement of Pentasodium Triphosphate (STP) in Detergents

Na_3NTA can be used to replace STP in the formulation of detergents with a low STP content; on the basis of zeolite and auxiliary washing agents, such as polycarboxylates, it is also possible to formulate STP-free detergents that achieve the effect of STP-containing products. The supply of phosphorus from detergents and cleaning agents is one of the causes of eutrophication in stagnant and slowly flowing waters when phosphorus is the limiting factor for algae growth causing eutrophication.

In comparison with STP, Na_3NTA is resistant to hydrolysis, so that no drop in efficiency occurs in the course of detergent production. According to the ratio of the complexation capabilities of STP and Na_3NTA, approximately 0.6 part Na_3NTA is required for 1 part STP.

3. Replacement of Tetrapotassium Diphosphate (TKPP) in Cleaning Agents

Difficulties are frequently encountered with the solubility of polyphosphates in the formulation of liquid cleaners. Na_3NTA is better soluble in water and provides advantages as a result of its threefold higher efficiency.

4. Masking of Heavy Metal Ions (Cu^{2+}, Mn^{2+}, and $Fe^{2+/3+}$)

NTA is very often used to mask disturbing heavy metal ions; its stability constants for the ions are lower by several orders of magnitude than those of EDTA (ethylenediaminetetraacetic acid). This may be of advantage if excessively stable complexes are undesirable.

III. CONSEQUENCES OF THE PHYSICOCHEMICAL PROPERTIES OF NITRILOTRIACETIC ACID (NTA) FOR ITS BEHAVIOR IN THE AQUATIC ENVIRONMENT

Essential physicochemical properties that decisively determine the environmental behavior of NTA are the complexation capacity and the high water solubility of the NTA salts and metal complexes. They influence the level of real NTA concentrations in the environment, the knowledge of which is indispensible to the proper discussion of the subject matter.

A. Water Solubility

1. Distribution in Environmental Compartments

The concentration and distribution of substances in the different environmental compartments is above all determined by their water solubility and adsorption properties on solid surfaces. Variables are the adsorption coefficient K_{oc} and the partition coefficient of the substance in *n*-octanol and water, P_{ow}. According to the empirical findings of Kenaga and Goring [7], the adsorption coefficient K_{oc} and thus the adsorption propensity decrease with increasing water solubility, whereas K_{oc} increases with rising partition coefficient P_{ow}:

$$\log K_{oc} = 3.64 - 0.55 \ (\log \text{WL}) \qquad \text{WL} = \text{water solubility, mg/liter}$$

$$\log K_{oc} = 1.377 + 0.544 \ (\log P_{ow})$$

The water solubility of NTA salts and NTA metal complexes is high [8d] and ranges between 500 and 1000 g/liter.

The partition coefficient P_{ow} or $\log P_{ow}$ cannot be experimentally determined for such polar and dissociating substances as NTA according to OECD methods 107 and 117, but a calculation according to the incremental method of Rekker [11] for H_3NTA yielded a value of -2.62 for $\log P_{ow}$. At the same time, this negative value shows a very low affinity for lipophilic media and also a very low propensity for bioaccumulation [8c,8d,13].

Any relevant propensity for adsorption on solid surfaces is not expected in view of these data [8d].

(a) Adsorption. Experimentally determined adsorption values on activated sludge and sediment were communicated by Alder and confirm other findings [14] according to which the fraction adsorbed per gram dry sludge solids is about

5% of the dissolved NTA concentration [15]. Trapp et al. [20] find no relevant NTA concentration differences between filtered and unfiltered water from the River Main. Adsorption on soils and active carbon filters is also correspondingly low, as shown by various results [8b,26,27].

(b) Bioaccumulation.   No data are available as yet revealing a particular bio-accumulation of NTA in fresh water or marine organisms. The environmental behavior of NTA is thus practically determined exclusively by the concentrations and effects in the homogeneous aqueous phase.

2. Environmental Concentrations

Because NTA is passed directly into the waste water, if properly applied, this path is of particular significance for evaluating environmental behavior. The concentrations in water then depend decisively on the degradability under various conditions.

(a) Biodegradation.   NTA is easily biodegradable, as confirmed by many in-vestigations in sewage plants, surface waters, and drinking water treatment. Degradation takes place without the accumulation of intermediates until com-plete mineralization. The average degradation in sewage plants amounts to 95% and more and is practically independent of the usual fluctuations of many pa-rameters, such as influent concentration ($\sim$1 to $\sim$30 mg/liter), low temperature in winter (6–16°C), sewage plant shock load, great short-term concentration fluctuations, oxygen content, and excess of easily degradable competitive nutri-ents [17].

NTA is also degraded in marine and estuarine regions. The rate of degradation increases with rising temperature and decreasing salinity and thus also with increasing bacterial cell density [17].

(b) Water Concentrations, Monitoring.   The following recent data (average values in µg/liter, Table 1) serve as examples of the large variety of NTA concentrations determined in numerous and, in part, extensive monitoring pro-grams. The Rhine, the largest central German river, is of particular significance at the German/Dutch frontier at Bimmen/Lobith and Wesel. At this location, the Rhine flows through densely populated and highly industrialized regions of Germany and its neighboring countries and contains the sum of all discharges (clarified and unclarified effluents) over its entire reach, including all its tribu-taries, such as the Neckar, Main, Mosel, Ruhr, and Emscher.

The NTA concentrations in surface waters are thus between 0.1 and 5 µg/liter, whereas values in the range of 10 to >1000 µg/liter are found for sewage plant influents and values of 1–100 µg/liter for sewage plant effluents, which docu-ments good degradability. This is also confirmed by the Zurich sewage plant, where the influent concentrations increased by a factor of 4 between 1984 and 1987, whereas the effluent concentrations rose only insignificantly by a factor of 1.5 [21].

It is also remarkable to note the small concentration difference in the Rhine between Village-Neuf (at the German/Swiss border) and Bimmen/Lobith (at the German/Dutch border), which, despite different loads, indicates a practically constant degradation potential of the Rhine.

A comparison with the WHO standard for drinking water of 150 μg/liter, moreover, shows that there is a high safety margin [34].

(c) Findings on Environmental Concentration. The acute and chronic toxicity of NTA for aquatic organisms is very low. The lowest LC_{50} for fresh water organisms is 98 mg/liter; for marine organisms it is considerably higher, in the range of ≥3000 mg/liter. This is attributed to the increasing complexation of free NTA with calcium in salt water, because increasing water hardness reduces the aquatic toxicity [8f,14].

Effects below 10 mg/liter for chronic toxicity are not known to date [8f]. In comparison with the concentrations encountered in waters, ecotoxic concentrations are thus higher by at least a factor of 1000, which indicates a high safety margin.

The opposite effect of inhibiting toxicity is eutrophication, that is, the enhancing effect of a substance (especially with respect to alga growth). As far as a direct eutrophic effect is concerned, a candidate primary nutrient for algae could be the nitrogen in NTA, which accounts for only 7% of the molecule, however, and is thus negligible in comparison with the nitrogen concentrations normally present in rivers (Rhine, 2–10 mg/liter). NTA could trigger a hypothetical eutrophic effect according to two mechanisms based on its central property, the capability for complexation.

B. Complexation

1. Eutrophication

NTA could make trace elements bioavailable by complexation (e.g. remobilization).

NTA could reduce or eliminate the toxic, that is, growth-inhibiting effects of trace elements and heavy metals by complexation.

(a) Fresh Water. Because the literature and laboratory experiments have so far provided only contradictory results on eutrophication, water ecology tests were carried out in a field ecosystem composed of several shallow ponds (surface area 2000 m^2 each; depth 0.7 m) in a 2 year experiment with 100 and 500 μg/liter of NTA [17]. The aim was above all to reproduce the conditions prevailing in impounded rivers, because elevated NTA concentrations are expected, in principle, in such waters as a result of increased sewage effects and, on the other hand, such waters are very likely to be eutrophic.

The extensive data on the different alga species observed, as well as on zooplankton and from physicochemical analyses, were statistically evaluated. The species composition of phytoplankton and zooplankton in the different ponds was examined for similarity by cluster analysis.

A comparison of roughly 100 phytoplankton species from each pond revealed that the similarity was greatest between the NTA-containing and NTA-free (control) ponds. No special selection of certain algae that might be undesirable was observed. NTA had no influence on zooplankton development or macrophyte inventory [17].

Interactions with trace elements (remobilization of metals) were not found, not even for iron as a possible algae-stimulating element contained in large quantities of 25–39 g/kg in the sediment [17].

The conclusion is that no relevant effects on the biocoenosis are expected in shallow waters and impounded rivers at NTA concentrations of up to 500 µg/liter.

(b) Seawater. The discussion of a eutrophic effect of NTA in coastal waters and in the sea ("alga bloom") must be based on concentrations clearly below 1 µg/liter of NTA (see the concentrations in the Rhine, Elbe, and Oder, Table 1). An influence on alga growth is therefore extremely improbable. Nevertheless, calculations were performed to clarify whether, under worst-case conditions, 1 µg/liter of NTA can significantly influence the iron concentration in seawater.

The calculation showed that 1 µg/liter of NTA increases the iron concentration by approximately 3% as a result of remobilization from iron-containing suspended matter at pH eight, which is representative of marine conditions.

This effect is considerably below the local and time-dependent fluctuations in the iron concentrations in seawater. Changes in the pH value, especially pH reductions, can have a much greater effect.

The cause for the small effect of 1 µg/liter of NTA on the iron concentration in seawater must be that the complexing agent is largely bound by alkaline earths, which are present in high concentrations. The species distribution results from the preceding calculation as follows: approximately 18% FeNTA, approximately 25% MgNTA, and approximately 56% CaNTA.

The calculated effect on the iron(III) concentration would be even further reduced if the concentrations of relevant bivalent heavy metals present in seawater, such as Cu, Ni, and Zn, were included. The probability of a significant effect of the NTA microconcentrations contained in seawater is additionally reduced by the fact that the presence and effect of natural complexing agents has not been taken into account.

A final evaluation under worst-case assumptions shows that NTA concentrations of up to 3 µg/liter of NTA cannot cause a significant increase in the concentration of dissolved iron in estuarine and marine regions [17].

TABLE 1 NTA Concentrations in Water

Water, location, and year	Concentration (μg/liter)	Source
Lake Constance, 1990	0.1	Rossknecht, 1991 [25]
Rhine at Village-Neuf, 1991	~2	Giger et al., 1991 [28]
Rhine at Wesel, 1992	2.9	ARW, 1992 [10]
Rhine at Bimmen/Lobith, 1991	3.9	ARW, 1991 [10]
	1.8	RIWA, 1992 [9]
Rhine at Lobith, 1992	1.48	RIWA, 1994 [37]
Main, 1991	1.93	Trapp et al., 1992 [20]
Elbe, 1991/92	3.5	Guderitz et al., 1993 [24]
Bank filtrate	<0.5	[24]
Oder/Frankfurt, 1991/92	3.0	[24]
Drinking water, 1993		DVGW [19]
Lake Constance,		
Wahnbachtalsperren verband,		
Hanover, Magdeburg,		
Wiesbaden, Regensburg,		
Düsseldorf	<0.5	
Mainz	0.6	[19]
Dortmund	1.4	[19]
Essen	~2.4	[19]
Drinking water (Canada)		
1976–1977	2.8	Malaiyandi et al., 1979 [33]
Sewage plant Zurich, Glatt		
Influent, winter 1984	40–380	Alder et al., 1990 [21]
Effluent,	3–30	
Influent, winter 1987	330–1490	
Effluent,	5–50	
Sewage plants in Hesse, 1987		
Influent	100–300	Kröber and Häckl, 1987 [29]
Effluent	<2–23	
Sewage plant Bielefeld-Heepen		
Influent	64–68	Lahl and Burbaum, 1988 [38]
Effluent	8–16	

2. Influence of Heavy Metals and Water Hardness

(a) Biodegradation. Some heavy metal complexes (Cd, Cr, Cu, Hg, Ni, Pb, and Zn) examined separately under laboratory conditions can show reduced degradation rates, with Hg possibly because of toxic effects on the degrading microorganisms [8a,30].

According to other observations, the presence of Al, Ca, Cd, Co, Cr, Cu, Fe, Mg, Mn, and Pb does not influence NTA degradation [8a,29,33], and NTA complexes with Ca, Cu, Fe, Mn, Pb, and Zn are degraded like NTA alone in adapted and nonadapted soils [30].

Under practical conditions in waste water, no significant influence of heavy metals on the degradability of NTA is recognized [14,27,35].

(b) *Photodegradation.* It is reported that iron and copper complexes can be degraded by photolysis, but this does not apply to complexes with lead, cadmium, calcium, or magnesium [8e]. Photodegradation as an elimination path is of relatively little significance on the whole because of easy biodegradability.

3. Influence of NTA on Heavy Metal Elimination in Sewage Plants

A regularly expressed concern about complexing agents like NTA is also based on the suspicion that they might impair the elimination of heavy metals from the (raw) waste water by adsorption on sewage sludge.

All the studies carried out in this field are based on the supply of complexing agents in the free form as (sodium) salt, although complexing agents are very likely to be bound as metal complexes in the inflow of sewage plants after proper application and thus cannot readily act as solubilizing complexing agents on free metal ions. Moreover, unrealistically high complexing agent concentrations were used in many investigations.

More recent results were obtained in semitechnical activated sludge plants using municipal waste water [17]. It was found, for example, that the elimination of heavy metals from waste water by adsorption on sewage sludge of up to approximately 20,000 µg/liter of NTA is not impaired in winter or summer. According to these sewage plant investigations by Bernhard [17], moreover, the outflow concentrations of Pb, Cd, Cr, Fe, Cu, Mn, and Zn were completely uninfluenced by the presence of NTA. The concentrations were insignificantly elevated for nickel, but it was expressly stated that no adverse effects on the waters can be deduced from these results [17].

Alder et al. [21] also arrive at the conclusion that NTA does not exert any influence on heavy metal elimination at the concentrations normally present in sewage plants (in comparison to the usual heavy metal concentrations in sewage plant outflow).

4. Remobilization of Heavy Metals

(a) *Basic Aspects of Remobilization.* Besides the interaction of NTA with heavy metals in sewage plants, the remobilization of heavy metals from water sediments is still of great interest in the literature.

What does this remobilization depend on? Anderson et al. [14] distinguish the following factors of influence:

Bonding strength between complexing agent and metal
Complexing agent concentration
Concentration of competing complexing agents
Bonding strength of adsorbed or precipitated metals
Concentration of adsorbed or precipitated metal ions
Concentration on adsorption sites and precipitative anions
pH value of the solution
Kinetics of the processes involved

Although there are a great many experiments and results, two essential prereq-
uisites for realistic reproduction in practice are not listed for most findings:

1. The concentration of complexing agents used is frequently much too high.
2. The complexing agent is used in the free form as a salt, although complexing
    agents in reality are bound as metal complexes in raw waste water, sewage
    plants, and surface waters, for example.

Furthermore, most of the discussions do not take account of the stoichiometric
mass difference between metal atoms and complexing agent molecules. It should
be remembered that about 3 µg/liter of NTA is required for binding 1 µg/liter of
Cr, Cu, Fe, Mn, Ni, or Zn each and that free, unbound NTA can no longer be
expected in the system if the metal concentration exeeds one-third the NTA
concentration.

Moreover, the metal concentrations in the system investigated are rarely specified, so that, in general, no relation can be established between the complexing
agent and the metal concentrations in the environmental compartments affected.
Proper evaluation of the results thus becomes difficult or impossible.

In principle, remobilization can only be expected if the dynamic equilibrium
between liquid phase and solid is changed, for example by concentration changes
or increases, because no remobilization can take place in thermodynamic equilibrium at constant concentrations.

In view of the use of synthetic complexing agents for decades in varied fields,
a practical equilibrium situation must be assumed today in waters, especially
because the NTA and heavy metal concentrations and loads in the Rhine have been
found to be approximately constant since the beginning of systematic measurements and, in part, show a rather clearly decreasing trend [9,10]; see Table 3.

Heavy metal remobilization is possible, in principle, if one of the following
steps are taken:

A free complexing agent is intermittently added directly to the water
A free complexing agent is used in stoichiometric excess
Recomplexation reactions occur, that is, an exchange of the central metal atom
    enclosed in the complex for metal ions in the sediment

Practically all experimental evidence of heavy metal remobilization by complexing agents is based on the first case. Moreover, very high input concentrations (mg/liter) of complexing agents were frequently used that do not occur under practical conditions. In German rivers, complexing agents are found at largely constant, somewhat decreasing concentrations, at most in the ppb range (µg/liter) [9,10,20,21,24].

The second case is hypothetical: in practice, the input quantity of complexing agents is proportional for both economic and ecological reasons so that an excess cannot occur if properly applied. If an NTA excess situation should arise, however, this is locally and temporarily limited because the excess amount of free complexing agent is bound by heavy metals from the aqueous phase in the sewer or in the sewage plant and is further diluted in heavy metal-containing waters. Moreover, NTA is rapidly biodegraded and thus is no longer available for further reactions.

For the third case it should be noted that for reasons of thermodynamic equilibrium in the aqueous phase dissolved metal ions are always present in other bond forms in addition to metal complexes. This can lead to an exchange of the central metal atoms in the complexing agent molecules ("recomplexation") in the case of changes in the equilibrium position, for example concentration changes. However, the balance of dissolved, homogeneously distributed metals is not basically influenced by such recomplexation reactions.

To understand recomplexation or remobilization reactions of heavy metals bound on sediment or other solid surfaces, the following aspects must be taken into account in addition to a concentration gradient:

There is generally a resting sediment over which water flows. The pore water of the sediment surface, which contains complexing agents, also already rests at a depth of a few millimeters. It is no longer exchanged for flowing water and is thus in equilibrium with its environment [17,18].

For remobilization (recomplexation) of metals from the water sediments, it is necessary, besides a (rather improbable) increase in the complexing agent concentration in the (pore) water, that metal complexes remain in contact with the metal species bound in the sediment or on solid particles until the recomplexation reaction has taken place. As already mentioned, metal complexes have a low propensity for adsorption on solid particles because of their good water solubility and high polarity. They are therefore preferentially contained in the "flowing wave," not in the sediment, in contrast to heavy metals. Under real conditions, recomplexation/remobilization processes in the sediment are very protracted for kinetic reasons and not verifiable to a measurable extent [17].

Experimental studies that do not take account of these facts (for example, (model) sediment specimens in the laboratory are presupplied with metal ions, which bears an unclear relation to real conditions, or solid matter (sediment) and test solution are intimately mixed over prolonged periods of time, which can only very rarely happen in practice) may provide reproducible results. Their relation to practice is unclear, however, and they thus involve a considerable risk of error in conjunction with correspondingly variable and problematic interpretation possibilities.

(b) Realistic Model Investigations. To obtain realistic results, two approaches were selected in Germany within the framework of the NTA research program [17,18]. On the one hand, the diffusion of heavy metals from an artificial sediment was determined in the presence and absence of free NTA in laboratory experiments. On the other hand, the concentrations of dissolved copper, nickel, and zinc were determined in a 1000 m flume filled with river sediment over which water with or without the addition of free NTA flowed at a slow rate.

These investigations led to the following findings:

From resting kaolinite as the model sediment approximately 10% (approximately 25 µg/liter of Zn) of the initial zinc concentration can just be significantly remobilized by 200 µg/liter of free NTA not yet bound on heavy metals within a contact period of 250 h.

In the flume experiment, a definite remobilization of the widely distributed and easily remobilizable zinc began at a concentration of 50 µg/liter of free NTA not yet bound on heavy metals. Other heavy metals, such as copper and nickel, are only remobilized at clearly higher NTA concentrations.

In this flume experiment, an NTA end concentration of less than 20 µg/liter of NTA was always obtained because of biodegradation on a flow section of approximately 900 m above a temperature of 10°C and independent of the initial NTA concentration (50–500 µg/liter of NTA).

In practice, therefore, remobilization effects are ruled out because NTA is not present in the free, unbound form in waters and concentrations above 50 µg/liter of NTA are expected only in waters in exceptional cases as a result of the good biodegradability.

It can be seen from the results of the NTA research program [17,32] that in bank filtration as an essential process for drinking water preparation from surface waters NTA concentrations of up to 200 µg/liter of NTA did not influence the behavior of the heavy metals during underground passage in connection with the low heavy metal concentrations in the groundwater of the Ruhr Valley under aerobic and anoxic conditions (e.g., in the presence of nitrate).

At elevated heavy metal contents, after adding 5 µg/liter of Cd, 50 µg/liter of Cu, 50 µg/liter of Ni, and 500 µg/liter of Zn to this system, a reduced sorption/ increased remobilization especially of nickel and zinc may occur under anaerobic conditions from 50 µg/liter of NTA upward. Concentrations above 50 µg/liter of NTA are not expected here, either.

(c) NTA and Heavy Metal Concentrations in the Environment. Results from laboratory and model experiments clearly gain in validity and acceptance if they are measured and evaluated in comparison with reality.

TABLE 2 NTA and Heavy Metal Concentrations in Water

	Concentration		
Origin of the water	µg/liter	µmol/liter	Source
Drinking water from Lake Constance			[36]
NTA	<0.5	<0.003	
Cd + Cr + Cu + Hg + Ni + Pb + Zn	2.97	0.0455	
Fe + Mn	2.04	0.0365	
Rhine at Lobith			[37]
NTA	1.48	0.0079	
Cd + Cr + Cu + Hg + Ni + Pb + Zn	41.68	0.62	
Fe + Mn	10.40	18.64	
Bielefeld-Heepen sewage plant			[38]
Outflow			
NTA	11	0.057	
Cd + Cr + Cu + Hg + Ni + Pb + Zn	332	4.88	
Fe + Mn	—	—	
Inflow			
NTA	66	0.344	
Cd + Cr + Cu + Hg + Ni + Pb + Zn	1064	16.36	
Fe + Mn	—	—	

It is therefore helpful to compare metal and NTA concentrations in water of different origins using a few examples (Table 2).

These data clearly show a pronounced stoichiometric excess of heavy metals over NTA, so that according to thermodynamic equilibrium all the NTA present is bound as a metal complex. Moreover, the difference between inflow and outflow of the sewage plant (Bielefeld) confirms a good elimination of the heavy metals over and above the degradation of NTA, which does not suggest any disturbance by NTA.

As already mentioned, the Rhine at the German/Dutch border at Lobith is a good example of the development of the situation under practical and environmental conditions, because the Rhine at this location has traversed densely populated and highly industrialized regions in Germany and its neighboring countries and thus contains the sum of all discharges (clarified and unclarified effluents) over its entire reach. The time-dependent development of molar heavy metal and NTA loads since 1987 (Table 3) documents a clearly decreasing tendency for a high stoichiometric heavy metal surplus.

The heavy metal fraction bound by NTA is obviously only a very small amount of the entire heavy metal quantity present and thus represents the low ecological relevance of the NTA concentrations.

The decisive finding for a realistic assessment of the remobilization signifi-

TABLE 3 Molar Annual Loads of Heavy Metals and NTA in the Rhine at Lobith Since 1987 (Mmol/a)

	1987	1988	1989	1990	1991	1992
NTA[a]	1.13	1.11	1.34	1.15	0.70	0.72
Metals[b]						
Cd + Cr + Cu + Hg + Ni + Pb + Zn	74	74	46	58	43	39
Fe, Mn	2402	2307	1212	1233	1069	1182

[a]Average values from Refs. 9 and 10.
[b]From annual averages [9].

cance of NTA can be explained, for example, using the data of RIWA, 1994 in Die Beschaffenheit des Rheinwassers bei Lobith im Jahre 1992 [37] (Table 4).

These data show that the concentration of heavy metals is not only higher but also shows greater fluctuations than the NTA concentration and that the span (as a measure of *scattering*) of the molar heavy metal concentration of Cd + Cr + Cu + Hg + Ni + Pb + Zn alone, without Fe and Mn, is higher by one order of magnitude than the (absolute) NTA concentration and its scattering.

Hence it follows that the heavy metal balance cannot be significantly influenced by fluctuations in the NTA concentration. (No remobilization is possible anyway at constant heavy metal and NTA concentrations in thermodynamic equilibrium.)

The proportion of the metal complexes formed depends on various factors and can be calculated theoretically [12,16,21,22]. Complete experimental data confirming the theoretically calculated metal/NTA species distributions are not known, because no direct detection methods exist for metal complexes, but only for either complexing agents or metals. The theoretically calculated proportion can deviate from the measured ratio of "pure" metals. For example, the calculated distribution of NTA-metal species at 10 µg/liter of H_3NTA is listed here for the composition of Rhine water [8g]:

CaNTA	~3%
CuNTA	~65%
NiNTA	~23%
ZnNTA	~7%
PbNTA + MgNTA + CaHNTA + $HNTA^{2-}$ + $Fe(NTA)(OH)^-$ + $Fe(NTA)(OH)_2^{2-}$ ~2%.	

TABLE 4 Scattering of Heavy Metal and NTA Concentrations in the Rhine

	Heavy metal concentration				NTA concentration	
	Cd + Cr + Cu + Hg + Ni + Pb + Zn		Fe + Mn			
	µM	µg/liter	µM	µg/liter	µM	µg/liter
Annual average	0.62	41.68	18.6	1040	0.0079	1.48
Maximum value	0.99	66.55	31.9	1780	0.0234	4.4
Minimum value	0.42	28.15	8.8	490	<0.0053	<1.0
Span	0.57	38.40	23.1	1290	~0.02 3.4	

The calculation for 1 µg/liter of H_3NTA in seawater (see Sec. III.B.1) provided:

CaNTA	~56%
FeNTA	~18%
MgNTA	~25%

If properly applied, synthetic complexing agents are passed in a bound form as metal complexes with the waste water into the sewage plants and, unless they are completely degraded there, also into surface waters, where they are detectable in constant concentrations in the range of µg/liter, which shows a rather decreasing trend and is ecotoxicologically harmless (see Tables 1–4).

Consider the following:

Presence in a bound form as metal complexes
Comparatively low metal complex concentrations in surface waters compared with high metal concentrations,
Higher scattering of the heavy metal concentration by one order of magnitude compared with the absolute metal complex concentration and its scattering
Good water solubility, high polarity, and thus low adsorption propensity of the metal complexes
Low exchange rate of the flowing water with the pore water of sediments

It must be assumed that complexing agents can neither change the heavy metal balance in surface waters in the homogeneous aqueous phase nor remobilize heavy metals from sediments.

Any data confirming an exchange of the central metal atoms in dissolved metal complexes for metals bound in the sediment (remobilization) are not experimentally verifiable for practice-relevant conditions but are not expected, either.

Natural changes in the pH value and the redox potential, as well as competition by natural complexing agents and the presence of precipitative anions, have a stronger influence on the heavy metal balance in surface waters (e.g., through solubility) than changes in the low metal complex concentrations [12,17,23].

Heavy metal remobilization by synthetic complexing agents, such as NTA, is neither of theoretical nor of practical or ecological relevance under practical and environmental conditions.

REFERENCES

1. G. Schwarzenbach and H. Flaschka, Die komplexometrische Titration, 5th edition, Enke Verlag, Stuttgart, 1965; S. Siggia, D. Eichlin, and R. Rheinhart, Anal. Chem. 27, 1745–1749 (1955).

2. A.E. Martell, and R.M. Smith, *Critical Stability Constants*, Vol. 1, Plenum Press, New York, 1974.

3. BASF, Technische Information Trilon Marken, Ludwigshafen, 1989.

4. J. Weiss, Handbuch der Ionenchromatographie, Dionex GmbH, Weiterstadt, 1985, p. 253.

5. N.T. de Oude, Wasser 64, 283–292 (1985).

6. DIN 38 413, Part 5.

7. E.E. Kenaga, and C.A.I. Goring, Relationship between water solubility, soil sorption, octanol-water partitioning, and concentration of chemicals in biota, in *Aquatic Toxicology*, (J.G. Eaton et al., eds.), ASTM STP 707, 1980, pp. 78–115.

8. H. Bernhardt, (ed.) "NTA: Studie über die aquatische Umweltverträglichkeit von Nitrilotriacetat (NTA)", Hauptausschuß Phosphate und Wasser der Fachgruppe Wasserchemie in der Gesellschaft Deutscher Chemiker, Verlag Hans Richarz, Sankt Augustin, 1984, ISBN 3-88345-376-5, 1984: (a) pp. 29,143,144; (b) pp. 30,152; (c) pp. 45,222; (d) pp. 48,243; (e) p. 153; (f) pp. 211,244; (g) p. 293;

9. RIWA, Samenwerkende Rijn- en Maaswaterleidingsbedrijven, Annual Reports, 1987–1992, Part A: Der Rhein, Sekretariat RIWA Postf. 8169, 1005 AD Amsterdam.

10. ARW 48th Report of the Arbeitsgemeinschaft Rheinwasserwerke e.V. 1991 Part 2, Entwicklung der EDTA- und NTA-Belastungssituation am Rhein, pp. 31–47; ARW, 48th Report of the Arbeitsgemeinschaft Rheinwasserwerke e.V. 1992, Part 1, Organische Komplexbildner, pp. 43–54.

11. R.H. Rekker, *The Hydrophobic Fragmental Constant*, Pharmacochemistry Library, Vol. 1, Elsevier, New York, 1977; R.F. Rekker and H.M. De Kort, Eur. J. Med. Chem.-Chim. Ther. *14*, 479 (1979).

12. A. Alder et al., Verhalten von EDTA in der aquatischen Umwelt, EAWAG, Annual Report 1992, pp. 29–38.

13. BUA, Beratergremium für umweltrelevante Altstoffe der Gesellschaft deutscher Chemiker, Nitrilotriessigsäure, BUA-Stoffbericht 5 (October 1986), VCH, Weinheim, 1987, ISBN 3-527-26680-1.

14. R.L. Anderson, W.E. Bishop, and R.L. Campbell, A review of the environmental and mammalian toxicology of nitrilotriacetic acid., Crit. Rev. Toxicol. *15*(1), 1–102 (1985).

15. H.A. Siegrist, W. Alder, W. Gujer, and W. Giger, "Verhalten der organischen Komplexbildner NTA und EDTA in Belebungsanlagen", Wasser/Abwasser, *68*, 101–109 (1988); H.A. Siegrist, W. Alder, W. Gujer, and W. Giger, Behaviour and modelling of NTA degradation in activated sludge systems, Water Sci. Technol. *21*, 315–324 (1989).

16. L. Sigg, Schwermetalle in Fliessgewässern communications from, EAWAG, December 1991; 32–35.

17. H. Bernhardt, Ergebnisse der Sonderforschungsvorhaben zu Fragen der aquatischen Umweltverträglichkeit von Nitrilotriacetat (NTA). Zusammenfassung und Bewertung der Ergebnisse, Folgerungen, An den Bundesminister für Forschung und Technologie und den Bundesminister für Umweltschutz, Reaktorsicherheit, September 1991.

18. D. Donnert, J. Horst, and S.H. Eberle, Ermittlung der Freisetzungsraten von Schwermetallen aus Sedimenten durch Nitrilotriessigsäure, ??? read at the AGF conference on November 28–29, 1991, Belastung von Böden und Gewassern, Bonn-Bad Godesberg.

19. DVGW, Deutscher Verein des Gas- und Wasserfaches e.V., EDTA/NTA-Konzentrationen in Trinkwässern (1993), Anlage zum Ergebnisprotokoll über das 7, Gespräch zur Verringerung der Gewässerbelastung durch EDTA, 31.03.1994, Umweltbundesamt, 14191 Berlin, III 3.6—20 113-6/23, 31.03.1994/Zan.

20. S. Trapp, R. Brüggemann, W. Kalbfus, and S. Frey, Organische und anorganische Stoffe im Main, GWF, Wasser, Abwasser, *133*, 495–504 (1992).

21. A.C. Alder, H. Siegrist, W. Gujer, and W. Giger, Behaviour of NTA and EDTA in biological wastewater treatment, Wat. Res., *24*, 733–742 (1990).

22. J.G., Hering, and F.M.M. Morel, Kinetics of trace Metal Complexation: ligand-exchange reactions, Environ. Sci. Technol., *24*, 242–252 (1990).

23. A.L. Bryce, W.A. Kornicker, A.W. Elzerman, and S.B. Clark, Metal desorption kinetics in a metal/ligand/sorbent System Natl. Meet.-Am. Chem. Soc., Environ. Chem., *33*, 84–86 1993.

24. T. Guderitz, W. Schmidt, and H.-J. Brauch, Die organische Belastung der oberen Elbe vor dem Hintergrund der Trinkwassergewinnung aus Ufer-filtrat, Vom Wasser *81*, 315–326 (1993).

25. H. Rossknecht, Die Entwicklung der NTA- und EDTA-Konzentrationen im Bodensee von 1985 bis 1990, Ber. Int. Gewässerschutzkomm. Bodensee, *41*, 1991, ISSN 1011-1263.

26. S. Schullerer and H.J. Brauch, Oxidative und adsorptive Behandlung EDTA- und NTA-haltiger Wässer, Vom Wasser, 72, 21–29 1989.

27. A. van Spaendonck, Schoner wassen met NTA, Chemiewinkel-CMCV, Universiteit van Amsterdam, 1989 Amsterdam, in Brouwer, N.M.; Ecological and toxicological

properties of NTA as detergent builder, (N.M. Brouwer and P.M.J. Terpstra, eds.), Report WAU, dept. Household and Consumer Studies, Wageningen, 1994.

28. W. Giger, C. Schaffner, F.G. Kari, H. Ponusz, P. Reichert, and O. Wanner, Auftreten und Verhalten von NTA und EDTA in schweizerischen Flüssen, Communications from EAWAG, December 1991, pp. 27–31.

29. B. Kröber and M. Häckl, Orientierende Messungen der Komplexbildner NTA und EDTA in Fliessgewässern und in ausgewählten kommunalen Kläranlagen in Hessen, in Annual Report 1987 Hessische Landesanstalt für Umwelt, pp. 66–86.

30. P.A. Thayer and C.J. Kensler, Current status of the environmental and human safety aspects of NTA, in CRC Crit. Rev. Environ. Control 3, 375–404 (1973).

31. O. van't Hof, T.J. Nieuwstad, and H.J. Pöpel, De invloed van NTA op de verwijdering van zware metalen in het actief-slibproces, H_2O, 17, (2), 26–29 (1984).

32. B. Kuhlmann and U. Schöttler, Behaviour and effects of NTA during anaerobic bank filtration, Water Supply 11, 119–128 (1993).

33. M. Malaiyandi, D.T. Williams, and R. O'Grady, A national survey of nitrilotriacetic acid in Canadian drinking water, Environ. Sci. Technol. 13(1), 59–62 (1979).

34. WHO (World Health Organization) EUR/ICP/CWS 032, 5882n.p.6, NTA, Copenhagen, 1991.

35. Hauptausschuss Phosphate und Gewässer in der Fachgruppe Wasserchemie, represented by Reg. Dir. Dr. A. Hamm, Beitrag zur Diskussion über Phosphatersatzstoffe in Waschmitteln, gwf Wasser, Abwasser, 132, 491–499 (1991).

36. Frey (Ed.), Wassergüte-Parameter des Trinkwassers aus dem Bodensee, Jahresmittelwerte 1993, Zweckverband Bodensee-Wasserversorgung, Postf. 801180, 70511 Stuttgart.

37. RIWA, Samenwerkende Rijn- en Maaswaterleidingsbedrijven, Annual Report, 1994, Die Beschaffenheit des Rheinwassers bei Lobith im Jahre 1992, Sekretariat RIWA Postf. 8169, 1005 AD Amsterdam.

38. U. Lahl and H. Burbaum, Einzelstoffanalysen im Zu- und Ablauf einer kommunalen Kläranlage, Korrespondenz Abwasser 35, 360–364 (1988).

10

Physicochemical Properties of Ethylene Dinitrilotetraacetic Acid and Consequences for Its Distribution in the Aquatic Environment

FRITZ H. FRIMMEL   Engler-Bunte-Institute, University of Karlsruhe, Karlsruhe, Germany

I. INTRODUCTION

Synthetic chelators have a broad technical application worldwide. Because of their ability to form stable complexes with multivalent cations, they can change the effective chemical and physical properties of metal ions. This effect of ethylene dinitrilotetraacetate (EDTA) has been used for numerous purposes, including analytical procedures, medical detoxification, industrial production, food processing, and water softening. EDTA has also found its way into the environment [1]. Natural aquatic systems have especially been polluted by this poorly biodegradable compound and its metal complexes. As a consequence, the fate of heavy metals in the environment must be considered. In particular, changes in sediment-water distribution, transport characteristics, and the bioavailability of EDTA-complexed heavy metals must be expected.

The aim of this chapter is to contribute to the understanding of the role of EDTA in the aquatic environment:

Describe the chemical and physical properties of EDTA and its metal complexes
Summarize the production and application of EDTA
Outline some analytical determination methods
Discuss the distribution of EDTA in natural water bodies
Discuss the distribution of EDTA in the course of water treatment steps

II. CHEMICAL AND PHYSICAL PROPERTIES

According to the IUPAC nomenclature, EDTA is ethylene dinitrilotetraacetate or acid. In *Chemical Abstracts* EDTA is listed as glycine, N,N'-1,2-ethanediylbis-[N-(carboxymethyl)-(CAS No. 60-00-4) and the EDTA tetrasodium salt is listed as glycine, N,N'-1,2-ethanediylbis[N-(carboxymethyl)-, tetrasodium salt (CAS No. 64-02-8). The elemental composition of EDTA is $C_{10}H_{16}N_2O_8$ (acid, H_4EDTA) and $C_{10}H_{12}N_2O_8Na_4$ (tetrasodium salt, Na_4EDTA). The molar mass of H_4EDTA is 292.3 g/mol, the molar mass of Na_4EDTA is 380.2 g/mol. The formal structure of H_4EDTA is shown in Fig. 1.

The tetrasodium salt of EDTA is the main component in many commercially available products, such as Hampene 100, Nervanaid B 65, Sequestrene ST, Trilon B, and Versene 100. The colorless and crystalline acid (H_4EDTA) has a relatively low water solubility of about 0.5 g/liter at 25°C, whereas the tetrasodium salt (Na_4EDTA) has a high water solubility of around 1030 g/liter at 25°C [2]. EDTA is relatively stable toward hydrolysis and strong acids and bases [3]. However, it can be oxidized easily, particularly in the form of some metal complexes.

A. Protonation Constants

Because it is a multifunctional acid, EDTA has several dissociation constants. In addition to the four carboxylic acid groups, protonation of the two amino groups

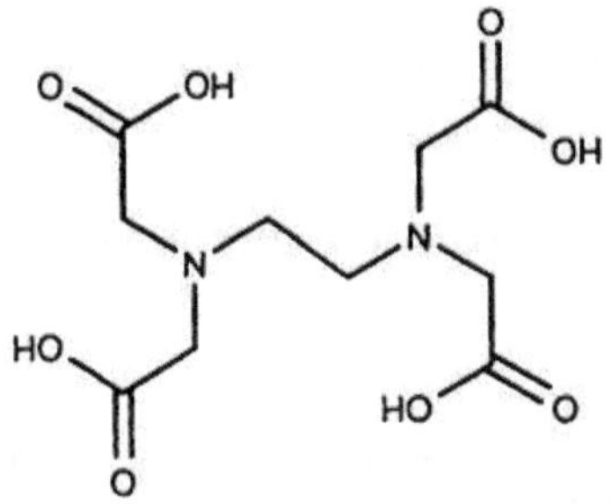

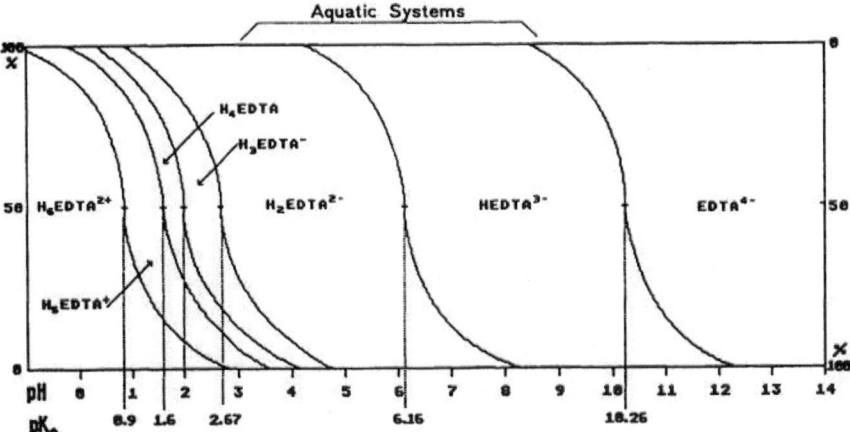

FIG. 1 Chemical structure of EDTA.

FIG. 2 Acid-base distributions for the different protonated EDTA species.

must be considered. Figure 2 shows the distribution of the different EDTA species and their dissociation constants pK_a [4].

From Fig. 2 it can be deduced that at pH values commonly found in natural aquatic systems H_2EDTA^{2-} and $HEDTA^{3-}$ must be expected. For the lowest solubility of EDTA, the range between pH 1.6 and 2.0 (H_4EDTA) is important.

B.   Metal Complexation and Equilibrium Constants

The formation of complexes of EDTA with metal ions can be described on the basis of electron donor-acceptor interactions. Being a multidentate ligand, EDTA can offer six donor positions per molecule. This leads to the formation of very stable chelate complexes [5].

The stabilities of the complexes for different metals can be characterized by individual stability constants. In Eqs. (1) through (9), the tetravalent EDTA anion ($EDTA^{4-}$) has been abbreviated as L^{4-}. The stability constants K and β were defined according to the equations given. In addition to the individual concentrations of the reacting partners the stability constants are dependent on the ionic strength and the temperature.

$$H^+ + L^{4-} \rightleftharpoons HL^{3-} \qquad K_1^H = \frac{c(HL^{3-})}{c(H^+)\,c(L^{4-})} = \beta_{HL} \qquad (1)$$

$$H^+ + HL^{3-} \rightleftharpoons H_2L^{2-} \qquad K_2^H = \frac{c(H_2L^{2-})}{c(H^+)\,c(HL^{3-})} \qquad (2)$$

$$H^+ + H_2L^{2-} \rightleftharpoons H_3L^- \qquad K_3^H = \frac{c(H_3L^-)}{c(H^+)\,c(H_2L^{2-})} \qquad (3)$$

$$H^+ + H_3L^- \rightleftharpoons H_4L \qquad K_4^H = \frac{c(H_4L)}{c(H^+)\,c(H_3L^-)} \qquad (4)$$

$$2H^+ + L^{4-} \rightleftharpoons H_2L^{2-} \qquad \beta_{H_2L} = \frac{c(H_2L^{2-})}{c(H^+)^2\,c(L^{4-})} \qquad (5)$$

$$3H^+ + L^{4-} \rightleftharpoons H_3L^- \qquad \beta_{H_3L} = \frac{c(H_3L^-)}{c(H^+)^3\,c(L^{4-})} \qquad (6)$$

$$Me^{2+} + L^{4-} \rightleftharpoons MeL^{2-} \qquad K_{MeL}^{Me} = \frac{c(MeL^{2-})}{c(Me^{2+})\,c(L^{4-})} = \beta_{MeL} \qquad (7)$$

$$Me^{2+} + HL^{3-} \rightleftharpoons MeHL^- \qquad K_{MeHL}^{Me} = \frac{c(MeHL^-)}{c(Me^{2+})\,c(HL^{3-})} \qquad (8)$$

$$H^+ + Me^{2+} + L^{4-} \rightleftharpoons MeHL^- \qquad \beta_{MeHL} = \frac{c(MeHL^-)}{c(H^+)\,c(Me^{2+})\,c(L^{4-})} \qquad (9)$$

The rapid and accurate assessment of the EDTA species present in natural aquatic environments and their concentrations as a function of pH can be powerfully visualized by species distribution diagrams. For the total concentration of

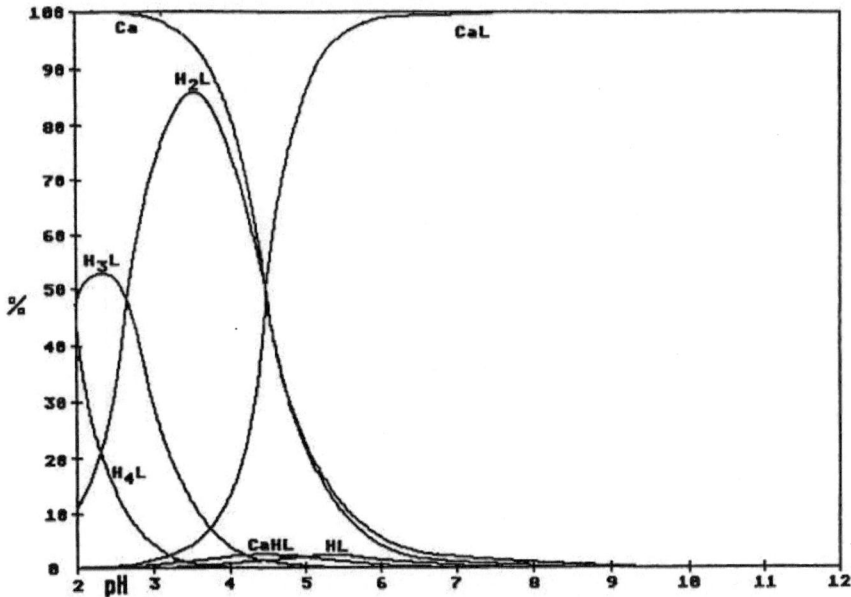

FIG. 3 Species distribution for the aqueous EDTA/Ca^{2+} system (L, ligand).

1.0 mM EDTA and 1.0 mM Ca(II) at an ionic strength of 0.1 M and a temperature of 25°C, the EDTA species distribution diagram is given in Fig. 3.

At a pH value of about 2, all Ca^{2+} in the system is in the uncomplexed form, accompanied by H_4L, H_3L^-, and H_2L^{2-}. As the pH increases, the 1:1 complex CaL^{2-} begins to dominate the system and finally becomes the only species present. Some critical stability constants [6] for 1:1 metal-EDTA complexes are given in Table 1. They were determined in K^+ media as supporting electrolyte.

TABLE 1 Selected Critical Stability Constants for Metal/EDTA Complexes at 0.1 M Ionic Strength and 25°C

Metal ion	Equilibrium quotient	Log K (with K^+)	$\Delta H°$
Mg^{2+}	$M/(M \cdot L)$	8.85 ± 0.08	3.3 ± 0.2
Ca^{2+}	$M/(M \cdot L)$	10.65 ± 0.08	-6.5 ± 0.1
Cu^{2+}	$M/(M \cdot L)$	18.78 ± 0.08	-8.2 ± 0.1
Zn^{2+}	$M/(M \cdot L)$	16.5 ± 0.1	-4.7 ± 0.2

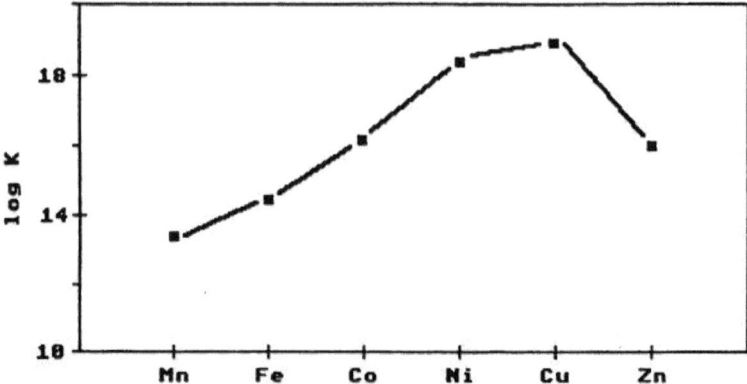

FIG. 4 Stability constants of EDTA chelates with divalent transition metal ions (increasing atomic number from left to right).

The values for Na^+ media are somewhat lower because of different stability constants. For KEDTA log K is 0.8 and for NaEDTA log K is 1.8.

The relative stability of EDTA complexes with some important divalent transition metal ions is given in Fig. 4. It is interesting that Cu(II) forms the most stable 1:1 chelate, followed by Ni(II), Zn(II), and Co(II).

For aquatic systems it is important to keep in mind that most transition metal ions occur only in the concentration range from 10^{-5} to 10^{-7} M [7]. An important exception is iron in the reduced oxidation state, which can occur in concentrations of up to 5×10^{-4} M. The most relevant alkaline-earth metals Ca^{2+} and Mg^{2+} are generally present in the 10^{-3} M concentration range. Because of the relatively high stability constants of Fe (for Fe^{2+}, see Fig. 4), practically all EDTA found in natural aquatic systems is associated with iron in various forms.

III. ANALYSIS OF EDTA

There are several procedures to identify and quantify EDTA. According to the analytical protocol, the methods are suited to different concentration ranges.

A. Gas Chromatography

For a sensitive gas chromatographic determination, EDTA must be derivatized. This can be done by bringing the aqueous sample to dryness [8] or by isolating EDTA from the aqueous solution by an anion exchange resin [9]. This sample preparation procedure is on its way to become a standardized method [10]. EDTA is derivatized to its tetra-*n*-butylester, which can easily be separated from other polyesters (e.g., from NTA, nitrilotriacetic acid) by capillary gas chromatography. Nitrogen-specific detection (NPD) allows the quantification down to 1 μg/liter concentrations. In samples with a high salt content the use of an internal standard (e.g., 1,2-diaminopropane-N,N,N',N'-tetraacetic acid, DPTA) is advisable to gain results with relatively small standard deviations.

The method has been applied to surface waters [1,11,12], drinking waters, and waters in water treatment plants [13,14]. An outline of the procedure is shown in Fig. 5.

B. Liquid Chromatography

High-performance liquid chromatography (HPLC) is well suited to the fast and reliable determination of EDTA. Relatively little sample pretreatment is necessary when primary EDTA concentrations are relatively high. Also, a wider range of detection systems (e.g., amperometric detection) [15] and stationary phases can be used. Other authors separated organic chelators like EDTA by ion chromatography [16] or applied ion pair chromatography to determine several aminopolycarboxylic acids [17]. The detection limit for the liquid chromatographic methods is around 0.2 mg/liter.

C. Polarography

Polarographic determination has become one of the official standard methods in Germany [10]. The pretreated aqueous sample is measured by differential pulse polarography (DPP) using a dropping mercury electrode (DME). The method is suited to the concentration range of 0.1–25 mg/liter of EDTA (H_4EDTA) in water and waste water. To minimize disturbance it is advisable to remove particulate matter from the original sample by membrane filtration and to use similar water without EDTA for system control. Nitrite, sulfite, and sulfide ions in the excess of 20 mg/liter must be eliminated by oxidation or purging after acidification. According to Fig. 6, disturbances by cations, detergents, and other surface-active substances can be minimized by adsorption on cation-exchange resins or polystyrene resins. Care must be taken should other chelators be present. For example, some phosphoric acid derivatives and diethylene trinitrilopentaacetic acid (DTPA) can disturb the quantification of EDTA if they are present in more than 20-fold excess.

Round-robin tests for EDTA-spiked waste water (2.9–11.9 mg/liter) yielded recoveries greater than 90% and a variability of less than 10% [10].

IV. PRODUCTION

The most widely used method for EDTA synthesis is the reaction of ethylenediamine with sodium cyanide and formaldehyde [18]. According to an official inquiry [19], about 35,000 ton EDTA tetrasodium salt were produced and applied

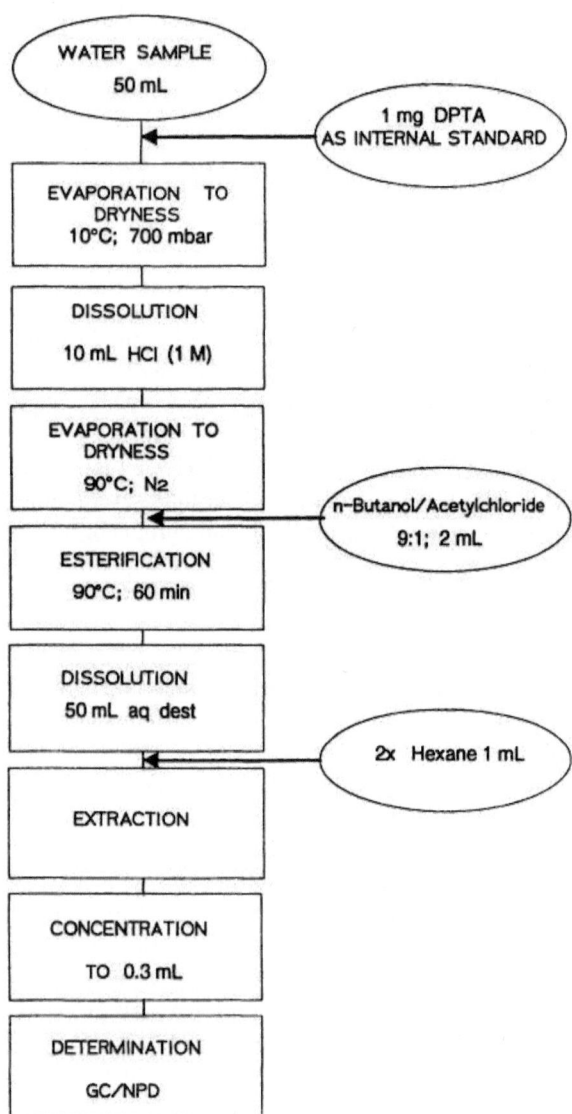

FIG. 5 Sample treatment for the determination of EDTA by gas chromatography.

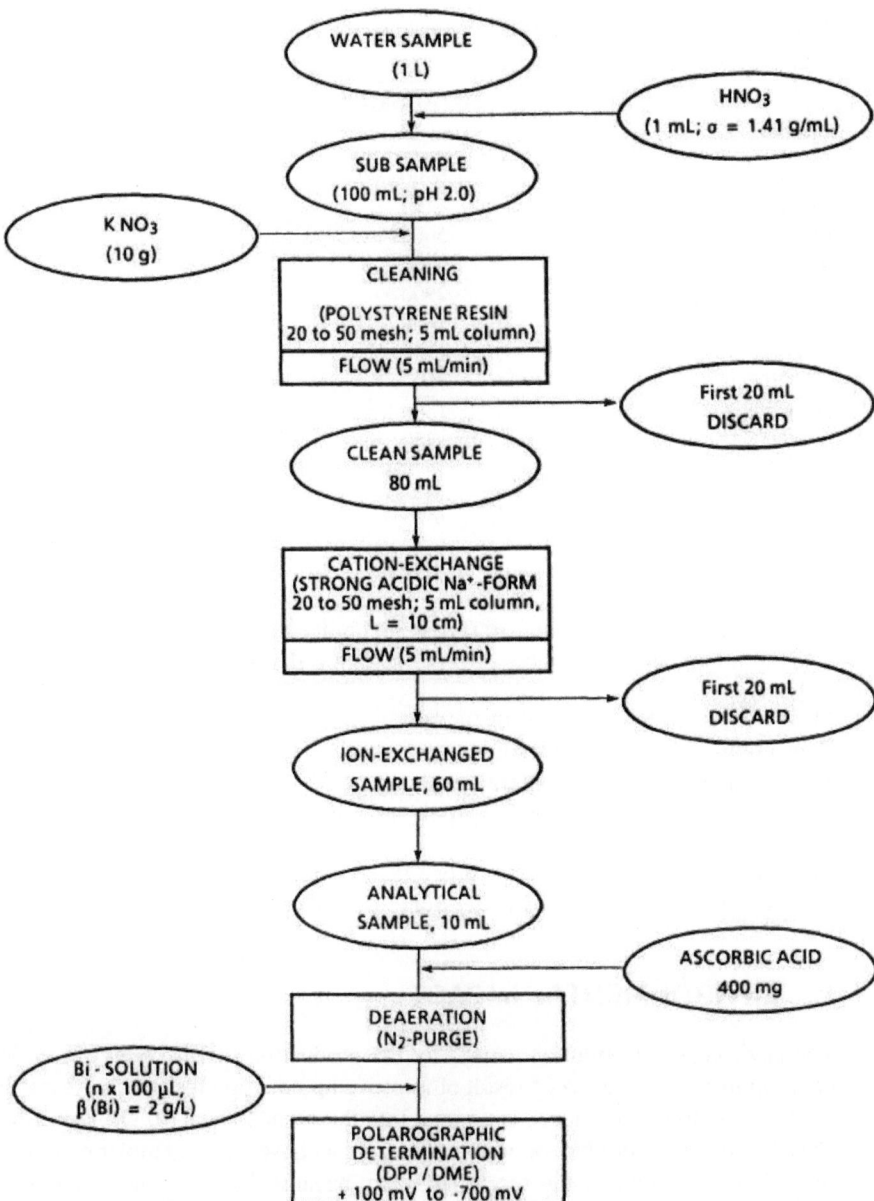

FIG. 6 Flow chart for sample treatment for the polarographic determination of EDTA.

TABLE 2 Estimation of EDTA Consumption
in Europe in 1990

Country	EDTA (Na$_4$) (t)
Germany	6900
Great Britain/Ireland	6400
France	5600
Italy	4400
Belgium/Luxembourg	2900
Spain	1800
Finland	1700
Netherlands	1600
Denmark	1600
Sweden	850

in Europe in 1990. The 10 leading countries in EDTA consumption are listed in Table 2.

The main reason for using EDTA is its ability to complex metal ions. This leads to a number of favorable effects, including the following:

Prevention of precipitation of metal salts
Prevention of catalyzed oxidation
Minimization of toxic and allergic reactions
Handling of trace metal elements in higher concentrations.

These technologically important properties led to a widespread application of EDTA in products of the daily life. Some examples are given in Fig. 7.

V. ENVIRONMENTAL IMPACT

The environmental pollution caused by the production of EDTA has been decreasing in recent years. As a result of improved production techniques, less than 2 kg per ton EDTA produced is emitted into the air and about 10–15 kg per ton EDTA ends in the production waste water. As a consequence, similar quantities of EDTA are found in the receiving rivers because EDTA is not effectively retained in waste water treatment plants [20]. The treatment plant of the main producer of EDTA in Germany emits about 380 kg EDTA each day. The effluent EDTA concentration is around 0.5–1.0 mg/liter.

More serious are diffuse EDTA emissions into the environment. According to the designation and application of EDTA, nearly all the EDTA used finds its way

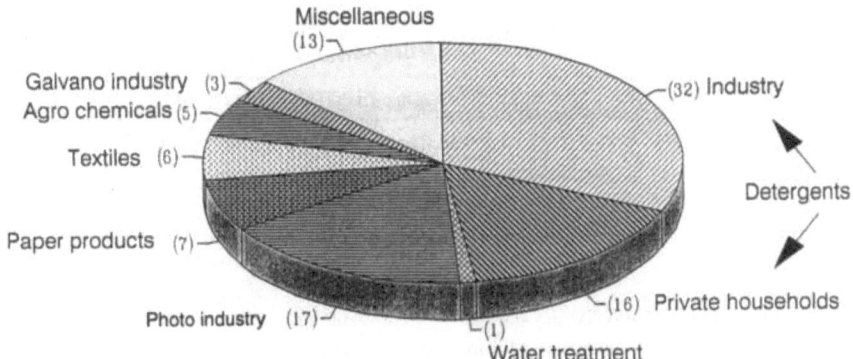

FIG. 7 Distribution of EDTA in products and application fields in Europe (1990).

into waste waters. As a consequence, a large fraction of this EDTA ends up in natural aquatic systems.

Starting from early investigations on the effect of EDTA in biological waste water treatment plants [21], there have been many studies focusing on the influence on flocculation and sedimentation [22–24]. The mobility and fate of trace metals, such as cadmium [25] or copper [26], and anaerobic sludge putrefaction have especially been investigated [27–30]. Also, the behavior of EDTA in many physical and chemical treatment steps has been extensively studied. A selection of studies and patents is listed with references in Table 3.

A. Rivers

Awareness of the EDTA problem has led to an investigation of EDTA distributions in the Rhine. In 1986–1991, a research project sponsored by the ARW (*Arbeitsgemeinschaft Rhein-Wasserwerke e.V.*) and the AWBR (*Arbeitsgemeinschaft Wasserwerke Bodensee-Rhein*) conducted a survey starting with 5 sampling points within the German territory [42]. From 1989 onward the survey was extended to 10 sampling points (including *Bischofsheim* at the River Main). The location of the sampling points for the extended survey is shown in Fig. 8. The data for 1986–1988 are shown in Table 4 (sampling on weekly basis) and for 1989–1991 in Table 5 (sampling on daily basis).

Because the data collected at individual stations are not distributed evenly around a mean value, median values (percentiles) were determined: P_{-50} designates the median of a given data population and indicates that 50% of the data are below that figure. Consequently, P_{-90} indicates that 90% of the data are below that figure.

There is a general trend for the EDTA concentrations increasing along the

TABLE 3 Influence of EDTA in Technical Steps Typically Used in Waste Water Treatment

Treatment steps	Reported effect	Reference	
Membrane separation	Metal (Hg) complex elimination	Bloch and Ausländer, 1976;	[31]
		Hopfenberg et al., 1978	[32]
	Ni complex elimination	Wen and Hamil, 1981	[33]
Adsorption on oxides (flocculation)	EDTA and metal complex elimination	Rubio and Matijevic, 1979;	[34]
		Huang and Bowers, 1980;	[35]
		Chang et al., 1983	[36]
Activated carbon adsorption	EDTA and metal complex elimination	Huang and Kao, 1981;	[37]
		Bhattacharyya and Cheng, 1987;	[38]
		Brauch and Schullerer, 1987;	[13]
		Rubin and Mercer, 1987	[39]
Anion exchanger	Cu complex elimination	Vignola, 1986	[40]
Heating (150–300°C)	EDTA and metal complex degradation	Martell et al., 1975;	[41]
		Booy and Swaddle, 1977;	[42]
		Motekaitis et al., 1986;	[43]
		Boles et al., 1987	[44]
Acidification	EDTA elimination	Katoh, 1986;	[45]
		Mitulla et al., 1987	[46]

Rhine's journey through Germany, starting with about 3 mg/m^3 (P_{-50}) at Öhningen and increasing to about 20 mg/m^3 (P_{-50}) at Bimmen in 1991. Very striking is the drastic increase between Karlsruhe and Mainz. Between these two sampling points, near Worms (Fig. 8), is the location of the largest EDTA-producing plant in Germany. The homogenization of the production waste water with the river water takes more than 100 km traveling distance. Because of this, representative concentration figures cannot be obtained from this location. Therefore, Worms was listed separately in Table 5. The high concentrations found in other rivers, such as the Main near Bischofsheim, the Neckar, and the Ruhr (not shown), are caused by the widespread application of EDTA in these areas.

As a second important result it becomes evident that from 1990 onward concentrations of EDTA decreased significantly. This reflects the generally better industrial waste water treatment in many larger and smaller industries using EDTA. However, the impact of these improvements is still relatively moderate because of EDTA emissions from municipal waste waters originating to a large extent from the domestic use of EDTA.

Other relevant information on EDTA pollution in rivers can be gained when the EDTA loads are considered instead of EDTA concentrations. Determination of the EDTA load F takes the concentration of EDTA β(EDTA) and the water

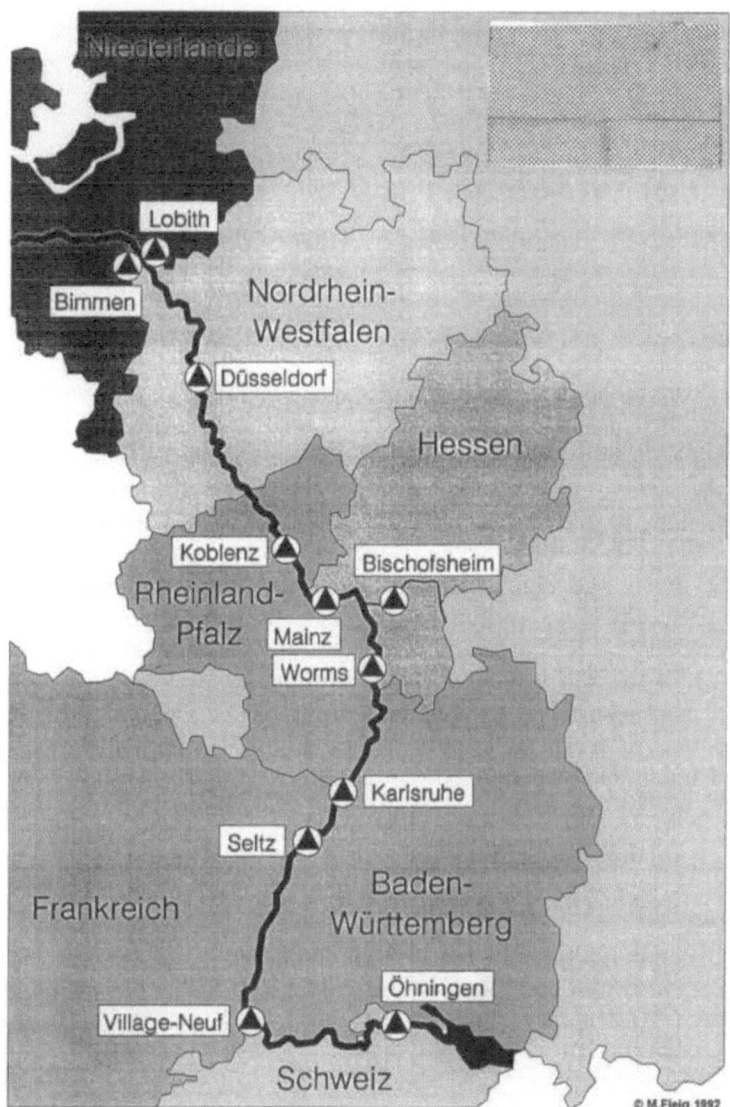

FIG. 8 Position of sampling points for the EDTA survey 1986–1991.

runoff data Q into account, according to Eq. (10). Absolute quantities of EDTA can be assessed when corrections for meteorological influences and waste water effluent strategies have been made.

$$F = \beta(\text{EDTA})\, Q \qquad \text{kg/day} \tag{10}$$

TABLE 4 P_{-50} and P_{-90} (mg/m^3) for EDTA in the Rhine, 1986–1988

	1986		1987		1988	
	P_{-50}	P_{-90}	P_{-50}	P_{-90}	P_{-50}	P_{-90}
Karlsruhe	8.4	18	9.6	12	11	16
Wiesbaden	20	50	14	32	20	26
Köln	18	31	16	26	24	38
Düsseldorf	21	58	20	24	23	39
Wesel	22	51	21	27	27	46

The EDTA loads at the different locations are shown in three-dimensional view for better visualization. Again, percentiles were used to summarize all the data obtained on a daily basis. As can be seen from Fig. 9, the downstream increase in the EDTA load is very drastic and far more pronounced than the increase in the corresponding concentrations. This reflects the widespread and similar distribution of EDTA in the contributing rivers.

In smaller rivers, variations in EDTA concentrations are much more visible because the effect of local influents on river water composition is more dominant. For example the EDTA concentrations in a medium-sized river (Neckar) are shown in Fig. 10. The variability is high but no clear trend can be deduced. The highest concentrations are found near Aldingen, behind a large sewage plant near Mülhausen. At Mannheim near the confluence of the Neckar with the Rhine, the concentrations are comparatively low [48].

For the Ruhr, another medium-sized tributary of the Rhine, EDTA concen-

TABLE 5 P_{-50} and P_{-90} (mg/m^3) for EDTA in the Rhine, 1989–1991

	1989		1990		1991	
	P_{-50}	P_{-90}	P_{-50}	P_{-90}	P_{-50}	P_{-90}
Öhningen	3.5	3.8	3.0	3.8	2.9	3.2
Village-Neuf	4.4	5.5	4.8	9.5	4.0	4.7
Seltz	8.5	9.9	7.6	13	6.6	8.4
Karlsruhe	9.8	12	7.5	12	6.6	9.7
Mainz	21	29	14	20	14	21
Koblenz	23	28	16	21	16	23
Düsseldorf	23	31	15	23	19	24
Lobith	25	29	16	24	21	25
Bimmen	24	28	17	26	20	22
Worms	39	56	21	40	26	39
Bischofsheim/Main	61	86	34	64	47	67

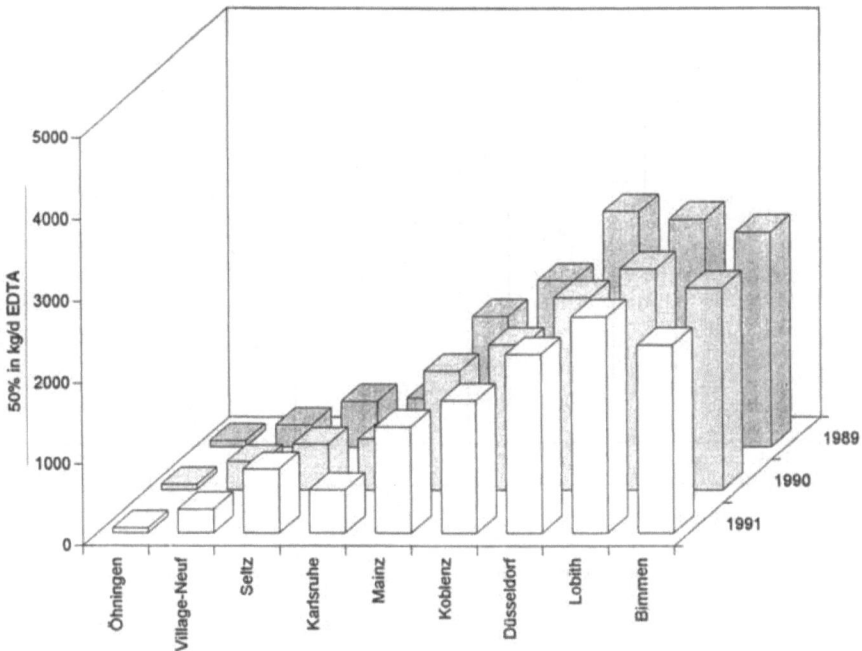

FIG. 9   EDTA loads along the Rhine from 1989 to 1991.

trations were reported to be around 18 µg/liter near the town of Essen. For drinking water (bank filtrate), 13 µg/liter was found [14].

Because of the undesirable occurrence of EDTA in rivers and its questionable ecological relevance, an agreement has been initiated by the German Ministry of Environment in 1991, according to which a decrease in EDTA emission from the industrial side [49] by 50% within the next 5 years can be expected. This will further reduce EDTA concentrations and EDTA loads in recipient rivers.

B.   Surface Lakes

Not many data on the distribution of EDTA in surface water bodies are available. One example, which illustrates well the stability of EDTA to chemical or biological degradation, is illustrated in Fig. 11. Here the vertical distribution of EDTA in Lake Geneva, a large stillwater body, in October 1987 was investigated [50].

In Lake Geneva, the overall very low concentrations of EDTA of about 1 µg/liter change very little with depth. This shows that over the course of several years a complete homogenization of the dissolved EDTA has taken place. Within

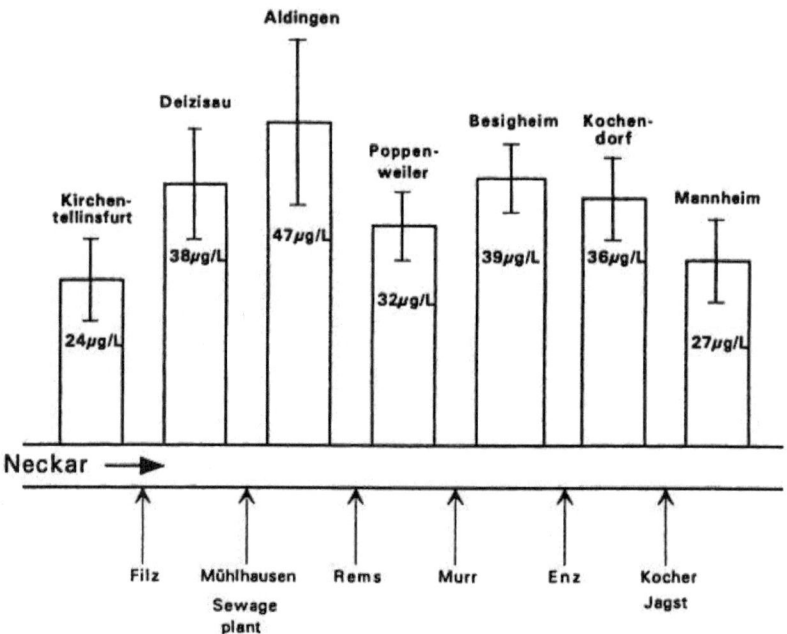

FIG. 10 Concentrations of EDTA in the downstream Neckar, 1989–1990.

this time no significant degradation or destabilization of EDTA occurred because this would have led to a gradual decrease in EDTA concentrations with depth.

A similar situation was found for Lake Constance, with slightly higher average EDTA concentrations of about 3 μg/liter [50]. With the water volume of the lake at 50×10^9 m³, a depot of 150 ton EDTA can be estimated [13].

C. Groundwater

There is little information about EDTA concentrations in groundwater. Well-protected aquifers can be expected to be practically free of this chemical. Only with broken waste water pipes or waste water infiltration will local concentrations above the analytical detection limit occur. A contribution of EDTA to groundwater may also come from surface water infiltration.

The effectiveness of bank filtration in the elimination of EDTA has been studied in a detailed survey [50]. At a testing site, groundwater was collected from wells with increasing distance from a small river (Glatt near Glattfelden, Switzerland). The initial concentrations of EDTA in the river were around 20 μg/liter. With increasing distance from the river bank, the concentration de-

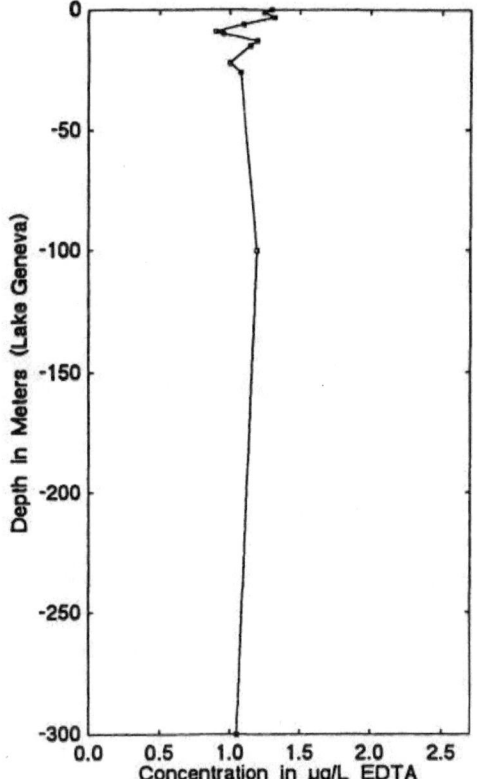

FIG. 11 Vertical distribution of EDTA in Lake Geneva.

creased slightly but increased again at larger distances (Fig. 12). It is assumed that the observed changes in concentrations were not affected by adsorption or biological degradation processes. More likely, a lateral flow of EDTA-free groundwater resulted in a slight dilution of the filtrate water [50]. However, the lack of knowledge of the specific hydrodynamics limits interpretation of the results.

D. Behavior During Water Treatment

EDTA is expected in many groundwater regimes that contain more or less bank filtrate from river waters. This is because of its poor biodegradability, good solubility in water, and inefficient adsorbability in natural aquifers.

From the point of view of drinking water supply, it is important to know how

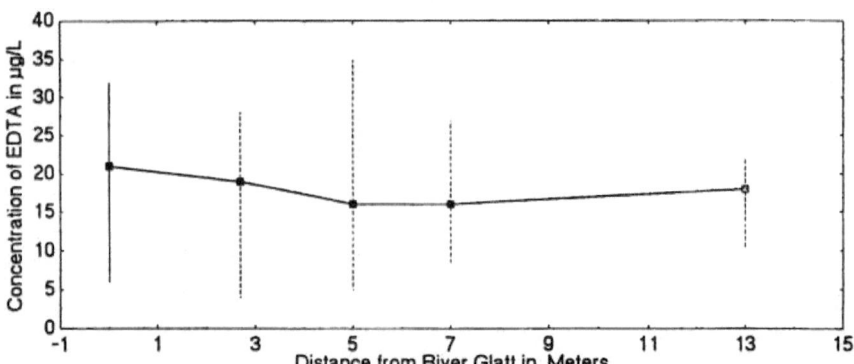

FIG. 12 Distribution of EDTA in river bank water with increasing distance from a small river (Glatt).

EDTA in the raw water behaves during the different drinking water treatment steps. The behavior of EDTA in the Rhine near Düsseldorf can be regarded as a typical example [13]. The analysis (see Fig. 13) includes (1) the river itself, (2) the corresponding bank filtrate, (3) the bank filtrate after ozonation, (4) sand filtration, and (5) after passage of an activated carbon filter. The concentrations of EDTA decreased from about 30 µg/liter (river water) to 20 µg/liter for the

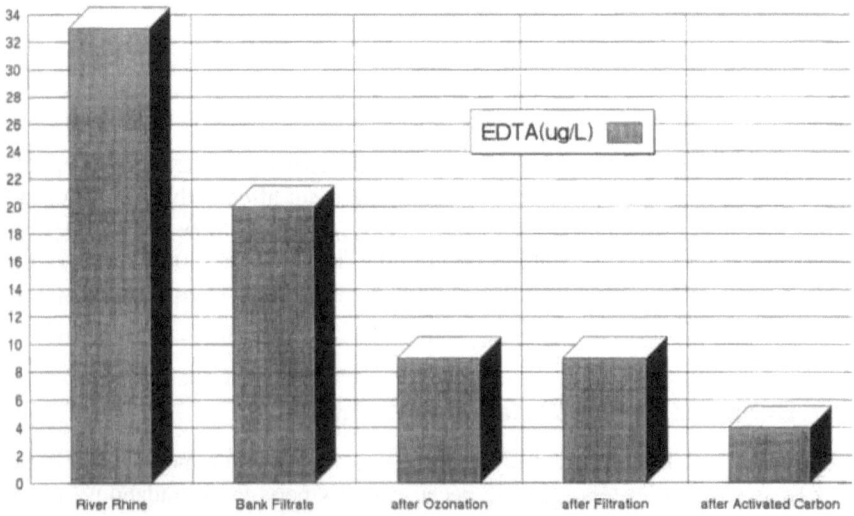

FIG. 13 Behavior of EDTA in a drinking water treatment plant.

bank filtrate. After ozonation and sand filtration, about 10 µg/liter of EDTA was found. Activated carbon filtration led to a further decrease by about 50%.

It becomes obvious that the oxidation step was fairly effective in decreasing the concentrations, whereas sand filtration did not change the EDTA content significantly. Even activated carbon did not remove much of the substance. From this result it can be deduced that drinking water in that area finally contains a few µg/liter of EDTA. The study also revealed that the addition of chlorine or chlorine dioxide for disinfection purposes had no influence on the EDTA concentration [13].

E. Oxidative Degradation

As mentioned earlier, EDTA is easily oxidizable. Oxidative degradation studies have been carried out with a variety of oxidants. A review of some important work is given in Table 6.

Most of the authors studied solutions with relatively high EDTA concentrations as they are found in the waste water of many metallic processes. For drinking water treatment, many of these methods cannot be applied because stringent regulations on the nature and concentration of the oxidizing agent must be observed. Model experiments [13] with ozone, chlorine, and chlorine dioxide and aqueous solutions containing 200 µg/liter of EDTA have shown that EDTA is readily oxidized with ozone but not with chlorine or chlorine dioxide (Fig. 14). In the experiment [13], buffered distilled water (pH 7.0) was used and the contact time was 10 minutes. At the end of the reaction time excess ozone was destroyed

TABLE 6 Oxidative Degradation Methods for EDTA

Oxidant	Reference	
$KMnO_4$	Feikes et al., 1985	[51]
Cl_2 and NaOCl	Bober et al., 1973;	[52]
	Liebgott and Fischer, 1986	[53]
H_2O_2	Gilbert, 1984;	[54]
	Sörensen and Frimmel, 1995	[55]
O_3	Fabjahn et al., 1976;	[56]
	Clem and Hodgson, 1978;	[57]
	Hartinger, 1986	[58]
UV radiation	Schneider and Rump, 1981	[59]
UV and oxidants	Schullerer and Brauch, 1989;	[60]
(e.g., H_2O_2, O_3)	Sörensen and Frimmel 1995	[55]
Anodic oxidation	Bollhalder and Sova, 1978;	[61]
	Hartinger, 1986;	[58]
	Müller et al., 1988	[62]

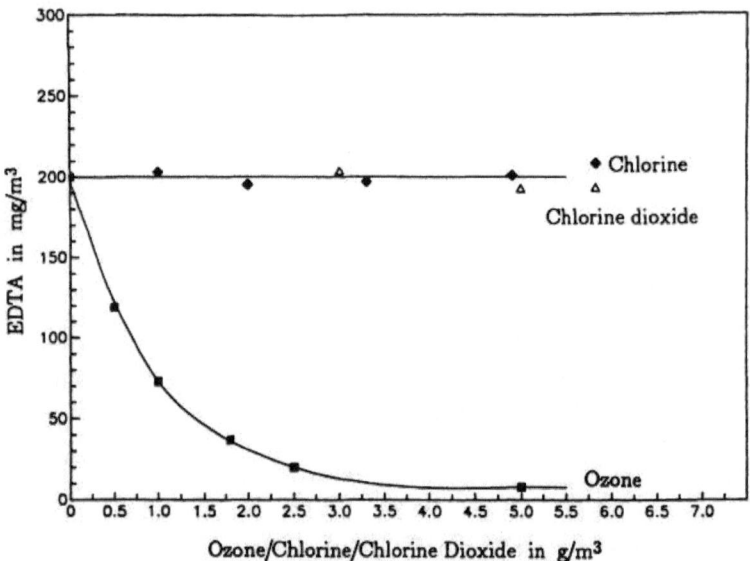

FIG. 14 Oxidation of EDTA with ozone, chlorine, and chlorine dioxide.

with thiosulfate. In natural aquatic systems, the yield is somewhat lower because of the competition effect of other dissolved organic matter, such as humic substances.

F. Photochemical Degradation

Because of its low volatility and high water solubility, EDTA occurs in the environment mainly in aquatic systems. Therefore, natural photochemical reactions are limited to the photic zone of surface water. In one study, the degradation of EDTA in water under sunlight conditions was investigated [63]. In solutions containing 1.25 g/liter of EDTA, more than 50% of the chelator was degradated within 10 minutes.

In another study the environmentally relevant ferric EDTA complex was investigated under sunlight and ultraviolet (UV)-conditions at 254 nm wavelength [64]. The yield of the reaction decreased with increasing pH (4.5–8.5). As stable reaction products, CO_2, formaldehyde, N-aminoethylglycerine, and glycine were detected. Irradiation with a wavelength of 350 nm led to the formation of CO_2, formaldehyde, and ethylenediaminetriacetic acid production [65].

Photochemical oxidation methods are much more effective when UV irradiation is used as catalyst for the rapid activation of oxidizing agents. For example,

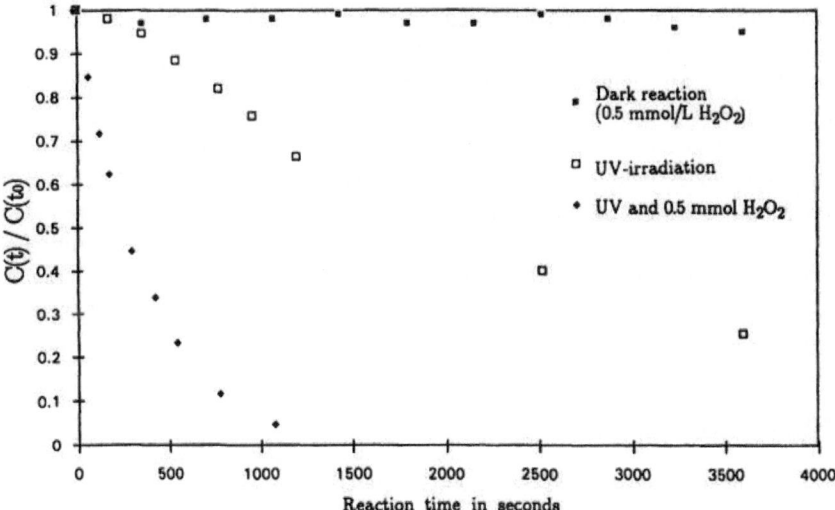

FIG. 15 Oxidation of EDTA with hydrogen peroxide and UV irradiation.

1 mol hydrogen peroxide can be decomposed into 2 mol hydroxyl radicals, which react nonspecifically with most organic compounds. This combined oxidation method can also be used to oxidize EDTA in highly diluted solutions within very short time. For illustration, the destruction of EDTA in model experiment [55] with hydrogen peroxide alone, UV irradiation alone, and hydrogen peroxide in combination with UV irradiation is shown in Fig. 15. Hydrogen peroxide decomposes in water very slowly; hence the dark reaction did not significantly affect the EDTA concentration. UV irradiation led to a clear and almost linear decrease in the acid with time. However, it is striking to see how the decomposition rate increases when UV irradiation was used in combination with hydrogen peroxide. The decomposition kinetics increased by a factor of about 8.

VI. CONCLUSIONS

EDTA is a well-known and widely used chemical. It is obviously not toxic to mammals. As a xenobiotic compound, however, its fate in the environment must be critically controlled. The kind of application, hydrophilic character, poor biodegradability, and the ability to form most stable complexes with transition metal ions suggest a broad distribution in aquatic systems. The results collected so far show that EDTA belongs to the organic chemicals with the highest mass concentrations measured in surface waters. In addition to the few rivers and lakes that have been investigated, more data on the occurrence of EDTA in surface and brown waters must be collected.

There is a severe lack of knowledge about the ecological importance of EDTA and its metal complexes in aquatic systems. Furthermore, the chemical, photo-chemical, and biological degradation of EDTA and its products must be inves-tigated to understand the function of these compounds in the environment and to introduce measures for their elimination, for example in water treatment. The interaction of the metal chelates with biota under environmental conditions is another field of research with many open questions. This has led to a remarkable agreement between German government and industry to reduce drastically emis-sions of this xenobiotic compound into the aquatic environment.

REFERENCES

1. F. Dietz, Korrespondenz Abwasser *32*, 988 (1985).
2. M. Windholz, S. Budavarri, R.F. Blumetti, and E.S. Otterbein (eds.), in *The Merck Index*, Merck&Co, Rathway, N.J., 1983.
3. J.-R. Hart, in *Ullmann's Encyclopedia of Industrial Chemistry* (W. Gerhartz, ed.), VCH Verlagsgesellschaft, Weinheim, Germany, 1987, Vol. A10.
4. G. Schwarzenbach and H. Flaschka, in *Die komplexometrische Titration*, F. Enke Verlage, Stuttgart, 1965, p. 8.
5. F.A. Cotton, G. Wilkinson, and P.L. Gaus, in *Grundlagen der anorganischen Chemie*, VCH Verlagsgesellschaft, Weinheim, Germany, 1990.
6. A.E. Martell and R.J. Motekaitis, in *Determination and Use of Stability Constants*, VCH Verlagsgesellschaft, Weinheim, Germany, 1990.
7. U. Förstner and G.I.W. Wittmann (eds.), *Metal Pollution in the Aquatic Environment* Springer, Berlin, 1983.
8. J.K. Reichert and A.H.M. Linckens, Environ. Technol. Lett. *1*, 42 (1980).
9. N.T. de Oude, Vom Wasser *64*, 283 (1985).
10. DIN 38 413, part 5 (German Industrial Standard), Beuth, Berlin.
11. F.H. Frimmel, R. Grenz, E. Kordik, and F. Dietz, Vom Wasser *72*, 175 (1989).
12. T. Wanke and S.H. Eberle, Acta Hydrochim. Hydrobiol. *20*, 192 (1992).
13. H.-J. Brauch and S. Schullerer, Vom Wasser *69*, 155 (1987).
14. E.A. Nusch, H.D. Eschke, and K.H. Kornatzki, Korrespondenz Abwasser *38*, 94 (1991).
15. J. Dai and G.R. Helz, Anal. Chem. *60*, 301 (1988).
16. J. Weiss and G. Hägele, Fresenius Z. Anal. Chem. *328*, 46 (1987).
17. W. Huber, Acta Hydrochim. Hydrobiol. *20*, 6 (1992).
18. F.C. Bersworth, U.S. Patent 2,387,735 for Martin Dennis Co. (1945).
19. *Conseil Europeen des federations de l'Industrie Chimique* (CEFIC), Chelating agents—questions and answers: EDTA, Brussels, 1990.
20. H.R. Siegrist, A. Alder, W. Gujer, and W. Giger, Gas, Wasser, Abwasser *68*, 101 (1988).
21. C. Potos., Journal WPCF *37*, 1247 (1965).

22. J.S. Balcerski and A.M. Schiller, U.S. Patent 4,224,149 (1980).

23. J. Bauer, D. Heathcote, and S. Krogh, J. Chromatogr. *369*, 422 (1986).

24. J.C. Le Bell, P. Stenius, and C. Axberg, Water Res. *17*, 1073 (1983).

25. J.F. Kao, L.P. Hsieh, S.S. Cheng, and C.P. Huang, Journal WPCF *54*, 1118 (1982).

26. M.H. Cheng, J.W. Patterson, and R.A. Minear, Journal WPCF *47*, 362 (1975).

27. J. Matsumoto and T. Noike., Technol. Rep. Tohoku Univ. *44*, 441 (1979).

28. P.J. Witkowski and J.S. Jeris, Proc. Ind. Waste Conf. *38*, 839 (1984).

29. K.R.K. Alibhai, I. Mehrotra, and C.F. Forster, Water Res. *19*, 1483 (1985).

30. C.-F. Forster, I. Mehrotra, and K.R.K. Alibhai, J. Chem. Tech. Biotechnol. *35b*, 145 (1985).

31. R. Bloch and J. Ausländer, German Patent DT 25 05 255 Al.29.07.1976 (1976).

32. H.B. Hopfenberg, K.L. Lee, and C.P. Wen, Desalination *24*, 175 (1978).

33. C.P. Wen and H.F. Hamil, J. Membrane Sci. *8*, 51 (1981).

34. J. Rubio and E. Matijevic, J. Coll. Interface Sci. *68*, 408 (1979).

35. C.P. Huang and A.R. Bowers, Environ. Eng. Div. Spec. Conf. 240 (1980).

36. H.-C. Chang, T.W. Healy, and E. Matijevic, Coll. Interface Sci. *92*, 469 (1983).

37. C.P. Huang and J.F. Kao, Environ. Int. Conf. 240 (1981).

38. D. Bhattacharyya and R.C.Y. Cheng, Environmental Progress *6*, 110 (1987).

39. A.J. Rubin and D.L. Mercer, Separation Sci. Technol. *22*, 1359 (1987).

40. M. Vignola, German Patent DE 36 14 061. Al. 30.10.1986 (1986).

41. A.E. Martell, R.J. Motekaitis, A.R. Fried, J.S. Wilson, and D.T. MacMillan, Can. J. Chem. *53*, 3471 (1975).

42. M. Booy and T.W. Swaddle, Can J. Chem. *55*, 1770 (1977).

43. R.J. Motekaitis, K. Ikemizu, H. Kamano, and Y. Kato, J. Chem. Eng. Japan *19*, 294 (1986).

44. J.S. Boles, K. Ritchie, and D.A. Crerar, Nuclear and chemical waste management *7*, 89 (1987).

45. S. Katoh, European Patent EP 0 168 752 A2. 22.01.1986 (1986).

46. K. Mitulla, J. Hambrecht, S. Marquardt, H. Brandt, B. Schmitt, H. Gausepohl, P. Siebel, and H. Dreher, German Patent DE 35 22 470 A1 (1987).

47. *Arbeitsgemeinschaft Rhein-Wasserwerke e.V.*, Report publ. by DVGW, University of Karlsruhe, Germany, 1991.

48. S. Schullerer-Iagiella, Internal Report, DVGW, University of Karlsruhe, 1992.

49. *Announcement of the German Ministry for Environment* (31.07.1991), WAI, 3. 23, 011/15, 1991.

50. *Untersuchungen über das Umweltverhalten des Phosphatersatzstoffes NTA und des organischen Komplexbildners EDTA*, Internal Annual Report, EAWAG, 1987.

51. L. Feikes, D. Schröder, and P. Reyer, German Patent DE 33 35 746. 11.04.1985, (1985).

52. T.W. Bober, T.J. Dagon, and I. Slovonsky, U.S. Patent 3, 767, 572, (1973).

53. H. Liebgott and F. Fischer, German Patent DE 35 01 932, (1986).

54. E. Gilbert, Vom Wasser *62*, 307 (1984).

55. M. Sörensen, F.H. Frimmel, Z. Naturforsch. *50b*, 1845 (1995).

56. C. Fabjahn, R. Davies, and K. Marschall, Galvanotechnik *67*, 643 (1976).

57. R.G. Clem and A.T. Hodgson, Anal. Chem. *50*, 102 (1978).

58. L. Hartinger, Galvanotechnik 77, 1814 (1986).
59. W. Schneider and H. Rump, in *Wasser 81*, Colloquium Verlag Otto Hess, Berlin, 1981, pp. 242–257.
60. S. Schullerer and H.-J. Brauch, Vom Wasser 72, 21 (1989).
61. H. Bollhalder and V. Sova, German Patent DE 27 21 994 A1 (1978).
62. K.-J. Müller, T. Bolch, and K. Mertz, Galvanotechnik 79, 172 (1988).
63. L.H. Hall and J.L. Lambert, J. Am. Chem. Soc. 90, 3036 (1968).
64. Y.K. Lee and S.J. Pirt, Microbiol. Lett. 6, 379 (1979).
65. J.H. Carey and C.H. Langford, Can. J. Chem. 51, 3665 (1973).

Index